Dispersal of Living Organisms into Aquatic Ecosystems

Dispersal of Living Organisms into Aquatic Ecosystems

Edited by

Aaron Rosenfield
National Marine Fisheries Service
National Oceanic and Atmospheric Adminstration

Roger Mann
Virginia Institute of Marine Science

A Maryland Sea Grant Publication
College Park, Maryland

Published by the Maryland Sea Grant College, University of Maryland, College Park.

Publication of this book is supported by grant NA86AA-D-SG007 from the National Oceanic and Atmospheric Administration to the Maryland Sea Grant College and by grant SES-8910789 from the National Science Foundation to the Center for Environmental and Estuarine Studies, University of Maryland System.

Book and cover design by Sandy Harpe

University of Maryland Publication UM-SG-TS-92-04
Library of Congress Card Catalog Number: 92-060531
ISBN 0-943676-56-8

For information on Maryland Sea Grant books, contact:

Maryland Sea Grant College
University of Maryland System
0112 Skinner Hall
College Park, Maryland 20742

Printed on recycled paper

It may have begun with Noah, but, wherever it started, the whole idea of rearranging the earth's wild creatures still seems irresistible. Man, the supreme meddler, has never been quite satisfied with the world as he found it, and as he has dabbled in rearranging it to his own design, he has frequently created surprising and frightening situations for himself.

George Laycock, *The Alien Animals*

Dedication

With gratitude to Carl J. Sindermann for opening wide the doors to a career filled with many opportunities and awesome challenges and to my family, Clarice and Sandra, for their devotion and the inspiration they provided.

—Aaron Rosenfield

With thanks to my wife Elaine Lynch for her love, support and tolerance of the evenings and weekends lost to work rather than to play.

—Roger Mann

Contents

List of Figures and Tables

Figures

Tables

Contributors

Standish K. Allen
Haskins Research Laboratory
Rutgers University
P.O. Box 687
Port Norris, New Jersey 08349

Charles D. Amsler
Department of Microbiology and Immunology
University of Illinois at Chicago
P.O. Box 6998 (M/C 790)
Chicago, Illinois 60680

Thomas A. Bell
Department of Veterinary Science
University of Arizona
Tucson, Arizona 85721

James A. Brock
Department of Land and Natural Resources
Area 4, Sand Island Parkway
Honolulu, Hawaii 96819

James T. Carlton
The Williams College — Mystic Seaport
Program in American Maritime Studies
Mystic Connecticut 06355-0990

Thomas T. Chen
Center of Marine Biotechnology
University of Maryland System
600 East Lombard Street
Baltimore, Maryland 21202

Robson A. Collins
Marine Resources Division
California Department of Fish and Games
Sacramento, California 85814

Rita R. Colwell
Department of Microbiology
University of Maryland
College Park, Maryland 20742

Walter R. Courtenay, Jr.
Department of Biological Sciences
Florida Atlantic University
Boca Raton, Florida 33431-0991

Jack R. Davidson
Hawaii Sea Grant College Program
University of Hawaii
1000 Pope Road
Honolulu, Hawaii 96822

Roy E. Drinnan
Department of Fisheries and Oceans
Halifax Fisheries Research Laboratory
Halifax, Nova Scotia B3J 2S7
Canada

Rex Dunham
Department of Fisheries and Allied Aquaculture
Auburn University
Auburn, Alabama

Ralph A. Elston
Battelle Marine Sciences
Laboratory
439 West Sequim Bay Road
Sequim, Washington 98382

C. Austin Farley
National Marine Fisheries
Service/NOAA
Cooperative Oxford Laboratory
904 S. Morris Street
Oxford, Maryland 21654

J.L. Fryer
Department of Microbiology
Oregon State University
Corvallis, Oregon 97331-3803

Jack Ganzhorn
OreAqua Inc.
88700 Marcola Road
Springfield, Oregon 97478

E. Spencer Garrett
NOAA/National Marine
Fisheries Service
National Seafood Inspection
Laboratory
3209 Frederick Street
Pascagoula, Mississippi 39568

Lucia Irene Gonzalez-Villaseñor
Bio Trax, Inc.
University of Maryland
Baltimore County
Baltimore, Maryland 21228

Robin Gregory
Decision Research
1201 Oak Street
Eugene, Oregon 97401

Phillip Hutton
Office of Pesticide Programs
U.S. Environmental Protection
Agency
401 M Street, S.W.
Washington, D.C. 20460

Frederick G. Kern
National Marine Fisheries
Service/NOAA
Cooperative Oxford Laboratory
904 S. Morris Street
Oxford, Maryland 21654

Christopher C. Kohler
Fisheries Research Laboratory
and Department of Zoology
Southern Illinois University —
Carbondale
Carbondale, Illinois 62901

Raymond J. Lewis
Harbor Branch Oceanographic
Institution
5600 Old Dixie Highway
Fort Pierce, Florida 34946

Donald V. Lightner
Department of Veterinary
Science
University of Arizona
Tucson, Arizona 85721

Chau-Min Lin
Center of Marine Biotechnology
University of Maryland System
600 East Lombard Street
Baltimore, Maryland 21202

David R. MacKenzie
U.S. Department of Agriculture
National Biological Impact
Assessment Program
901 D Street, N.W.
Washington, D.C. 20251-2200

James A. McCann
National Fisheries Research
Center — Gainesivlle
U.S. Fish and Wildlife Service
Gainesville, Florida 32606

James P. McVey
National Sea Grant College
1335 East West Highway
Silver Spring, Maryland 20910

Roger Mann
Virginia Institute of Marine
Sciences
Gloucester Point, Virginia 23062

G. Malcolm Meaburn
NOAA/National Marine
Fisheries Service
P.O. Box 12607
Charleston Laboratory
Charleston, South Carolina
29412

Michael Mendelsohn
Office of Pesticide Programs
U.S. Environmental Protection
Agency
401 M Street, S.W.
Washington, D.C. 20460

Mir S. Mulla
Department of Entomology
University of California
Riverside, California 92521

Michael Neushul
Marine Science Institute
University of California
Santa Barbara, California 93106

Nick C. Parker
U.S. Fish and Wildlife Service
Texas Cooperative Fish and
Wildlife Research Unit
Texas Tech University
Lubbock, Texas 79409-2125

Robert A. Peoples, Jr.
Division of Fish and Wildlife
Management Assistance
U.S. Fish and Wildlife Service
Arlington, Virginia 22203

Dennis A. Powers
Hopkins Marine Station
Stanford University
Pacific Grove, California 93950

Rita M. Redman
Department of Veterinary
Science
University of Arizona
Tucson, Arizona 85721

Daniel C. Reed
Marine Science Institute
University of California
Santa Barbara, California 93106

Lloyd W. Regier
NOAA/National Marine
Fisheries Service
Charleston Laboratory
P.O. Box 12607
Charleston, South Carolina
29412

Amy Rispin
Office of Pesticide Programs
U.S. Environmental Protection Agency
401 M Street, S.W.
Washington, D.C. 20460

J.S. Rohovec
Department of Microbiology
Oregon State University
Corvallis, Oregon 97331-380

Aaron Rosenfield
National Marine Fisheries Service/NOAA
Cooperative Oxford Laboratory
904 S. Morris Street
Oxford, Maryland 21654

Thomas K. Sawyer
Rescon Associates, Inc.
P.O. Box 206
Royal Oak, Maryland 21662

David J. Scarratt
Department of Fisheries and Oceans
Halifax Fisheries Research Laboratory
Halifax, Nova Scotia B3J 2S7
Canada

Carl J. Sindermann
National Marine Fisheries Service
Northeast Fisheries Center
Oxford Cooperative Laboratory
Oxford, Maryland 21654

Lynn B. Starnes
U.S. Fish and Wildlife Service
U.S. Courthouse & Federal Bldg.
500 Gold Avenue, S.W.
Albuquerque, New Mexico 87108

Gary H. Thorgaard
Department of Zoology and Program in Genetics and Cell Biology
Washington State University
Pullman, Washington 99164-4220

Robert B. Thurman
Department of Veterinary Science
University of Arizona
Tucson, Arizona 85721

William E. Walton
Center for Great Lake Studies
University of Wisconsin Milwaukee
600 East Greenfield
Milwaukee, Wisconsin 53204

James D. Williams
National Fisheries Research Center
U.S. Fish and Wildlife Service
7920 N.W. 71st Street
Gainesville, Florida 32606

Leonard G.L. Young
Department of Land and Natural Resources
335 Merchant Street
Honolulu, Hawaii 96813

Z. Zhu
Aberdeen University
Aberdeen, Scotland

Acknowledgments

We are grateful to the following offices within the National Oceanic and Atmospheric Administration for providing support for the symposium on which this volume is based: the National Ocean Pollution Program Office, the National Marine Fisheries Service's Northeast Regional Office, Office of Research and Environmental Information, Office of Protected Resources, and Office of Trade and Industry Services. We are also grateful to the National Science Foundation for its timely grant, No. SES-8910789, that helped support publication of this volume.

Robert Wildman, formerly of NOAA's National Office of Sea Grant, is especially deserving of our thanks as is the Maryland Sea Grant College. Special thanks are due Laura Gabanski of the National Ocean Pollution Program Office, for her part in coordinating the preparation of that section of the National Marine Pollution Program covering goal no. 3, "Understanding the Sources, Fates, and Effects on Marine Organisms of Biological Agents that Are Introduced or Influenced by Human Activities." This goal laid the foundation for the symposium and her advocacy was instrumental in bringing the symposium to fruition. We are particularly indebted to Merrill Leffler, Jack Greer, and Sandy Harpe of the Maryland Sea Grant College and Martin Wylie of the University of Maryland's Chesapeake Biological Laboratory for their attention to the details involved in publishing this volume. Mr. Leffler was generous of his own time and energy in rewrites of sections of some manuscripts, ensuring editorial instructions were observed and liaison activities carried out. Ms. Karen Hayman of the Northeast Fisheries Center Oxford Cooperative Laboratory was very helpful and merits our sincere thanks for preparing correspondence, communicating with appropriate parties, keeping records, and above all showing great patience and understanding when besieged with the many details of meeting time constraints and other demands associated with deadlines. Ms. J.B. Keller was exceptionally helpful in preparing manuscript transcriptions and contributing editorial-formatting suggestions and guidelines for which we acknowledge our sincere appreciation.

We should also like to thank Victor Kennedy, University of Maryland Center for Environmental and Estuarine Studies, and Sandra Shumway, Maine Department of Marine Resources, for their helpful role in the symposium's program development, including their reviews of the abstracts and their having the program printed.

Of greatest importance to this listing of acknowledgements is the wish to express our deepest gratitude to the contributors of this volume for their good will, cooperation and, above all, their patience as we progressed toward final completion of this document. Finally, one of us (A.R.) acknowledges his gratitude to the Northeast Fisheries Center and the University of Maryland Center for Environmental and Estuarine Studies (CEES) and the late Dr. Ian Morris for permitting him to participate in an Interagency Personnel Agreement as a visiting adjunct professor of biology at the CEES facility in Cambridge, Maryland, while this volume was in preparation.

Foreword

The papers in Part I of this volume discuss the risks associated with releases, introductions, and transfers of living organisms into aquatic ecosystems and resulting impacts. Part II focuses on management approaches relative to translocation of living organisms that can be applied for reduction of risk to aquatic biota and their habitats, enhancement of food production, and protection of human health.

The papers in *Dispersal of Living Organisms into Aquatic Ecosystems* contain many terms to connote the involvement of humans in the dispersal of organisms from one location to another, for example, translocation, movement, intrusion, incursion, transplanted, transported, transmitted, released, removed, shipped, exported, imported, relocated. The intent of these terms should be clearly understood according to their syntax. However, these definitions given below of introduced and transferred species, as applied to aquaculture activities, appear in a glossary prepared jointly by the working group on introductions and transfers of marine organisms of the International Council for the Exploration of the Sea and by the working party on introductions of the European Inland Fisheries Advisory Commission. Consequently, the expressions used most frequently throughout this volume are as follows:

- Introduced species (= non-indigenous species which includes exotic species). Any species intentionally or accidentally transported and released into an environment outside its present range.

- Transferred species (= transplanted species). Any species intentionally or accidentally transported and released within its present range.

Preface

The papers that appear in this volume were first presented as part of a National Shellfisheries Association symposium entitled "Human Influences on the Dispersal of Living Organisms and Genetic Materials into Aquatic Ecosystems." The symposium was held in February 1989 in Los Angeles, California, as part of the Aquaculture '89 meeting, which was sponsored jointly by several organizations with interests in aquaculture, fishery biology, conservation, pollution, and living resource/ecosystem management.[1] The rationale for the symposium was a consequence of efforts by the National Oceanic and Atmospheric Administration's National Ocean Pollution Program Office to coordinate and update the Interagency Federal Plan for Ocean Pollution Research, Development, and Monitoring FY 1988-1992. Operating under the aegis of the President's National Ocean Pollution Policy Board, representatives from government, industry and academic communities contributed to the preparation of an updated five-year plan, published in September 1988 (available from U.S. Department of Commerce, National Ocean Pollution Program Office, Rockwall Building, Rockville, Maryland 20852). The next update of the plan should be available in 1992.

---

[1] Sponsors: Catfish Farmers of America; Fish Culture & Bio-engineering Sections of the American Fisheries Society; National Shellfisheries Association; Shellfish Institute of North America; U.S. Trout Farmers Association; World Aquaculture Society. Associate Sponsors: Florida Aquaculture Association; Clemson/South Carolina Wildlife & Marine Resource Department Aquaculture Cooperative; International Association of Astacology; Louisiana Crawfish Farmers Association; Pacific Coast Oyster Growers Association; Washington Aquaculture Council; Washington Fish Growers Association.

The executive summary of the Plan lists six goals within the National Marine Pollution Program, with goal number 3, "Understanding the sources, fates and effects on marine organisms of biological agents that are introduced or influenced by human activities," having the most pertinence to the symposium. Thus, information provided through this symposium and publication of its papers should prove helpful in achieving the goal as stated, both for microorganisms and macroorganisms. In addition, other information needs on dispersal of living organisms into aquatic ecosystems must be satisfied for better understanding of national and international programs even though they are not necessarily associated with marine pollution-related activities, under the five-year National Marine Pollution Program.

Freshwater and marine aquaculture programs in the United States and those abroad are expanding. Several perplexing questions about the necessity and advisability of translocating plant and animal species, and their possible consequences, now confront the scientist, entrepreneur, and habitat and resource manager. Many exotic aquatic organisms with several life history stages are being transported with increasing frequency, via a multiplicity of pathways, to locales throughout the globe. While many of these transplantations are ostensibly for growth enhancement, cultivation and ocean ranching in aquaculture operations, other transplantations are for scientific experiments, resource restoration, depuration and aesthetic purposes. Inspection of imports and surveillance and bioassay systems to detect undesirable organisms that accompany transplanted species are seldom carried out effectively, largely because communications, information and planning are poor. Propagation and cultivation systems as well as holding facilities differ and breakdowns are always possible; water treatment systems also vary: many are poorly designed and inadequately built. Further contributing to ineffective controls over undesirable and potentially harmful imports are incomplete legislation, unenforceable regulations and a lack of sufficient funding. Accidental and deliberate imports of exotic species released at the conclusion of a scientific study, or cast off by collectors or discharged in ballast water, have already resulted in the establishment of exotic populations inimical to the health of ecosystems.

Programs in biotechnology, particularly those incorporating innovative biochemical and genetic approaches to develop useful products and organisms with "desirable characteristics," have been receiving great attention; thus, the need for more effective policy statements, protocols, standards and guidelines for product testing and application will continue to expand. The information in this volume should contribute to this need.

More recent publications will also be of importance. Tiedje et al. (1989) covers the release of genetically engineered organisms and includes ecological considerations and recommendations. Issued by the Ecological Society of America, it provides important information on the mechanisms, management and consequences on natural and human-assisted translocations and releases of living organisms into diverse ecosystems. A book edited by Drake et al. (1989) was initiated in mid-1982 by the Scientific Committee on Problems on the Environment (SCOPE) of the International Council of Scientific Unions (I.S.C.U.). While the book does not address human-assisted translocations of organisms into aquatic ecosystems, its primary focus is on biological agents that have successfully invaded non-agricultural regions of the globe. The emphasis is on those agents that have disrupted terrestrial ecosystem functions, including the consequent effects on human beings. Though several chapters discuss theoretical issues about the commonality, convergence and applicability of the principles governing plant and animal invasions, they are eminently applicable to marine, brackish and freshwater ecosystems as well. These recent publications as well as this book should have value for delineating the role human beings can play to promote, retard or at least partially control the impacts of ill-considered actions that affect their habitats and associated biota.

As a final note, the original title of this volume was "Biological Pollution by Aquatic Organisms"; it was provocatively titled to emphasize the negative aspects of human-assisted invasions of living organisms into bodies of water. Such a title was restrictive in that it excluded the highly beneficial effects of many bioremedial, biological control, biomedical and aquaculture programs. Another title, "Human Influences on the Dispersal of Genetic Information into Aquatic Ecosystems," which was the original title of the symposium this book derives from, was also restrictive, for it suggested the dispersal or release primarily of genetically manipulated microorganisms. The eventual title was agreed to because it embraces the varied possibilities when microorganisms and macroorganisms are dispersed into aquatic environments.

Literature Cited

Drake, J.A., H.A. Mooney, F. di Castri, R.H. Groves, F.J. Kruger, H. Réjmanek and M. Williamson, editors. 1989. Biological invasions: A global perspective. John Wiley & Sons, Chichester.

Tiedje, J.M., R.K. Colwell, Y.L. Grossman, R.E. Hodson, R.E. Lenski, R.N. Mack and P.J. Regal. 1989. The planned introduction of genetically engineered organisms: ecological considerations and recommendations. Ecology 70: 298-315.

PART I.
RISKS AND IMPACTS

Introduction to Part I

Risks Associated with Translocations of Biological Agents

AARON ROSENFIELD

Introduction

Coastal and inland waters throughout the world have long histories of exploitation for the production and harvest of natural resources. These waters are heavily used, for example, for crop irrigation and cooling of power plants, for transportation and recreation; they receive discharges of human, industrial and agricultural wastes. Such use has led to widespread introductions of substances, both biotic and abiotic. While some have been beneficial, many are also harmful to public health and are injurious to living resources and environmental health. Figure 1 is an attempt to visualize human resource and environmental health interrelationships resulting from the movements of biotic and abiotic agents into aquatic ecosystems. If the area of each circle represents the intensity that biotic and abiotic agents can have on health, then increases or decreases in any one area (for instance, resource health) suggest how other areas (in this case, human and environmental health) can be impacted.

Though numerous coastal states and communities are trying to limit the impact of pollution, indications are that human populations will continue to increase and aquatic ecosystems will continue to feel the impact of transportation, housing and factory construction, mining and other development. It appears obvious that there will continue to be significant involvement with the dispersal of chemical compounds and biotic agents into aquatic

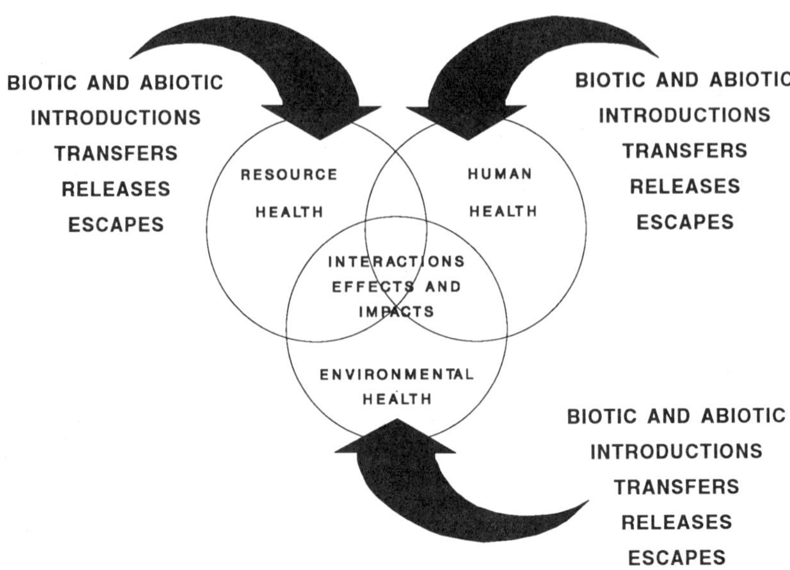

Figure 1. Ecological interrelationships: Humans, biota and environments.

ecosystems, ranging from microorganisms to more complex multicellular organisms, many of them pollutants. On the other hand, the dispersal of living agents into aquatic ecosystems could be highly beneficial to resource productivity, environmental quality, and to humankind. Examples of successful and adverse introductions are discussed in this book.

Living Organisms = Genetic Information

All living organisms by definition contain genetic information consisting of nucleic acids, that are capable of replicating under appropriate conditions. Some organisms contain very minute amounts of genetic material, for example viruses, while others contain massive amounts. Genetic material carrying heritable information has been inserted as genes into the germ plasm of some species to produce transgenic strains of those species. Most of the organisms under discussion have attained their genetic composition and ability to adapt to various environmental conditions, either through natural evolutionary mechanisms such as polyploidy, or through human intervention, via selective breeding. With advances in biotechnology and genetic engineering, it is now possible to hasten and augment such mechanisms, with the potential for creating assemblages of genetic material that would have otherwise failed to develop, except perhaps over evolutionary time. Some consider organisms whose genetic composition has been altered through such genetic engineering techniques, or even through traditional genetic breeding methods, to be exotic species and that for risk assessment and habitat management purposes, they must be treated as such. Consequently, the topics addressed in this collection were designed to cover broad issues and approaches to species management and to reach an audience with a wide spectrum of interests and backgrounds.

Dispersal of Genetic Material and Its Survival in Aquatic Ecosystems

The quantity, type, and arrangement of genetic material (genotype) present in organisms that are dispersed do not necessarily confer survival value on them. Nor does information about an organism's genotype necessarily provide indications of their ability to invade, adapt, propagate or sustain reproducing populations. Knowledge of the scale, type and mechanisms of movement and frequency of movement from one ecosystem to another is very helpful for planning purposes; still, even that knowledge does not necessarily provide for accurate predictions of successful population establishment or colonization. Obviously, successful invasion into another organism — for example, by a pathogen — or into a new environment by a biological agent, and its manifestation (phenotype), is most important and depends upon a number of factors. These include the invading organism's concentration, its virulence, its assiduousness to prevail, the presence of transmission mechanisms, enhancers, competitors, inhibitors, reproduction capabilities, effectiveness of host defense mechanisms, and the biological, chemical and physical influences of the environment.

Some thought has been given to the mechanisms and conditions organisms require to invade and successfully establish populations that can reproduce and grow in a particular ecological niche (see Mooney and Drake 1986). However, other than the case of well understood strains of plants and animals intended for controlled cultivation, few would be willing to give assurances that they can accurately predict whether an invasion by a biological agent will establish a self-sustaining population. Wild forms, however, in contrast to cultivated species, possess the full genetic potential to survive and grow without human assistance. Therefore, they can adapt to much wider ranges of environmental conditions than those that are genetically modified or bred selectively for specific characteristics.

Ecological Effects

Current studies have focused on purposeful movements of biotic and abiotic agents to aquatic environments, particularly with regard to species for aquaculture and habitat alteration. With the growing interest in aquaculture, introduced species which contain foreign segments of genetic material could bring about profound environmental change to alter habitat conditions if they become established as wild populations in their new environments. They could, for example, become fouling agents or competitors, or serve as vectors, alternate or reservoir hosts for parasites and other infectious agents. Likewise, predators, disease agents, and nuisance species that accompany organisms and materials translocated deliberately or accidentally for a variety of purposes, including aquaculture, could have profound unintended effects on resident biota and their environments. Some well known examples are the introductions of oyster drills onto the west coast of the United States and Canada, whirling disease into several fish hatcheries in the United States, and the zebra mussel into the Great Lakes. These introductions, respectively, were the accidental results of shipments of oysters from Japan and the United States east coast, salmonids from Europe (infected frozen fish used as fish food), and discharges of ballast water from ships originating in eastern Europe.

Many additions, both biotic and abiotic, could also greatly modify ecological conditions to influence reproduction, growth, and development of obscure resident forms. Thus, the sudden and abnormal appearance of large numbers of previously undetected or unrecognized endemic forms might be mistaken as the result of a "population explosion" of a newly introduced species. Arguments can be made defending the issue of enhancement versus establishment of a newly introduced species, for example, as in the increasing number of cases of sudden unusual algae dinoflagellate blooms in several locations throughout the world. Some additions, such as those that result from ocean dumping, could affect physiological, metabolic and genetic processes, i.e., gene exchanges via plasmids, that lead to a variety of effects, from the elaboration of toxins to antibiotic and heavy metal resistance, to depletion of dissolved oxygen. In turn, the final consequences of these additions to resident biota and food chain organisms could be impairment that results in abnormal morphology, function and behavior, decreased immunocompetency, and possible death. Environmental degradation could also result with possible devastating effects.

While studies are examining dispersal of agriculturally important biotic agents, including those that were genetically altered using gene splicing or transgenic techniques, most of this attention has been applied to terrestrial ecosystems and plant crops (MacKenzie et al. 1985). It is important to recognize that some genetically altered and unaltered organisms used as pest control agents in terrestrial environments may also reach aquatic environments to adversely affect non-target aquatic species. As in cases where pesticides and other toxic agents have been misapplied, their effects could lead to loss — perhaps even extinction — of biological-genetic diversity, at least in geographically confined populations. In addition to possible harm to non-target species, other non-advantageous or even advantageous effects could result from the introduction, transfer or release of genetically altered species into aquatic ecosystems. Examples might include the development of forms that have been genetically programmed to vary qualitatively and quantitatively in the following: resistance to diseases and toxins, temperature, pH, salt, oxygen tolerance, elaboration and possession of various chemical-physical-biological components and characteristics, and enhanced nutrient intake and metabolic activity. Whether introductions are compatible with the needs of society or harmful will depend on the number, rate, and method of application and their interaction with other biota and the environment. Of course, the perspectives and objectives of the individual group or organization undertaking an introduction, transfer, or release — as well as the socioeconomic, political or aesthetic evaluations by resource managers and interpretations by an informed public — will enter into judgments regarding benefits or harm.

Calculated Risks

Whereas the impression of the foregoing remarks may connote a degree of negativism about the impacts that might result from the dispersal of genetic material into aquatic environments, this is not the total intent of the papers gathered here. Just as proper use of some manufactured chemicals and other abiotic agents have been beneficial to society, so the introductions of natural and genetically altered living organisms to aquatic ecosystems can be beneficial as well. With judicious application and management. It should be possible to provide for continuity and even enhancement of genetic diversity of aquatic organisms for increased production of natural resources for food and energy sources, for manufactured products, and for help in rehabilitating despoiled and polluted environments. Indeed, several introductions can be documented as "successful," particularly in terms of establishing populations of commercially important aquatic resource species, for example, striped bass (*Morone saxatilis*) and the Japanese oyster (*Crassostrea gigas*) along the United States west coast, the European oyster (*Ostrea edulis*) on the coast of Maine, and brown trout (*Salmo trutta*) into Chile and other parts of the world.

Nothing can be accomplished without some risk; some translocations of living organisms are compelling, in terms of survival and economic necessity, regardless of the degree of risk involved. However, whether or not an organism intended for movement into an aquatic environment is an exotic or native species, genetically altered or not, considerable thought must be given to the final consequences of its translocation and establishment in a new ecosystem. Unlike terrestrial environments, field testing is not very feasible in aquatic environments and eradication programs at best are apt to be costly and difficult, if not impossible, to implement. Environmental impact evaluations and checklists comparing advantages against disadvantages or possible problems that might arise from translocations should be prepared in advance. Possible effects to the genetical, behavioral, pathological, and biochemical-physiological stability of resident biota and on ecological integrity resulting from the planned releases of biologic agents into aquatic systems must be subjected to some form of ecological risk assessment.

Ecological Risk Assessments

The scientific community, policymakers and resource managers have directed much of their attention toward predicting and understanding the fates and effects of the sources of anthropogenic toxic compounds with regard to human health. Until very recently, with a few exceptions — namely large disposal operations and contamination of food supplies — relatively little attention has been shown for assessing risks to resident biota and in determining the sources and ultimate fate and effects of living organisms introduced into aquatic ecosystems; that is, unless their perceived effects would be expected to be evident immediately, highly unusual, or clearly visible on resident resource biota and the environment. Generally speaking, however, assessments have been made on a retrospective rather than on a predictive basis. Examples of these post-event effects would be dolphin mortalities, which some have attributed to the disposal of medical and pharmaceutical wastes, and discharges of domestic sewage into areas where seafood is harvested or when dead fish wash up onto beaches.

In making risk assessments, scientific and technological information is used to estimate the ecological consequences that can result from the interaction of human activity with naturally-occurring conditions, for example, floods, storms, outages. When implementing risk management decisions, risk assessment information must be integrated to a greater or lesser degree into strategies that are designed to bring about, mitigate or avoid potential results, whether they be social, economic, political or aesthetic.

Predictive capability and risk reduction actions concerned with translocations would be improved significantly with more reliable information on the characteristics of particular organisms and the pathways they take in producing beneficial or adverse ecological effects and impacts. In particular, and as shown in Figure 2, their quantity, source and routes of entry, their processes, mechanisms, and mode of action, their interactions with other organisms and the environment, and their final effects on various species, their trophic levels and levels of organization. With this information, risk assessments resulting from deliberate and some accidental introductions and transfers of organisms from one ecosystem to another could be done with more reliable assurance, thus facilitating environmental and resource management decisions. As with chemical pollut-

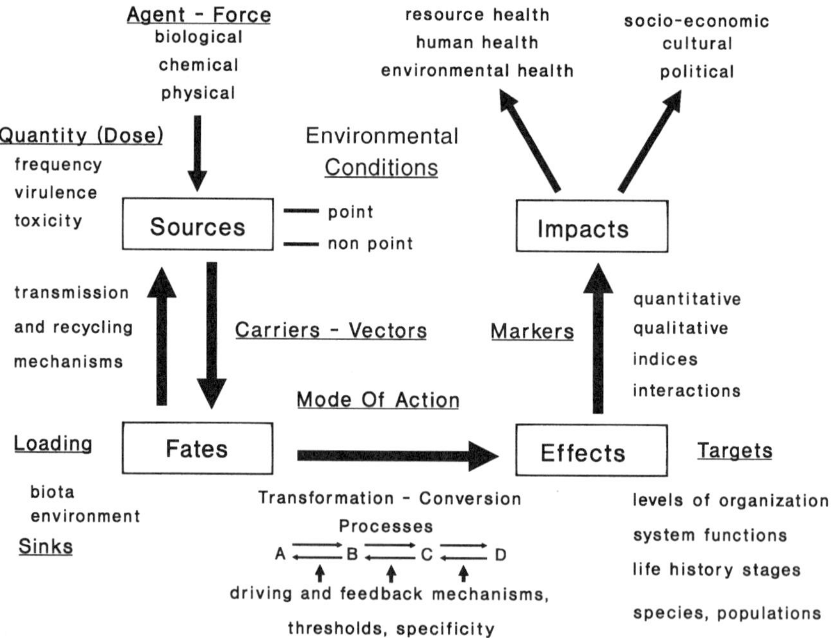

Figure 2. Information needs for ecological risk assessments.

ants, primarily toxic compounds, it seems entirely possible that parallel models can be developed more completely for dispersed living agents to assess their hazard and risk potential and for better managing their impacts on aquatic ecosystems.

In summary, seven major hypotheses are woven throughout the papers presented during the symposium and then expressed in this volume:

1. Parallels exist between biotic and abiotic contamination and pollution and in the way risks can be assessed and managed. Parallels are particularly close or overlapping in cases where introduced species have the potential to become nuisances or pathogens to humans and living resources, or adversely affect the environment.

2. Introduction of an exotic species (in some cases a transgenic form) is equivalent to translocating strange pieces of genetic information from one location to another.

3. Regardless of the amount of genetic material involved in species translocations, from molecular amounts (as with transplanted genes), virus particles and prokaryotes to much greater amounts (as present in eukaryotic, larger multicellular forms and polyploid species), the principles and factors involved in their successful adaptation and establishment in a new environment are the same.

4. The risk of successful establishment of populations byan exotic species is minimal if the introduced organisms are domesticated forms or are incapable of reproducing under the conditions that prevail in the new environment.

5. Invasions and colonization of aquatic (and terrestrial) ecosystems by exotic biological agents are continuous, naturally-occurring events. Human influences can profoundly speed up and retard the associated processes involved.

6. From the socioeconomic and cultural points of view, invasion and subsequent colonization of an aquatic ecosystem by an exotic biological agent can be dramatically disruptive. However, from a biological-evolutionary perspective ecosystems represent a continuum of changing conditions. Hence, an ecosystem will accommodate or assimilate introduced exotic biological agents and will continue to function in an altered or changed state.

7. Ecological risk assessment methodologies such as qualitative and quantitative models that estimate the degree of safety or harm to resident biota and their ecosystems require comprehensive and solidly-based scientific and technical information. Risk management approaches concerned with translocations of living agents will need to be integrated with risk assessment information if plans and actions to be taken are to culminate in desired socioeconomic, cultural and political results.

Acknowledgments

I thank Ms. J. Keller for her editorial comments and typing, Mr. J. Lewis for preparing the graphics of Figures 1 and 2, and Dr. C.J. Sindermann for helpful opinions and stimulating discussion.

Literature Cited

MacKenzie, D.R., C. S. Barfield, G. G. Kennedy, and R. D. Berger, editors, with D. J. Taranto. 1985. The movement and dispersal of agriculturally important biotic agents. Claitor's Publishing Division, Baton Rouge, Louisiana.

Mooney, H.A. and J.A. Drake, editors. 1986. Ecology of biological invasions of North America and Hawaii. Ecological Studies, Volume 58. Springer-Verlag, New York, London.

Turner, G.E., editor. 1988. Codes of practice and manual of procedures for consideration of introductions and transfers of marine and freshwater organisms. Cooperative Research Report No. 159. International Council for the Exploration of the Sea, Palaegade 2-4 1261 Copenhagen K, Denmark.

Dispersal Mechanisms: A Conceptual Framework

Dispersal of Living Organisms into Aquatic Ecosystems as Mediated by Aquaculture and Fisheries Activities

James T. Carlton

Abstract: The deliberate and accidental releases of marine, brackish, and freshwater organisms by aquaculture and fisheries industries have led to the introduction of a large but unknown number of exotic species around the world, often with profound ecological and economic consequences. Both *target* species (those intentionally transported and liberated) and *non-target* species (those that accompany target species and/or are found on or in the transported species or in the transport media) have been released in complex intercontinental patterns over hundreds of years. In addition to aquacultural and fisheries enhancement activities, the potential for species introductions by the aquarium industry and by other fishing activities (such as bait organisms and fishing vessel water wells and gear) is reviewed. These processes have led workers to underestimate the role of human-mediated dispersal, and in turn may lead to the belief that the distribution patterns of many species of animals and plants are natural. When the modern-day practices of aquaculture and fisheries industries are combined with the extent of the global movement of marine and freshwater organisms by ballast water, it is clear that the potential for exotic species to continue to invade and restructure most aquatic systems in the 1990s is staggering.

Introduction

The intercontinental dispersal of living organisms and their component genetic material into marine and freshwater ecosystems by human agency has been steadily increasing with global human migrations over the past five or more centuries. The intentional movement of human food species — and the unintentional movement of associated species — always accompany these migrations, a concomitant phenomenon far better known and documented for terrestrial than for aquatic ecosystems. While the movements of terrestrial domestic animals and plants are often recognized as part of "human geography," the movements of many aquatic species have gone, in large part, both undocumented and unrecognized, such that the significance of the alteration of natural species distributions in rivers, lakes, and coastal waters has frequently been underestimated (Carlton 1989).

We now recognize, at the close of the twentieth century, that there have been a very large number of human-mediated dispersal mechanisms affecting marine and freshwater organisms. These agencies have served to alter the natural distributions of what may be eventually recognized as thousands of species of plants and animals [see for example Dromgoole and Foster (1983); Zibrowius (1983); Leppakoski (1984); Carlton (1985, 1987, 1989); and Chapman (1988)]. The organisms accompanying human dispersal are not, of course, solely those intended for food. Transport vehicles (on land or in the water) have played a profound role in the largely accidental movement of these "shadow" species, ranging from barnacles to ants, and shipworms to rats. As cultures evolve, other species are moved for purposes of pleasure (hobbies, ornaments, non-critical food enhancement).

What specifically are these dispersal mechanisms in aquatic environments? I review here briefly most of these agencies. I then consider in detail the role of the aquaculture and fisheries industries in inter- (and in some cases intra-) continental dispersal.

Dispersal Mechanisms Other than Aquaculture-Fishery Related

Vessels (ships) have historically transported innumerable invertebrates and algae, and to a lesser extent fish, as fouling organisms on (or boring organisms in) their hulls. The role of modern oceangoing vessels in the dispersal of external fouling organisms is not clear, but is believed to be less than was historically the case for a number of reasons (summarized by Carlton and Scanlon 1985). Vessels, however, continue to play a significant role in the transoceanic and interoceanic dispersal of marine and freshwater organisms through the transport of planktonic and benthic organisms in their ballast (not bilge) water (Carlton 1985, 1989; Williams et al. 1988; Hebert et al. 1989). Red-tide causing dinoflagellates, for example, transported in ballast water, had devastating effects on aquaculture and fisheries industries in the 1980s in Australia (Hallegraeff and Sumner 1986; Hallegraeff et al. 1988).

Other maritime and aquatic activities that have served to transport species include the movements of semisubmersible exploratory drilling platforms, amphibious planes and seaplanes, recreational boats, and the release of ornamental plants and fish. Researchers have also released species for various purposes, occasionally without thought as to their potential for subsequent colonization. A dramatic example is the success of the southern Californian compound ascidian *Botrylloides diegensis,* released in 1972 in the Eel Pond in Woods Hole, Massachusetts (Carlton 1989; Carlton, in preparation). It is now one of the most abundant fouling organisms of southern New England.

Corridors built for inland and oceangoing vessels — the sea level and lock canals crossing continents and connecting previously isolated water bodies — quickly become biotic corridors as well. These have led to numerous biological invasions (Por 1978; Carlton 1985), although no global synthesis of this phenomenon is available.

Aquaculture-Fisheries Dispersal Mechanisms

The deliberate and accidental introductions of marine and freshwater organisms by fisheries and related aquaculture (mariculture) industries are second only to ships in the historical role that they may have played in the alteration of natural distribution patterns of marine and freshwater organisms. Aquaculture may now, however, be poised to rival shipping in its modern-day *potential* to accelerate the rate of introduction of exotic species.

There appears to be no formal classification available of those dispersal modes associated with the aquaculture-fisheries industries. Based upon a review of over 1,000 case histories and records, I have categorized the potential mechanisms of introduction of exotic species, as mediated by aquacultural and other fishery activities, by combining both type of release and the purpose or mechanism of introduction (Table 1). I consider and give examples of each of these here. In general no distinction is made between private (entrepreneurial), industry, and government actions.

Target Species

1. Deliberate Releases into the Environment

Target species are those intentionally moved from one locality to another. Species may be moved and then deliberately released into the environment for a number of reasons. These include placement in open waters for:

a. "Grow-out," that is, the growth of the target species to a marketable size. Depending upon the interests of the growers, there may be little or no concern relative to the potential of the species to reproduce.

b. Experimental studies, testing, for example, what the survival and growth rates of a species might be in a new geographical region.

c. Potential establishment (with or without prior experimentation to determine survival, growth, or reproductive potential). The intent of the introduction may be for human food, for forage stock for food species, for use as bait, or for biological control of pests. The introduction may represent an entirely *new spe-*

Table 1. Aquaculture and other fisheries activities: the mechanisms of introductions of exotic species.

Target Species

1. Deliberate Release into Environment
   a. Grow-out for marketing
   b. Experimental studies (survival, growth)
   c. Potential establishment (food, forage, bait, biocontrol)
      New species
      Replenishment (Restoration)
      Reestablishment ("Reintroduction")
   d. Stocking (continual)
   e. "Direct Consumption" discards
2. Accidental Escape

Non-Target Species

(Deliberate Releases and Accidental Escapes)
1. Associated Species
2. Biota on/in Target/Non-Target Species
   Epibota (epizoics, epiphytes), parasites, pathogens, diseases
3. Biota on/in Transport Media
   Transport (holding) water, detritus, sediment, algae and other dunnage (packaging materials), packing boxes

Aquarium Trade

1. Deliberate Releases

Fishing Activities

1. Movement of Bait Organisms
   a. Biota in/on transport media
2. Water Wells in Fishing Vessels
3. Fishing Gear
4. Movement of Algae as Fish Egg Substrate

cies (never before released), or the *replenishment (restoration)* of a previously existing population depressed for one reason or another (overfishing, pollution), or the *reestablishment* of a species now locally extinct (such reestablishments are frequently but erroneously referred to as "reintroductions," the species in question, almost always native, never having been introduced in the first place).

d. Stocking, usually continuous and of a high energy and economic investment. Stocking programs are generally of two categories: stocking (i) of a species known to have never reproduced in the target environment (although it originally may have been thought possible given the known biology of the species and the characteristics of the receiving environment) and (ii) of a species thought to be unable to reproduce, the management biologists believing, given the known biology of the target species and the conditions of the new environment, that it could never reproduce. Continual stocking of species (i) may be accompanied by the hope (often not officially expressed by the agency concerned) that "something" will change and the species will become established. Inversely, stocking of species (ii) is often accompanied by the assurances of the stocking agencies involved that the species poses no long-term threat to the environment, as cessation of stocking will lead to the demise of the population. Stocking programs exist for the purpose of enhancing sport fishing, or for bait or forage food stocks.

e. Disposal of unwanted food items. Species, brought in alive, and intended for direct consumption, may be subsequently released into the environment (reasons include the perception that the shipment was "spoiled," when in fact some of the individuals are alive; a stock surplus; sympathetic releases of captured animals; private curiosity as to whether a "desirable" species would live and reproduce).

Numerous workers have written extensively on the movement of target species, covering all of the above categories. Intentional introductions have been made by private individuals, by private industry, and by government agencies. Much of the available (published) literature focuses on commercial fish or invertebrate species, for example, Hanna 1966; Whitney 1967; Lachner et al. 1970; Cole 1972; Walford and Wicklund 1973; Moyle 1976a, b; Jhingran and Natarajan 1979; Mann 1979; Glude 1979; Andrews 1980; Hedgpeth 1980; Rosenthal 1980; 1985; de Groot 1985; Randall 1987; Welcomme 1988 [see also Kwain (1982) and Kwain and Lawrie (1981) relative to the introduction of the pink salmon *Oncorhynchus gorbuscha* into the Great Lakes, a species originally thought to be unable to reproduce successfully in that environment]. Far more "literature" exists in the form of unpublished documents, progress and annual reports, state fish and wildlife (game) agency memoranda, bulletins, newsletters, file reports, letters, and so forth. In short, most records of most introductions are not in the published literature.

Some species are so abundant today in the "natural environment" that it seems difficult to believe that one or a few individuals, acting on their own, could have introduced these species; again, there are generally no records of most such private introductions. One of the more spectacular marine invasions in North America is the common periwinkle *Littorina littorea,* occurring on rocky shores, in salt marshes, on mudflats, and in most other intertidal and shallow sublittoral habitats, from Labrador to (commonly) New Jersey (Carlton 1982; Vermeij 1982; Brenchley and Carlton 1983). It seems probable that British or French settlers in Eastern Canada introduced this periwinkle, with the hope of establishing it as a food item, in the early decades of the nineteenth century. It spread naturally thereafter down the Atlantic coast, and is today one of the ecologically most important intertidal species of the northwest Atlantic coast.

Monographic works on species of aquaculture importance (for example, Bardach et al. 1972; Korringa 1976a, b, c; Lutz 1980; New 1982; Kafuku and Ikenoue 1983; Morse et al. 1984; Hunter and Brown 1985; Tucker 1985; Manzi and Castagna 1986) intentionally or unintentionally provide extensive documentation of actual or potential species' movements. The species reviewed in these monographs either are, or often shortly become, those involved in international trade activities.

Perhaps the best sources, however, by which to assess modern-day movements of target species are the industry's own buyer guides. I know of no analysis of these data sets, which would provide valuable information on the historical development and dramatic increase in species availability over at least the past 20 years. In North America, *Aquaculture Magazine* issues an annual "Buyer's Guide"; for Europe, works such as those of Frimodt (1987) are available. The "1989 Buyer's Guide" lists over 25 species of marine and freshwater invertebrates, over 50 species of fish, four species of macroalgae, and three species of microalgae (a taxonomic list of species important in aquaculture lists over 50 invertebrate species). The Guide's numerous advertisements imply the availability of far more species of animals and plants "upon special request."

How many of these commercially sold species are from cultured stocks, and how many are wild-collected animals or plants resold on the aquaculture market as "seed" stock, have not, to my knowledge, been rigorously reviewed. Conversely, the existence of a rapidly increasing number of dedicated and controlled hatchery populations, creating unique genotypes, raises important questions relative to the release back into the environment of such manipulated stocks. One company, for example, "Buyer's Guide 1989," (page 135), advertises,

> "Quality Quahog Seed: Hatchery reared prime stock, *Mercenaria mercenaria*, genetically selected for fast growth, grown successfully from Texas to Maine."

Where have these "genetically selected for fast growth" seed of the clam *Mercenaria* been released back into the environment? (It would be difficult to imagine that such private releases have not occurred!) The genetic manipulation and genome-diversity reduction of aquatic food-stock species, especially fishes, dates back thousands of years (Ling 1977) and parallel the genetic domestication of land food stocks. Modern-day concerns with such practices and manipulations focus on the increasing sophistication of genetic engineering *combined with* the even more rapidly increasing ability of humans to disperse altered or modified genetic stocks globally.

2. *Accidental Escape*

Increasingly greater numbers of exotic species are held in confinement in a new region and not released directly into the environment. These confinement conditions may include some measure of quarantine (either no effluent or treated effluent released), or no quarantine at all (untreated effluent flows into the environment from the holding facility, but there is a perception that by keeping the target species confined there is less chance of escape and reproduction in the natural environment). Accidental escapes from these holding facilities are common but rarely documented (although records appear not infrequently in local newspapers).

Accidental escapes of fish species are reviewed in many of the works cited above under the releases of target species. Welcomme's (1988) monograph is a valuable synopsis. Randall (1987) notes the establishment of the silvery tilapia *(Tilapia melanotheron [= T. macrocephala])* in Oahu, Hawaii, which "somehow escaped — perhaps through a dislodged screen on an outflow pipe" (of a holding facility).

In 1988, the large penaeid shrimp *Penaeus monodon* ("giant tiger shrimp"), native to Southeast Asia, India, and Australia, accidentally escaped from the Waddell Mariculture Center, South Carolina. The Center, a state-operated research facility, imported 100,000 postlarvae in the spring of 1988 from an Hawaiian hatchery for stocking and grow-out. At some point thereafter, postlarvae may have been siphoned out passively through a drainpipe into a canal, which in turn leads to a river and thus to the Atlantic Ocean (the holding ponds have since been replumbed, and no further escapes are thought possible). The first reports of ocean-captured *Penaeus monodon* were in July 1988. Between July and

October 1988 less than 500 were captured from the open ocean over a range of 500 km extending from northern South Carolina (at Georgetown) to northern Florida (at St. Augustine). The largest ocean specimen captured was about 22 cm in length (125 g), or approximately seven months old. No ovigerous females were collected (Davis 1988; S. Hopkins, personal communication, May 1989; see also Anonymous 1988).

Other examples include the occasional appearance of young American lobsters *(Homarus americanus)* in the tidepools at Bodega Bay, California, adjacent to the Bodega Marine Laboratory (where long-term studies are undertaken on American lobsters) and of adult seahares *(Aplysia californica)* subtidally at Woods Hole, Massachusetts, adjacent to the Marine Biological Laboratory (where the species is frequently used as an experimental animal) (both cases: J. T. Carlton, personal observations). Both facilities have seawater systems draining to adjacent waters.

It would appear to be only a matter of time, given the massive movements of so many species, before accidental escapes lead to the establishment of reproducing populations of certain of the more widely transported aquaculture species of the 1980s and 1990s. High-risk candidates are found among the molluscs (particularly mussels and clams) and crustaceans (particularly shrimps and lobsters). While the above escapes have not established populations, the combination of the "right" species released at the "right" time in the "right" environment — that is, the realization of an invasion window — would appear to be inevitable.

Non-Target Species

Frequently accompanying target species are non-target species, of which three categories can be recognized. These non-target species, as with target taxa, may be deliberately or accidentally released, or accidentally escape, upon arrival at the new locality. In most cases, and in concert with target species, historical perspective on whether the introduction was deliberate or accidental is difficult to obtain (and in the long run may not matter).

1. Associated Species

Taxa others than those "requested" or "ordered" frequently accompany shipments of target species, having been added to the shipment either intentionally or unintentionally. There is little formal documentation of most such cases. Moyle (1976a, b) notes that bigscale logperch *(Percina macrolepida)* accidentally accompanied a shipment of a target species, largemouth bass *(Micropterus salmoides)*, from Texas to California, as did the rainwater killifish *(Lucania parva)* with a shipment of gamefish from New Mexico into Utah and California. Stickleback *(Gasteosteus aculeatus)* distributions in some regions may be more nearly related to the patterns of trout enhancement programs than to natural dispersal events (Moyle 1976b, p. 111). Hickey (1979) notes that the accidental inclusion of Pacific oysters *(Crassostrea gigas)* with a shipment of quahogs *(Mercenaria mercenaria)* from California to Massachusetts lead to subsequent experiments with the former species! Randall (1987) has documented the establishment of Marquesas Islands fishes, the kanda *(Valamugil engeli)* and the striped goatfish *(Upeneus vittatus)* in the Hawaiian Islands as a result of the State of Hawaii Division of Fish and Game mixing these species unintentionally with shipments of Marquesan sardines *(Sardinella marquesensis)*.

2. Biota on/in Target Species

It is typical to find an often impressive variety of smaller organisms, whose presence was either overlooked or thought to be inconsequential, on or in the target species. These include host specific or non-specific species attached to the organisms (epizoics or epiphytes), or epibiotic or endobiotic parasites, pathogens, and diseases. More species of aquatic organisms have been successfully introduced by this means than by any of the other aquaculture/fisheries mechanisms discussed here. The movement of commercial oysters around the world has been the most significant vector for hundreds of species. Pilgrim (1967), Hoffman (1970), Boschma (1972), Cole (1972), Walford and Wicklund (1973), Gruet et al. (1976), Shotts et al. (1976), Carlton (1979a, 1979b, 1985),

Rosenthal (1985) and McKenzie and Moroni (1986) provide oyster-related and many other examples. Oyster pathogens have frequently been transported with disastrous economic results (Andrews 1980). The spread of the oyster disease bonamiasis (caused by the protist *Bonamia ostreae*) with commercial oysters from the Pacific coast of the United States to Europe is documented by Bucke (1988), Elston et al. (1986) and Friedman et al. (1989). Steenbergen and Schapiro (1974) reported the "probable transplantation" of the bacterium *Aerococcus viridans* var. *homari* (formerly known as *Pediococcus homari* and before that as *Gaffkya homari*), the cause of the lobster disease gaffkemia, into at least one southern California estuary. Shipments of the American lobster *Homarus americanus* to the Pacific coast may have led to the introduction of this bacterium, which has been shown experimentally to be pathogenic to both native west American spiny lobsters *(Panulirus)* and to crabs *(Cancer)*. *A. viridans* was isolated from sediment samples adjacent to a commercial facility holding American lobsters and which had in the past used an open seawater system and outfall to an estuary (Steenbergen and Schapiro 1974). Cole (1972) cites a similar case of the introduction of *A. viridans* with American lobsters to Ireland. Lightner et al. (1983) and Rosenthal (1985) document the global, and almost instantaneous, dispersal of a lethal shrimp virus, infectious hypodermal and hematopoietic necrosis virus (IHHN), with the movements of commercial stocks of penaeid shrimp in the 1980s.

A little-known example of accidental introductions of non-target species occurred in the Salton Sea, a large inland body of salt water located in the Colorado Desert of southern California. In 1957 a Texas marine angiosperm, "shoal grass" *(Diplanthera wightii)*, was introduced into the Salton Sea to provide food for waterfowl, and along with it "an unknown number of many species of invertebrates were introduced unintentionally" (Linsley and Carpelan 1961). The successful introduction of a common Texas amphipod, *Gammarus mucronatus*, "which appears to be the principal food of the sargo" (the fish *Ancisotremus davidsoni*) in the

Salton Sea resulted from this action, as well as of another amphipod, *Corophium louisianum* (Barnard and Gray 1968). Intracontinental and global movements of maritime, emergent, and other coastal vegetation have been extensive; the number of small invertebrates simultaneously moved must be profound.

Djajasasmita (1982) and Taylor (1966) provide evidence for the dispersal of the glochidia larvae of freshwater mussels *(Anodonta)* with the dispersal of introduced fish stocks. This has been a little-examined dispersal mechanism to explain modern-day distribution patterns of these mussels.

There remains, however, no systematic study of the diversity or abundance of organisms now being dispersed with what may be, on a daily basis, thousands of shipments of invertebrates, fish, and plants, around the world. While many of these shipments now either leave the point of origin with "health and disease" certificates, or are examined at the receiving point, there remains a significant potential for the accidental transport of associated species, symbionts, parasites, and disease agents. There are at least five reasons for this situation: One, only a small subsample (ten percent or less) of any shipment is examined. Two, these subsamples are searched for only certain species which have been preidentified as those of special concern (usually known predators, parasites, or pathogens). Three, histochemical and histopathological examinations may not detect certain pathogens in seasonal resting stages (Hickey 1979). Four, as reviewed below (p. 28) [Note refer. to B-3], many species may occur in the transport media (water, packaging) which may not be subject to any level of inspection. And five, species that are believed to be of no concern, but are found with the target species upon inspection, may be seen but passed over.

Thus, for example, it is well known in the commercial oyster trade that movements of seed stock are still accompanied by species believed to be "harmless"—these may include herbivorous snails, hydroids, sponges, bryozoans, ascidians, and other small mobile, encrusting, or attached organisms. I have been shown living gastropods, including small trochids (*Calliostoma* sp.) and slipper shells (*Crepidula* sp.), that accompanied shipments of the oyster *Ostrea edulis* from Sendai, Japan, and Maine, respectively, received at Moss Landing, Monterey Bay, California (D. Shonman, personal communication).

Hickey (1979) noted that 100,000 cultchless, hatchery-reared oysters *(Crassostrea gigas)* were obtained from a hatchery at Moss Landing, California, "after being examined histologically and certified 'free of disease.'" These oysters were then placed in the open environment in a salt pond on Cape Cod, Massachusetts. What might have accompanied these oysters?

In 1979 R. Mann received a shipment from Moss Landing of 3,000 hatchery-reared *Crassostrea gigas* (25 to 35 mm in length) at the Woods Hole Oceanographic Institution. Shipping time was 48 hours. These seed were accompanied by two government inspection certificates: one stating that ten percent of the shipment had been inspected and "no oyster drilling snails, egg cases, or portions thereof" were found, and a second indicating that (based upon an unstated number of specimens examined), "no evidence of parasites or diseases was observed in these specimens." I used three methods to examine this shipment of seed oysters: 100 oysters each from (1) the top and (2) the bottom of the container were thoroughly rinsed in filtered seawater, and any associated debris or species retained. In addition, (3) the surfaces of the shells of 25 oysters from the bottom of the container were examined microscopically. The entire container was inspected in general for any other associated species that might have been missed by the first three sampling methods. The results of this inspection are shown in Table 2. The most common organism was a small amphipod of the genus *Corophium* (up to 3 mm in length) represented by juveniles, males, and ovigerous females; many of the latter appeared freshly "spent," and many of the juveniles were very small (less than one mm in length). If samples (1) and (2) are representative, more than 1,000 specimens of this amphipod accompanied this shipment of oyster seed. *Corophium* also occurred in mud tubes under flutes on the oyster valves. The remaining species were represented in samples (1) and (2) by relatively few specimens of the mussel *Mytilus "edulis"* (5 to 10 mm in length), an additional gammarid amphipod, two species of polychaete worms (5 to 10 mm in length), and fragments of the green alga *Enteromorpha* and of a small hydroid. [I refer to *Mytilus "edulis"* in quotation marks at some points in this paper (and not at others) depending upon the geographic region involved: see McDonald and Koehn 1988]. No species occurred in sample (3), nor in the general inspection, that were not found in samples (1) and (2). In all, seven species of invertebrates and algae were found. Russell (see below) has found 29 additional species of algae, diatoms, protozoans, and invertebrates in water accompanying oyster and clam shipments from Moss Landing.

Table 2. Species and numbers of individuals associated with hatchery-reared Pacific oysters *(Crassostrea gigas)* shipped from California to Massachusetts (+ = present).

Species	*Sample (see text)* (1)	(2)	(3)
Mollusca:Bivalvia			
Mytilus "edulis"	2	2	+
Annelida:Polychaeta			
Opheliidae			
Armandia brevis	4	2	+
Nereidae			
Undetermined sp.	1	—	1
Crustacea:Amphipoda			
Corophium sp.	31	16	1
Undetermined sp.	—	1	—
Coelenterata:Hydrozoa			
Undetermined hydroids	—	+	+
Chlorophyta			
Enteromorpha sp., cf. *E. intestinalis*	+	+	+

The release of 100,000 California-reared Japanese oysters on Cape Cod thus also released thousands of small protists, invertebrates and algae, and perhaps endobiotic organisms as well. Ten months after the oysters were placed in this tidal pond, the outlet was blocked to prevent the escape of oyster larvae (Hickey (1979, p.133)) in anticipation of the spawning season—too late, however, for the escape of associated species.

It is thus clear that hatchery-reared stocks of species shipped with "quarantine" or "health" certificates are not necessarily free of other associated organisms. The ease of availability of so many species underscores the potential breadth of this phenomenon. What species accompany the shipments of the algae *Laminaria, Porphyra,* and *Macrocystis,* that can now be easily ordered? How many epizoic, microscopic algae, peritrichous protozoans, and rotifers, or endobiotic commensal protozoans, can be found on or in the Australian crayfishes (Astacidae) *Cherax destructor* ("yabbies") and *Cherax tenuimanus* ("marrons"), now shipped all over the world?

A very large number of species are transported around the world alive, intended for direct human consumption and not for introduction. Once imported there are few subsequent controls; these species can also be easily released in a new region. Living New Zealand "green mussels," *Perna canaliculus,* have been shipped on a weekly, if not daily, basis from New Zealand to California since at least 1983. The shells and the dense byssal mats of these mytilids provide excellent surfaces for epizoic, nestling, and even boring species. I have examined freshly killed specimens of these mussels (provided to me by Michael Graybill; material shipped to a Eugene, Oregon, restaurant), upon which were recently living balanoid barnacles (*Balanus trigorus*), hydroids, serpulid polychaete tubeworms, boring spionid polychaetes (*Polydora* sp.), and folliculinid protozoans. The widespread availability of this mussel in the restaurant and seafood industry suggests that little could prevent the release of these mussels (and their epibionts) in, for example, southern California bays. Oyster farmers in Oregon have inquired about the feasibility of open release or grow-out of *Perna canaliculus* in southern Oregon estuaries.

3. Biota on/in Transport Media

A wide variety of non-target organisms may be associated with the transport media (such as water, packaging [algae or other dunnage], detritus), or with the shipping container itself. If the transport medium includes algae or other marine plants, these species are themselves potential introductions, often being discarded into the new environment. An example of the latter may be the successful establishment in Washington and Oregon of saltmeadow hay *(Spartina alterniflora)* used as packing material with Atlantic oysters *(Crassostrea virginica)* (Frenkel and Boss 1988, and references therein). As discussed below the New England seaweed *Ascophyllum* is regularly discarded into San Francisco Bay, California.

Many species of freshwater, estuarine, and marine fish have been transported within and between all continents (Bardach et al. 1972; Ling 1977; Moyle 1976a, b; McNeil 1979; Vaini 1985; Welcomme 1988). These fish were transported in barrels, tanks, and aquarium railroad cars. There is no doubt that the water in which these fish were transported carried protists, zooplankton and phytoplankton. Stone (1876) noted the presence of "minute forms of life" (i.e., zooplankton) as fish food occurring in the water used to transport fish from the American eastern seaboard to California. Hazel (1966) has suggested that the eastern American freshwater polychaete *Manayunkia speciosa* may owe its presence in Oregon and in the Sacramento-San Joaquin River delta of California to transport in water associated with catfish *(Ictalurus)* introductions from the northeastern United States, including catfish (and water) taken directly from *Manayunkia*'s type locality in Pennsylvania.

Belk (1973) found the freshwater waterflea (cladoceran) *Latonopsis australis* in a man-made reservoir in Guam, and noted its probable introduction with fish from Hawaii. The presence of the marine isopod *Sphaeroma serratum* and the gammarid amphipod *Gammarus aequicauda* in an Egyptian inland salt lake is believed by Holdich and Tolba (1985) to be related to a fish restocking program, and the concomitant movement of water and algae with these fish from the Mediterranean. Shotts et al. (1976) found 14 genera of bacteria associated with the shipping water of aquarium fishes imported from Southeast Asia to Georgia. These cases must represent a very minor fraction of actual introductions and transportation events.

Russell (1981, and personal communication, 1989) cultured 18 species of macroalgae and microalgae, seven species of protozoans, and copepods from water from oyster and clam shipments received in Hawaii from Moss Landing, California. In addition, he collected rotifers, nematodes, and isopods from this water. Russell further cultured six species of red, green, and brown algae from water accompanying tropical fish and coral shipments from Canton Island in the central Pacific Ocean, East Southeast of the Gilbert Islands. For many years it was common practice in Hawaii to place shipments of fish and coral received from throughout the Indo or South Pacific in holding tanks whose non-treated effluent flowed out onto adjacent reefs (D. Russell, personal communication, 1989).

Given the widespread movement of freshwater fish across the continental United States, the modern-day distributions of many invertebrates — such as the waterfleas *Daphnia* spp. and gammarid amphipods — should be examined in light of nineteenth-century railroad routes, as well as relative to natural means of dispersal. How many estuarine and marine organisms may have been distributed in shipping waters is equally unknown.

Living organisms intended for deliberate release or direct consumption are often packed in seaweeds or other plants. In North America beginning in 1888, for example, shipments of Atlantic lobsters *(Homarus americanus)* for release on the Pacific coast were packed in "rockweed," the brown alga *Fucus* (Rathbun 1890).

The transcontinental shipping of lobsters from New England to California resumed in the 1960s, with air flights of lobsters for the restaurant trade (Carlton 1979a). These lobsters are packed in the brown algae *Ascophyllum* and *Fucus.* Miller (1969) and Dawson and Foster (1982) have documented that such algae may be discarded into San Francisco Bay. Miller lists more than 20 species of invertebrates, including sponges, hydroids, flatworms, spirorbid polychaetes, barnacles, amphipods, snails, mussels, bryozoans, and seastars, associated with these imported algae. The common Atlantic periwinkle *Littorina littorea* has appeared on occasion in San Francisco Bay (Carlton 1969), and may still be there, as a result of these algal discards. Intertidal brown seaweeds are also used as packing for baitworms (p. 32).

Packing containers themselves are unique mechanisms of dispersal in the aquaculture and commercial fisheries trades, although this possibility has rarely been explored. Quayle (1964) suggested that the wood-boring gribble (isopod) *Limnoria tripunctata* may have been introduced from Japan to British Columbia in the wooden boxes in which Japanese seed oysters *(Crassostrea gigas)* are packed and shipped. Popham (1983) has made the same suggestion for the establishment of a Japanese species of shipworm *(Lyrodus takanoshimensis)* in British Columbia.

Aquarium Trade

Hundreds of species (there are no definitive lists) of invertebrates, fish, other vertebrates, seaweeds, and other plants, have been and are now moved in complex patterns through the commercial aquarium industry. The increasing ability to move these species at faster speeds around the world has resulted in the survival of a far greater variety of transported species than ever before. Exotic taxa can now arrive alive (and potentially be released) within 12 to 24 hours at virtually any point in the world. Few countries have established adequate controls over such movements. In turn, there is no control over the fate of such species once sold on the public market. The public may purchase such species with the direct desire to release them (in some cases out of curiosity as to the likelihood of success of such species in establishing a population, and in other cases out of sympathy, the more so in the 1980s and 1990s as a result of animal rights and welfare concerns). More commonly such purchases are made with the desire of holding the organisms as pets; pet owners may then later release their pets for one reason or another.

Aquarium-released fish are considered by Moyle (1976a, 1976b), de Groot (1985), and Welcomme (1988), among others. Courtenay (1978) describes the remarkable case of the introduction of the Southeast Asian walking catfish *(Clarias batrachus)* into Florida by the fish escaping from the transport truck as it was driven from the airport to the aquarium farm: "(the driver) proceeded on his journey north while Walking Catfish were falling and jumping off the truck onto the road next to a major drainage canal."

Fishing Activities

The overall picture of fishery and related activities in the movement of fish, shellfish, and other species must include four additional categories of dispersal mechanisms, some rarely considered but potentially of equal importance to some of the other mechanisms discussed here.

1. Movement of Bait Organisms

The collection, transport, and release of invertebrates and small fish for bait have doubtless resulted in the redistribution of far more species than has been realized. There are virtually no controls on such movements. The private fisherman, purchasing or collecting organisms at a "bait shop" or in one lake or river, can release the same species in a lake or river scores or hundreds of kilometers away within hours or days. One of the best-known modern examples is the distribution of the Asian freshwater clam *Corbicula fluminea* throughout the United States; its rapid dispersal was due, at least in part, to its spread by fishermen as a bait organism (Counts 1986). The introduction of the goldspot herring *(Herklotsichthys quadrimaculatus)* from the Marshall Islands to Hawaii was apparently the result of these fish having been taken aboard as bait for tuna fishing (Randall 1987). Clark (1932) recorded six species of fish found alive in bait tanks of vessels returning from tuna fishing off Mexico, all of which were released into Los Angeles Harbor or nearby waters.

A corollary to such movement is the dispersal of organisms in the transport medium (usually water [which may contain zooplankton and phytoplankton] or packing materials) from the source region. The practices of the marine worm bait industry, such as those located in Maine (Sandrof 1946; Pettibone 1963; Dow and Creaser 1970; Creaser and Clifford 1986) provide an excellent example. Worms as bait (family Nereidae, "sandworms" or "pileworms," and family Glyceridae, "bloodworms") are regularly shipped from Maine to many locations around the country. I have examined samples of brown algae *(Fucus vesiculosus* and *Ascophyllum nodosum)* shipped with worm bait from Maine to Newport Bay in southern California (material provided by M. E. Smith and W. Wagg). In the algae were three common New England periwinkle species *(Littorina saxatilis, Littorina obtusata,* and *Littorina littorea),* the mussel *Mytilus edulis,* isopods, and gammarid amphipods. A marine ascomycete (fungus), *Pleospora* sp., also occurred on the algae (J. J. Kohlmeyer *in litt.* to R. B. Setzer). These algae are commonly discarded in the bay. In turn, living *Littorina littorea* have on occasion been found in Newport Bay (Carlton 1979a). Carlton and Scanlon (1985) and Dawson and Foster (1982) suggest that the green alga *Codium fragile tomentosoides* may have been transported to Virginia and to San Francisco Bay, California, respectively, in worm bait packing.

2. *Water Wells in Fishing Vessels*

Fishing vessels have long moved living organisms in "live wells" in their holds. Some species are brought back to land for the purpose of live-selling at market; others, as described above, are held as live bait for fishing (and later released when no longer wanted). Species taken up with "target" species are deliberately or inadvertently often moved as well (examples cited above). These wells may contain plankton from the source region, or a mixture of plankton taken up en route from different sites. Wolff (1977) documented the transport to and release in Europe of horseshoe crabs *(Limulus polyphemus)*, held in live wells, by East European fishermen returning from fishing grounds off Atlantic North America en route back to their Baltic home ports.

3. Fishing Gear

Fishing vessels' nets are an obvious means for transportation and subsequent release of living organisms over long distances. Nets piled on deck and later streamed out for cleaning in distant waters may release a wide variety of species, especially many small invertebrates and algae. Bottom otter trawls frequently come aboard with a broad array of invertebrates entangled in the netting, ranging from bits of sponge and hydroid colonies to bryozoans and ophiuroids. Some species may survive in these nets (and in the vessel's scuppers) for at least 48 hours (J. T. Carlton, personal observations, Atlantic coast of North America), sufficient time for a modern fishing vessel to make considerable progress along a coastline. Carlton and Scanlon (1985) propose that the green alga *Codium fragile tomentosoides* may have been transported around Cape Cod by this (among other) means. The seagrass *Halophila stipulacea*, a euryhaline subtropical species native to the Indo-Pacific Ocean, may similarly have been transported through the Suez Canal to the Mediterranean Sea on fishing nets (Lipkin 1972).

The fishing gear of sport fishermen, including tackle, poles, hand nets, containers, boots, waders, and so forth, are ideal sites for the potential entrainment of many small organisms, from protozoans and cladoceran ephippia to bryozoan statoblasts and algae. These may impinge upon local, lake-to-lake distributions, rather than long-distance dispersal. Lange and Cap (1986) suggested that a European freshwater cladoceran *(Bythotrephes)* may have been introduced to the Great Lakes with such private fishing equipment; in this case, however, *Bythotrephes* was more likely introduced by ballast water.

4. Movement of Algae as Substrate for Fish Eggs

Certain egg fisheries are sufficiently lucrative to have led to the experimental placement of substrates for the enhancement of egg laying. An example is the herring egg industry of Alaska and the Pacific Northwest coast of North America (Krakauer 1986).

As an example, in 1986 the Oregon Department of Fish and Wildlife issued "experimental fishing gear" permits to several private individuals to place the brown kelp *Macrocystis pyrifera* on racks in Coos Bay, Oregon, as a potential substrate for egg deposition by herring (the kelp and eggs then to be harvested together). The *Macrocystis* was brought from Avila Beach (San Luis Obispo County), California, just north of Point Conception, being kept cold during the 24 hour transit. I sampled the kelp upon its arrival in Coos Bay, and recovered over 40 species of invertebrates and algae, including alloeocoel flatworms on the eggs of the kelp crab *Pugettia producta*, the bryozoan *Membranipora membranacea*, several species of shelled gastropods and nudibranchs, copepods, barnacles (on crabs), amphipods, isopods, shrimps, and mites. All of these species were released directly into Coos Bay.

Discussion

The uncontrolled release of exotic species into the open environment, by individuals involved in aquaculture and other fisheries industries, is realized on a daily basis around the world. An individual or organization can today request 50,000 individuals of "Target Species A," to be shipped, for example, from the Philippines to Hawaii. The container arrives by air, assuring maximum survival of all taxa. In the container is the target species — but perhaps only 49,000 are the requested species and, by mistake or inattention, 1,000 are another similar (non-target) species. On these taxa, which may or may not have passed through inspection or quarantine, may be several species of epibionts; several hundred specimens of "Target Species A" may have a small symbiont or parasite as well, easily overlooked. In the water in the container may be several dozen species of microscopic zooplankton and phytoplankton. The contents of the entire container, target species, non-target species, epibiota, and biota in the transport medium, can be emptied into the open sea.

It is difficult to judge the immediate significance of such an event, or to place it fully in the context of the many other mechanisms that are now in place and that serve instantaneously to alter hundreds of thousands to millions of years of genetic isolation. The release of a single target species would appear to be a potentially minor occurrence, on simple numerical grounds (although perhaps not on ecological grounds), in light of the thousands of species that likely have been and are now being released through the discarding of live food, the release and escape of non-target biota, or the movement of untold quantities of bait organisms, and the organisms with which they are packed. These difficulties, and appearances, are exacerbated by the lack of sufficient data sets by which to judge the sizes of these waves upon waves of exotic species. When one combines the dispersal events I have outlined here with the global movement of ballast water, and the vast numbers of planktonic and benthic organisms being transported, released, and inoculated by ships on a daily basis in bays, estuaries, and harbors, the potential for exotic species to continue to invade and restructure most aquatic systems in the 1990s is staggering.

Evolutionary biologists, biogeographers, and natural historians have, if not a genetic, at least a well-ingrained belief that the distributions of most organisms on the Earth's surface are reflections of long-term, natural processes. This well-anchored assumption is clearly reflected in most biogeography texts and monographs and in most discussions of animal or plant dispersal mechanisms in ecology texts. And yet it would appear difficult to understand the patterns of distribution of many continental aquatic organisms without understanding the patterns of movement of the vast amounts of water used to transport the fish species and stocks that were endlessly manipulated "to improve nature" over the past several centuries. It would appear equally difficult to understand the patterns of distribution of many coastal marine and estuarine organisms without understanding the historical and modern-day traffic patterns of the small fishing vessels that have plied from harbor to harbor over the centuries, with fish, algae, and invertebrates in their holds, nets, and on their hulls.

The ecological and genetic ramifications of the introductions of exotic species have been considered in recent years by many authors (for example, Hornberg and Williamson 1986; Mooney and Drake 1986; Drake et al. 1989), although most of these works largely focus upon terrestrial communities. Seemingly innocuous movements of young hatchery oysters, or seaweed with bait, could lead to profound ecological effects. The transcontinental movements of the amphipod *Corophium*, noted above, could predictably lead to the establishment of new species or new populations; *Corophium* are critical components in many estuarine food chains, including those involving salmonid fishes (J. Chapman, personal communication, 1989). The wide geographic transfers of predators and competitors, pathogens, parasites and pests, by the ten or so now active aquaculture and fisheries related dispersal mechanisms reviewed here, have a predictably significant statistical chance of leading to the establishment of yet more exotic species that could have profound biological and ecological effects on the invaded communities.

On a more specific level, quantitative documentation of the many ways in which certain species are transported are fundamental to understanding the population biology and genetics of such taxa. How the mussel *Mytilus edulis* has been and is now being transported around the world would appear to be a critical foundation to the many genetic studies now being undertaken with this species group (Edwards and Skibinksi 1987; McDonald and Koehn 1988; Varvio et al. 1988; Gosling 1989). Atlantic *Mytilus* are released into California, and California *Mytilus* are shipped to the Atlantic (records above). Japanese *Mytilus* are released, probably on a daily basis, as veliger larvae in ballast water, on the North American Pacific coast (Carlton et al. 1990; J. Geller and J. Carlton, in preparation), if not throughout the Pacific Rim. Farm-raised, hatchery-raised, or wild-collected *Mytilus* are presumably available for purchase in many areas of the world on the open market, shippable (and releasable) anywhere. *Mytilus* is also one of the world's most common ship-fouling organisms (Woods Hole Oceanographic Institution 1952). Most harbor populations of *Mytilus* around the world remain to be examined genetically, either for their amount and nature of allozyme divergence, or for their variation in mitochondrial DNA. It is predictable that introduced populations of the mussels *Mytilus trossulus* and the Atlantic *Mytilus edulis* (both species as defined by McDonald and Koehn 1988 and Varvio et al. 1988) eventually will be recognized at widespread localities around the world, as have introduced populations of the Mediterrranean mussel *Mytilus galloprovincialis* (McDonald and Koehn (1988)).

"In contrast to land and fresh waters," wrote Charles Elton (1958) concerning biological invasions, "the sea seems still almost inviolate." It is becoming increasingly clear, some 30 and more years later, that that contrast has all but faded: the human role in the alteration of species compositions, and of the evolutionary role of gene flow, in the ocean — and in all aquatic ecosystems — has been, and is, as pervasive as on land. The creation and release of genetically manipulated species have and will fundamentally contribute to this pervasiveness. Strategies for reducing the rate of alteration due to invasions in both natural environments and in those environments already invaded (Hubbs 1977; Courtenay and Taylor 1986; Sindermann 1986; Welcomme 1986; Carlton 1989; and the chapters in the present volume) will require the same national and international cooperative efforts that have led to the partial reductions in the discharge of organic and inorganic pollutants into water bodies or into the atmosphere. Perhaps such a goal can be achieved by the beginning of the twenty-first century.

Acknowledgments

I am grateful to Aaron Rosenfield for the invitation to participate in the "Aquaculture 89" symposium held in Los Angeles, California, in February 1989. The work reported here was begun while I was a Postdoctoral Scholar in Biology at the Woods Hole Oceanographic Institution, Woods Hole, Massachusetts. For valuable information and materials I am indebted to John Chapman, Douglas Conklin, Steve Hopkins, Roger Mann, Aaron Rosenfield, Dennis Russell, Robert Setzer, David Shonman, Carl Sindermann, and my colleagues of the International Council for the Exploration of the Sea's (ICES) "Working Group on Introductions and Transfers of Marine Organisms." The oyster-shipment intercept data were originally presented in an unpublished ICES paper at Woods Hole in 1981. The kelp-shipment intercepts in Oregon were made possible by the field assistance of Jim Reed, Lisa Haggblom, Jan Hodder, and Debby Carlton. The mussel-shipment intercepts were made possible by the inspiration of Michael Graybill; the bait-kelp intercepts were facilitated by Mary Smith and Winifred Wagg. I am indebted to Dr. William Newman, Scripps Institution of Oceanography, for identifying some New Zealand barnacles.

This paper is dedicated to the late Robert C. Terwilliger, Professor of Biology, University of Oregon, and Director, Oregon Institute of Marine Biology. Bob, who was a colleague, co-teacher and friend, died while I was away attending this symposium.

Literature Cited

Andrews, J.D. 1980. A review of introductions of exotic oysters and biological planning for new importations. Mar. Fish. Rev. 42:1-11.

Anonymous. 1988. Waddell Center: emphasis on research and service to fish farmers in state. Aquacult. Mag. 14:49-59.

Bardach, J.E., J.H. Ryther and W.O. McLarney. 1972. Aquaculture. The farming and husbandry of freshwater and marine organisms. Wiley-Interscience, New York.

Barnard, J.L. and W.S. Gray. 1968. Introduction of an amphipod crustacean into the Salton Sea, California. Bull. So. Calif. Acad. Sci. 67:219-232.

Belk, D. 1973. The Cladocera of Guam. Crustaceana 24:146-147.

Boschma, H. 1972. On the occurrence of *Carcinus maenas* (Linnaeus) and its parasite *Sacculina carcini* Thompson in Burma, with notes on the transport of crabs to new localities. Zool. Meded. 47:145-155.

Brenchley, G.A. and J.T. Carlton. 1983. Competitive displacement of native mud snails by introduced periwinkles in the New England intertidal zone. Biol. Bull. 165: 543-558.

Bucke, D. 1988. Pathology of bonamiasis. Parasitol. Today 4:174-176.

Carlton, J.T. 1969. *Littorina littorea* in California (San Francisco and Trinidad Bays). The Veliger 11: 283-284.

Carlton, J.T. 1979a. History, biogeography, and ecology of the introduced marine and estuarine invertebrates of the Pacific coast of North America. Ph.D. dissertation, University of California, Davis.

Carlton, J.T. 1979b. Introduced invertebrates of San Francisco Bay, p. 427-444. *In* T.J. Conomos (ed.) San Francisco Bay: The urbanized estuary. American Association for the Advancement of Science, Pacific Division, California Academy of Sciences, San Francisco.

Carlton, J.T. 1982. The historical biogeography of *Littorina littorea* on the Atlantic coast of North America, and implications for the interpretation of the structure of New England intertidal communities. Malacolog. Rev. 15:146.

Carlton, J.T. 1985. Transoceanic and interoceanic dispersal of coastal marine organisms: the biology of ballast water. Oceanogr. Mar. Biol. Annu. Rev. 23: 313-371.

Carlton, J.T. 1987. Patterns of transoceanic marine biological invasions in the Pacific Ocean. Bull. Mar. Sci. 41: 452-465.

Carlton, J.T. 1989. Man's role in changing the face of the ocean: biological invasions and implications for conservation of near-shore environments. Conserv. Biol. 3: 265-273.

Carlton, J.T. and J.A. Scanlon. 1985. Progression and dispersal of an introduced alga: *Codium fragile* ssp. *tomentosoides* (Chlorophyta) on the Atlantic coast of North America. Bot. Mar. 28:155-165.

Carlton, J.T., J.K. Thompson, L.E. Schemel and F.H. Nichols. 1990. The remarkable invasion of San Francisco Bay (California, USA) by the

Asian clam *Potamocorbula amurensis* (Mollusca). I. Introduction and dispersal. Mar. Ecol. Prog. Ser. 66:81-94.

Chapman, J.W. 1988. Invasions of the Northeast Pacific by Asian and Atlantic gammaridean amphipod crustaceans, including a new species of *Corophium*. J. Crust. Biol. 8:364-382.

Clark, F.N. 1932. Introduction of Mexican fishes into Southern California waters. Calif. Fish Game 18:61-62.

Cole, H.A. 1972. Report of the working group on introduction of non-indigenous marine organisms. International Council for the Exploration of the Sea, Cooperative Research Report 32, 59 pp.

Counts, C.L. 1986. The zoogeography and history of the invasion of the United States by *Corbicula fluminea* (Bivalvia: Corbiculidae). Am. Malacolog. Bull. Spec. Ed. 2:7-39.

Courtenay, W.R. 1978. The introduction of exotic organisms, p. 237-252. *In* H.P. Brokaw, ed., Wildlife and America. Council on Environmental Quality, U.S. Government Printing Office, Washington, D.C.

Courtenay, W.R. and J.N. Taylor. 1986. Strategies for reducing risks from introductions of aquatic organisms: a philosophical perspective. Fisheries 11:30-33.

Creaser, E.P. and D.A. Clifford. 1986. The size frequency and abundance of subtidal bloodworms (*Glycera dibranchiata* Ehlers) in Montsweag Bay, Woolwich-Wiscasset, Maine. Estuaries 9:200-207.

Davis, J. 1988. Escapes by sea and air: giant shrimp loose along coast. The Atlanta Journal and Constitution, Atlanta, Georgia, October 18, 1988, p. 81.

Dawson, E.Y. and M.S. Foster. 1982. Seashore plants of California. University of California Press, Berkeley.

de Groot, S.J. 1985. Introductions of non-indigenous fish species for release and culture in the Netherlands. Aquaculture 46:237-257.

Djajasasmita, M. 1982. The occurrence of *Anodonta woodiana* Lea, 1837 in Indonesia (Pelecypoda: Unionidae). The Veliger 25:175.

Dow, R.L. and E.P.Creaser. 1970. Marine bait worm, a valuable inshore resource. Marine Resources of the Atlantic Coast, Atlantic States Marine Fisheries Commission, Leaflet 12, 4 pp.

Drake, J.A., H.A. Mooney, F. di Castri, R.H. Groves, F.J. Kruger, M. Rejmanek and M. Williamson. 1989. Biological invasions, a global perspective. John Wiley and Sons.

Dromgoole, F.I. and B.A. Foster. 1983. Changes to the marine biota of the Auckland harbour. Tane 29:79-96.

Edwards, C.A. and D.O.F. Skibinksi. 1987. Genetic variation of mitochondrial DNA in mussel (*Mytilus edulis* and *Mytilus gallopro-*

vincialis) populations from southwest England and south Wales. Mar. Biol. 94:547-556.

Elston, R.A., C.A. Farley and M.L. Kent. 1986. Occurrence and significance of bonamiasis in European flat oysters, *Ostrea edulis* in North America. Dis. Aquat. Organ. 2: 49-54.

Elton, C.S. 1958. The ecology of invasions by animals and plants. Methuen and Co., Ltd., London.

Frenkel, R.E. and T.R. Boss. 1988. Introduction, establishment and spread of *Spartina patens* on Cox Island, Siuslaw Estuary, Oregon. Wetlands 8:33-49.

Friedman, C.S., T. McDowell, J.M. Groff, J.T. Hollibaugh, D. Manzer and R.P. Hedrick. 1989. Presence of *Bonamia ostreae* among populations of the European flat oyster, *Ostrea edulis* Linne, in California, USA. J. Shellfish Res. 8:133-137.

Frimodt, C. 1987. The European fishing handbook. Directory of the European fish trade. Osprey Books, Huntington, New York.

Glude, J.B. 1979. Oyster culture — a world review, p. 325-332. *In* T.V.R. Pillay and W.A. Dill (eds.), Advances in aquaculture. FAO, Fishing News Books Ltd., England.

Gosling, E.M. 1989. Genetic heterozygosity and growth rate in a cohort of *Mytilus edulis* from the Irish coast. Mar. Biol. 100:211-215.

Gruet, Y., M. Heral and J.-M. Robert. 1976. Premieres observations sur l'introduction de la fauna assocée au nassain d'huitres japonaises *Crassostrea gigas* (Thunberg), importe sur la cote Atlantique française. Cah. Biol. Mar. 17: 173-184.

Hallegraeff, G.M., C. Bolch, B. Koerbin and J. Bryan. 1988. Ballast water a danger to aquaculture. Aust. Fish. 47:32-34.

Hallegraeff, G.M. and C.E. Sumner. 1986. Toxic plankton blooms affect shellfish farms. Aust. Fish. 45:15-18.

Hanna, GD. 1966. Introduced mollusks of western North America. Occ. Pap. Calif. Acad. Sci. 48, 108 pp.

Hazel, C.R. 1966. A note on the freshwater polychaete, *Manayunkia speciosa* Leidy, from California and Oregon. Ohio J. Sci. 66:533-535.

Hebert, P.D. N., B.W. Muncaster and G.L. Mackie. 1989. Ecological and genetic studies on *Dreissena polymorpha* (Pallas): a new mollusc in the Great Lakes. Can. J. Fish. Aquat. Sci. 46:1587-1591.

Hedgpeth, J.W. 1980. The problem of introduced species in management and mitigation. Helgol. Meeresunters. 33:662-673.

Hickey, J.M. 1979. Culture of the Pacific oyster, *Crassostrea gigas,* in Massachusetts waters, p. 129-148. *In* R. Mann (ed.), Exotic species in mariculture. MIT Press, Cambridge, Massachusetts.

Hoffman, G.L. 1970. Intercontinental and transcontinental dissemination and transfaunation of fish parasites with emphasis on whirling disease *(Myxosoma cerebralis).* Amer. Fish. Soc. Spec. Publ. 5:69-81.

Holdich, D.M., and M.R. Tolba. 1985. On the occurrence of *Sphaeroma serratum* (Isopoda, Sphaeromatidae) in an Egyptian inland salt lake. Crustaceana 49:211-214.

Hornberg, H. and M.H. Williamson, editors. 1986. Quantitative aspects of the ecology of biological invasions. Philos. Trans. R. Soc. Lond. B, Biol. Sci. 314:501-746.

Hubbs, C. 1977. Possible rationale and protocol for faunal supplementations. Fisheries 2:12-14.

Hunter, J.V. and E.E. Brown, editors. 1985. Crustacea and mollusk aquaculture in the United States. AVI Publishing Company, Inc., Westport, Connecticut.

Jhingran, V.G. and A.V. Natarajan. 1979. Improvement of fishery resources in inland waters through stocking, p. 532-541. *In* T.V.R. Pillay and W.A. Dill (eds.), Advances in Aquaculture. FAO, Fishing News Books, Ltd., London.

Kafuku, T. and H. Ikenoue, editors. 1983. Modern methods of aquaculture in Japan. Elsevier, Amsterdam, The Netherlands.

Korringa, P. 1976a. Farming the cupped oysters of the genus *Crassostrea.* Elsevier, Amsterdam, The Netherlands.

Korringa, P. 1976b. Farming the flat oysters of the genus *Ostrea.* Elsevier, Amsterdam, The Netherlands.

Korringa, P. 1976c. Farming marine fishes and shrimps. Elsevier, Amsterdam, The Netherlands.

Krakauer, J. 1986. A fishing frenzy strikes on Sitka when herring run. Smithsonian 17:97-109.

Kwain, W. 1982. Spawning behavior and early life history of pink salmon *(Oncorhynchus gorbuscha)* in the Great Lakes. Can. J. Fish. Aquat. Sci. 39:1353-1360.

Kwain, W. and A.H. Lawrie. 1981. Pink salmon in the Great Lakes. Fisheries 6:2-6.

Lachner, E.A., C.R. Robins and W.R. Courtenay. 1970. Exotic fishes and other aquatic organisms introduced into North America. Smithsonian Contr. Zool. 59.

Lange, C. and R. Cap. 1986. *Bythotrephes cederstroemi* (Schodler) (Cercopagidae: Cladocera): a new record for Lake Ontario. J. Great Lakes Res. 12:142-143.

Leppakoski, E. 1984. Introduced species in the Baltic Sea and its coastal ecosystems. Ophelia (Supplement) 3:123-135.

Lightner, D.V., R.M. Redman and T.A. Bell. 1983. Infectious hypodermal and hematopoietic necrosis (IHHN), a newly recognized virus disease of penaeid shrimp. J. Invertebr. Pathol. 42:62-70.

Ling, S.-W. 1977. Aquaculture in southeast Asia. A historical overview. University of Washington Press, Seattle.

Linsley, R.H. and L.H. Carpelan. 1961. Invertebrate fauna, p. 43-47. *In* B.W. Walker (ed.) The ecology of the Salton Sea, California, in relation to the sportfishery. California Department of Fish and Game Fish Bulletin 113.

Lipkin, Y. 1972. Marine algal and sea-grass flora of the Suez Canal. J. Zool. 21:405-446.

Lutz, R.A., editor. 1980. Mussel culture and harvest: a North American perspective. Elsevier, Amsterdam, The Netherlands.

Mann, R., editor. 1979. Exotic species in mariculture. MIT Press, Cambridge, Massachusetts.

Manzi, J.J. and M. Castagna, editors. 1986. Clam mariculture in North America. Elsevier, Amsterdam, The Netherlands.

McDonald, J.H. and R.K. Koehn. 1988. The mussels *Mytilus galloprovincialis* and *M. trossulus* on the Pacific coast of North America. Mar. Biol. 99:111-118.

McKenzie, K.G. and A. Moroni. 1986. Man as an agent of crustacean passive dispersal via useful plants — exemplified by Ostracoda *ospiti esteri* of the Italian ricefields ecosystem — and implications arising therefrom. J. Crust. Biol. 6:181-198.

McNeil, W.J. 1979. Review of transplantation and artificial recruitment of anadromous species, p. 547-554. *In* T.V.R. Pillay and W.A. Dill (eds.), Advances in aquaculture. FAO Fishing News Books, Ltd., London.

Miller, R.L. 1969. *Ascophyllum nodosum:* a source of exotic invertebrates introduced into west coast near-shore marine waters. The Veliger 12:230-231.

Mooney, H.A. and J.A. Drake, editors. 1986. Ecology of biological invasions of North America and Hawaii. Springer-Verlag, New York.

Morse, D.E., K.K. Chew and R. Mann, editors. 1984. Recent innovations in cultivation of Pacific molluscs. Elsevier, Amsterdam, The Netherlands.

Moyle, P.B. 1976a. Inland fishes of California. University of California Press, Berkeley, California.

Moyle, P.B. 1976b. Fish introductions in California: history and impact on native fishes. Biological Conservation 9:101-118.

New, M.B., editor. 1982. Giant prawn farming. Elsevier, Amsterdam, The Netherlands.

Pettibone, M.H. 1963. Marine polychaete worms of the New England region. I. Aphroditidae through Trochochaetidae. U. S. Nat. Mus. Bull. 227:1-356.

Pilgrim, R.L.C. 1967. *Argulus japonicus* Thiele, 1900 (Crustacea: Branchiura) — a new record for New Zealand. N. Z. J. Mar. Freshwater Res. 1:395-398.

Popham, J.D. 1983. The occurrence of the shipworm, *Lyrodus* sp. in Ladysmith Harbour, British Columbia. Can. J. Zool. 61:2021-2023.

Por, F.D. 1978. Lessepsian migration. Springer-Verlag, Berlin.

Quayle, D.B. 1964. Dispersal of introduced marine woodborers in British Columbia waters. Comptes Rendus de la Congrès Internationale de la Corrosion Marine et des Salissures, Cannes, France. Centre de recherches et d'études Océanographiques: 407-412.

Randall, J.E. 1987. Introduction of marine fishes to the Hawaiian Islands. Bull. Mar. Sci. 41:490-502.

Rathbun, R. 1890. The transplanting of lobsters to the Pacific coast of the United States. Bull. U. S. Fish. Comm. 1888, 8:453-472.

Rosenthal, H. 1980. Implications of transplantations to aquaculture and ecosystems. Mar. Fish. Rev. 42:1-14.

Rosenthal, H. 1985. Constraints and perspectives in aquaculture development. GeoJournal 10:305-324.

Russell, D.J. 1981. Introduction of alien seaweeds to Hawaii. Phycologia 20:112.

Sandrof, S. 1946. The worm turns. Nat. Geogr. Mag. 89:775-786.

Shotts, E.B., A.L. Kleckner, J.B. Gratzek and J.L. Blue. 1976. Bacterial flora of aquarium fishes and their shipping waters imported from Southeast Asia. J. Fish. Res. Board Can. 33: 732-735.

Sindermann, C.J. 1986. Strategies for reducing risks from introductions of aquatic organisms: a marine perspective. Fisheries 11:10-15.

Steenbergen, J.F., and H.C. Schapiro. 1974. *Gaffkemia* in California spiny lobsters. Proceedings of the Fifth Annual Meeting of the World Mariculture Society, 5: 139-147.

Stone, L. 1876. Report of the operations in California in 1873. Rep. U. S. Comm. Fis. 3:377-429.

Taylor, D.W. 1966. An eastern American freshwater mussel, *Anodonta,* introduced into Arizona. The Veliger 8:197-198.

Tucker, C.S., editor. 1985. Channel catfish culture. Elsevier, Amsterdam, The Netherlands.

Vaini, F.A. 1985. Introduzione di specie ittiche esotiche nelle acque interne: storia, motivazioni, aspetti ecologici e sanitari. Riv. Ital. Piscicol. Ittiopatol. 20:87-97, 118-126.

Varvio, S.-L., R.K. Koehn, and R. Vainola. 1988. Evolutionary genetics of the *Mytilus edulis* complex in the North Atlantic region. Mar. Biol. 98:51-60.

Vermeij, G.J. 1982. Environmental change and the evolutionary history of the periwinkle *Littorina littorea* in North America. Evolution 36:561-580.

Walford, L., and R. Wicklund. 1973. Contribution to a world-wide inventory of exotic marine and anadramous organisms. FAO Fisheries Technical Paper 121.

Welcomme, R.L. 1986. International measures for the control of introductions of aquatic organisms. Fisheries 11:4-9.

Welcomme, R.L. 1988. International introductions of inland aquatic species. FAO Fisheries Technical Paper 294.

Whitney, R.R. 1967. Introduction of commercially important species into inland marine waters, a review. Contrib. Mar. Sci. 12:262-280.

Williams, R.J., F. B. Griffiths, E.J. Van der Wal, and J. Kelly. 1988. Cargo vessel ballast water as a vector for the transport of non-indigenous marine species. Estuarine Coastal Shelf Sci. 26:409-420.

Wolff, T. 1977. The horseshoe crab *(Limulus polyphemus)* in North European waters. Vidensk. Medd. Dan. Naturhist. Foren. 140:39-52.

Woods Hole Oceanographic Institution. 1952. Marine fouling and its prevention. United States Naval Institute, Annapolis, Maryland.

Zibrowius, H. 1983. Extension de l'aire de repartition favorisée par l'homme chez les invertebres marins. Oceanis 9:337-353.

CHAPTER 1

Movement and Dispersal of Exotic Species

Dispersal of Exotic Species from Aquaculture Sources, with Emphasis on Freshwater Fishes

WALTER R. COURTENAY, JR.
JAMES D. WILLIAMS

Abstract: Since the beginning of translocations of aquatic species beyond their historical ranges, there have been escapes and releases from culture facilities. Most escapes have resulted from carelessness in construction and operation of these facilities. In some instances, there appear to have been deliberate releases of stocks. To date, pet industry culture facilities have been the source of more introductions in the United States than has the culture of fishes for food or other purposes. Of the 46 established exotic fishes in the waters of the contiguous U.S., 22 are the result of aquaculture activities. Future development of aquaculture of food resources, however, promises to become a major source of introductions unless precautions are taken early. Recognition that introduced exotic aquatic species have been, or have the potential to be, detrimental to native species and ecosystems, and can create negative economic impacts, is reason for concern and caution. Guidelines for safety in importation and culture practices must be developed that will enhance both the future of aquaculture and protection of irreplaceable native natural resources.

Introduction

Aquaculture is a relatively new term for a very old practice — the culture of aquatic organisms. Bardach et al. (1972) stated that pond culture of carps is known from the fifth century B.C. in China. In Europe, culture of common carp, *Cyprinus carpio* Linnaeus, apparently began during the time of the Roman Empire as part of the medieval monastic pond fish culture practice (Balon 1974; Welcomme 1984, 1988). The source of wild common carp for use in monastic fish culture is presumed to have been the Danube, where this species is native. From there, this fish was distributed widely throughout Europe, with subsequent escapes into natural waters where it became established as reproducing populations.

Primarily because of its long history as a popular aquaculture species in Europe, emigrants to other continents were in large part responsible for movement of common carp into continents where the species was not native. These transfers began on a massive scale in the latter half of the 1800s. Welcomme (1988) noted that the common carp now has a nearly global distribution because of these transfers, subsequent escapes from culture facilities, and deliberate introductions into natural waters. As an introduced exotic (of foreign origin) species, it ranks third in international transfers, behind Mozambique tilapia, *Oreochromis mossambicus* (Peters) and rainbow trout, *Oncorhynchus mykiss* (Walbaum), and has been successfully released into 59 countries (Welcomme 1988).

Common carp provides an interesting contrast in opinions on its impact. It is a major species in inland aquaculture. Pullin (1986) noted that approximately 250,000 metric tons of carp are produced annually by aquaculture. In several countries, it is a subject of capture fisheries. Common carp has also been accused of being a detrimental species in many areas where it has found its way or was released into open waters, due to its habit of uprooting aquatic vegetation with subsequent increases in turbidity and lowering of oxygen levels. It is also known to feed on eggs of native fishes, including those of endangered and threatened species. Thus, it is a species that is praised in some aquaculture circles and simultaneously cursed by environmentally oriented groups (Cooper 1987).

Prior to 1900, international transfers of fishes for aquaculture purposes largely involved salmonids intended for introduction or repeated stockings as sport species, and some limited food production. Movements of common carp to North America began in the 1830s (DeKay 1842) and peaked in the last quarter of the 1800s (Smiley 1886; Courtenay et al. 1984; Crossman 1984; Contreras and Escalante 1984); southern Africa experienced a similar history with this species (de Moor and Bruton 1988). Welcomme (1988) reported that common carp reached its greatest popularity in international transfers between 1910 and 1940, being replaced by tilapias from 1950 to 1979 and Chinese carps from 1960 to 1980 "as preferred species." Present international interest in exotic species for aquaculture purposes includes a number of crustaceans, while in the United States tilapias and Chinese carps are of growing popularity among aquaculturists.

Similar opinions exist in relation to the second most widely transferred aquaculture fish, Mozambique tilapia, now found in 66 nations (Welcomme 1988). Its present distribution is almost pan-tropical. The first escape into waters where it is exotic may have been during the 1930s in Java, where it is thought to have been an aquarium fish release. This species is of great and increasing importance in modern aquaculture as a food fish. It has also been promoted for biological control purposes but its effectiveness is questionable. Like common carp, it is typically cultured in ponds, but in some nations it has been widely disseminated into natural waters. In several nations, the Mozambique tilapia is considered as a pest species due to its high fecundity, nature of producing stunted stocks, and its ability to displace native fishes (Bardach et al. 1972; Ling 1977; Shelton and Smitherman 1984).

Aquaculture and agriculture provide some important parallels and contrasts. The early development of agriculture in the Western Hemisphere was characterized by importations of exotic plant and animal species, just as there is now great interest in the use of exotic species in the developing aquaculture industry. One of the major contrasts is that agriculture involves mostly plants and animals that are so far removed genetically from their wild ancestors that they require care and husbandry to survive, with few persisting in a feral state. The few able to persist on their own include feral goats, pigs, donkeys, and horses, and these have created environmental management problems in several areas. Species employed in aquaculture are mostly feral stocks being reared artificially, and most have the capability to return to a feral state if released, within or outside their native ranges of distribution. A second contrast is that the technology of agriculture far outpaces that of aquaculture, largely due to the longer history of the former and its predominant importance to mankind. In such areas as nutrition, disease control, genetics, and husbandry, aquaculture has a lot of experience yet to be obtained and utilized.

In nearly every instance where an exotic aquatic species has been the subject of culture, escape into open waters has occurred. Shelton and Smitherman (1984) noted, "For whatever purpose an exotic fish is used, escape is virtually inevitable; thus, this eventuality should be considered." Welcomme (1988) added, "Species originally introduced for aquaculture eventually escape from the confinement of their ponds often but not always to colonize natural waters. Therefore any introduction made for aquaculture must be thought of as a potential addition to the wild fauna in the receiving country." Moreover, many introduced exotic species have had negative impacts on native organisms, habitat, and regional and national economies (Elton 1958; Laycock 1966; Courtenay 1979, in press a). It is these concerns, escape and impacts, that are the focus of this contribution.

Currently there are 46 species of exotic fishes established in open waters of the contiguous U.S. (Table 1). Additionally, another 14 species were established, but some were purposefully eradicated and others failed to survive, probably due to cold winter temperatures (Table 2). Fifty-three more, identified to species, and 7, identified only to genus (doubtless representing far more than 7 species), have been collected or reported from open waters; none is known to be established (Table 2). The majority of these introduced exotic fishes escaped from ornamental aquarium fish culture facilities or were released by aquarists (Courtenay et al. 1984, 1986b; Courtenay and Stauffer 1990). An unquantified number of alien aquatic invertebrate and plant species is also present, with several, particularly plants, having achieved status as major pests. Nearly all the introduced exotic fishes, a few invertebrates, and several aquatic plants were cultured at some time in the United States prior to their escape or intentional dissemination into open waters.

Bardach et al. (1972) defined aquaculture as "the growing of aquatic organisms under controlled conditions." Controlled conditions obviously does not mean "escape-proof." This definition is broad and could encompass activities from outdoor farming of aquatic organisms to maintaining aquarium species indoors. Indeed, it also encompasses the culture of species intended for stocking or introduction into natural waters. As used here, aquaculture will be confined to the culturing of aquatic organisms for food, sport, biological control, and aquarium purposes.

Species from Aquaculture for Food Resources

In many protein-deficient nations, governmental agencies regard the widespread introduction of exotic species to open waters as a form of aquaculture (Contreras and Escalante 1984; Erdman 1984; Maciolek 1984; Courtenay and Kohler 1986; Courtenay, in press a). Their philosophy appears to be that they have made these new resources available to the public. The assumption is that capture fisheries by individuals or commercial fishermen will solve a continuing and perhaps growing nutrition problem. While aquaculturists in the U.S. may disagree with or perhaps find humor in such an unfortunate approach, our own history includes similarities.

Of the 46 exotic fishes now established in waters of the contiguous U.S. (Table 1), several were imported during the latter part of the 1800s for the purpose of introducing new food resources. Most, however, were imported primarily as potential new sport species, with the food aspect of secondary importance. The major reason for introducing common carp was to provide a food resource that was highly prized by immigrants from Europe. Nevertheless, in contrast to European practices of culturing this species in ponds, the only culture activities this species experienced was prior to release into open waters (Baird 1879, 1893). Whatever positive traits pond culture in Europe provided this species

Table 1. List of exotic fishes established in open waters of the contiguous United States[1]. Abbreviations: A = from aquarium fish hobbyists; B = ship ballast release; C = from fish culture activities; E = release of experimental stock; F = released bait; P = introduction by governmental agency; Z = escape from zoo.

Family	Scientific Name	Common Name	Source
Cyprinidae	*Carassius auratus*	goldfish	A/C/F
	Ctenopharyngodon idella	grass carp	P
	Cyprinus carpio	common carp	P
	Hypophthalmichthys nobilis	bighead carp	C
	Leuciscus idus	ide	A/C/P
	Rhodeus sericeus	bitterling	A
	Scardinius erythrophthalmus	rudd	F/P
	Tinca tinca	tench	C/P
Cobitidae	*Misgurnus anguillicaudatus*	oriental weatherfish	A/C
Clariidae	*Clarias batrachus*	walking catfish	C
Loricariidae	*Hypostomus* sp.	armored catfish (FL)	A/C
	Hypostomus sp.	armored catfish (NV)	A
	Hypostomus sp.	armored catfish (TX)	Z
	Pterygoplichthys multiradiatus	sailfin catfish	A
Salmonidae	*Salmo trutta*	brown trout	P
Osmeridae	*Hypomesus nipponensis*	wakasagi	P
Poeciliidae	*Belonesox belizanus*	pike killifish	E
	Poecilia mexicana	shortfin molly	A/C
	Poecilia reticulata	guppy	A
	Poeciliopsis gracilis	porthole livebearer	C
	Xiphophorus helleri	green swordtail	A/C
	Xiphophorus maculatus	southern platyfish	A/C
	Xiphophorus variatus	variable platyfish	A/C
Percidae	*Gymnocephalus cernuus*	ruffe	B
Sciaenidae	*Bairdiella icistia*	bairdiella	P
	Cynoscion xanthulus	orangemouth corvina	P
Cichlidae	*Astronotus ocellatus*	oscar	C
	Cichla ocellaris	peacock cichlid	P
	Cichlasoma bimaculatum	black acara	C
	Cichlasoma citrinellum	Midas cichlid	C
	Cichlasoma managuense	jaguar guapote	A
	Cichlasoma meeki	firemouth cichlid	C
	Cichlasoma nigrofasciatum	convict cichlid	A

Table 1. Continued

Family	Scientific Name	Common Name	Source
Cichlidae	*Cichlasoma octofasciatum*	Jack Dempsey	A/C
	Cichlasoma urophthalmus	Mayan cichlid	A
	Geophagus surinamensis	redstriped eartheater	C
	Hemichromis bimaculatus	jewelfish	C
	Oreochromis aureus	blue tilapia	C/F/P
	Oreochromis mossambicus	Mozambique tilapia	A/C/P
	Oreochromis urolepis hornorum	Wami tilapia	P
	Sarotherodon melanotheron	blackchin tilapia	A/C
	Tilapia mariae	spotted tilapia	A/C
	Tilapia zilli	redbelly tilapia	P
Gobiidae	*Acanthogobius flavimanus*	yellowfin goby	B
	Tridentiger trigonocephalus	chameleon goby	B
Anabantidae	*Trichopsis vittata*	croaking gourami	C

[1] Postal abbreviations are used for the three distinct morphological species of *Hypostomus* as follows: FL = Florida, NV = Nevada, TX = Texas.

in terms of appearance, perhaps taste and reduced intramuscular bones, were doubtless lost when the fish was returned to feral conditions (Laycock 1966). Within the first two decades of the 1900s, the introduction of common carp was generally recognized as a mistake, particularly as it began to dominate many waters in the midwest and east (Courtenay et al. 1984, 1986b).

The common carp was not the only cyprinid to enter the U.S. during the late 1800s. Several shipments from Europe contained ide, *Leuciscus idus* (Linnaeus), and tench, *Tinca tinca* (Linnaeus) (Baird 1879, 1893). Reports of the U.S. Fish Commission, a predecessor agency of our current National Marine Fisheries Service and the U.S. Fish and Wildlife Service, indicated that these species were intentionally imported for distribution purposes. What "distribution" meant is unclear, but probably involved dissemination toward establishment for food and sport uses. Not surprisingly, both ide and tench were released accidentally into the Potomac River when a flood overflowed federal

Table 2. List of exotic fishes collected from, but not known to be established in, open waters of the contiguous United States. Species indicated by an asterisk (*) were previously established and were either eradicated or became extirpated naturally. Presumed source of introduction is indicated by: A = aquarium fish release; B = from culture for food, sport, or biological control purposes; C = from aquarium fish culture activities; D = ship ballast release; P = purposeful release by governmental agency; T = escape from tourist attraction; U = unknown.

Family	Scientific Name	Common Name	Source
Osteoglossidae	*Osteoglossum bicirrhosum*	arawana	A
Anguillidae	*Anguilla anguilla*	European eel	U
	Anguilla australis	shortfinned eel	U
Cyprinidae	*Barbodes schwanefeldi*	tinfoil barb	C
	Danio rerio	zebra danio	A/C
	Danio malabaricus	malabar danio	C
	Hypophthalmichthys molitrix	silver carp	B
	Labeo chrysophekadion	black sharkminnow	A
	Puntius conchonius	rosy barb	C
	Puntius gelius	golden barb	C
	Puntius tetrazona	tiger barb	A/C
Erythrinidae	*Hoplias malabaricus**	trahira	C
Characidae	*Colossoma* spp.	pacus	A
	Colossoma bidens	(no common name)	A
	Colossoma macropomum	tambaqui	A
	Gymnocorymbus ternetzi	black tetra	A
	Hemigrammus ocellifer	head and tail light	A
	Leporinus fasciatus	banded leporinus	A
	Metynnis sp.	(no common name)	A
	Paracheirodon innesi	neon tetra	C
	Pygocentrus nattereri	red piranha	A
	Serrasalmus humeralis *	pirambeba	T
	Serrasalmus rhombeus	redeye piranha	A
Doradidae	*Oxydoras niger*	ripsaw catfish	A
	Platydoras costatus	Raphael catfish	A
	Pterodoras granulosus	(no common name)	A
Pimelodidae	*Phractocephalus hemilopterus*	redtail catfish	A
Callichthyidae	*Callichthys callichthys*	cascarado	A
	Corydoras sp.	corydoras	A
Loricariidae	*Hypostomus* sp.	suckermouth catfish	A
	Otocinclus sp.	(no common name)	A

Table 2. Continued

Family	Scientific Name	Common Name	Source
Esocidae	*Esox reicherti*	Amur pike	B
Plecoglossidae	*Plecoglossus altivelis*	ayu	U
Salmonidae	*Coregonus maraena*	German whitefish	P
	Oncorhynchus masou	cherry salmon	P
	Salmo letnica	Ohrid trout	P
Adrianichthyidae	*Oryzias latipes* *	medaka	A
Aplocheilidae	*Cynolebias bellottii* *	Argentine pearlfish	P
	Cynolebias nigripinnis	blackfin pearlfish	P
	Cynolebias whitei	pearlfish	P
Cyprinodon-tidae	*Rivulus harti* *	giant rivulus	P
Goodeidae	*Ameca splendens*	butterfly splitfin	A
Poeciliidae	*Poecilia* hybrids		A/C
Atherinidae	*Chirostoma jordani*	charal	P
Centropomidae	*Lates mariae*	bigeye lates	P
	Lates nilotica	Nile perch	P
Percidae	*Stizostedion lucioperca*	zander	P
Cichlidae	*Aequidens pulcher* *	blue acara	C
	Cichla temensis	speckled pavon	P
	Cichlasoma beani *	green guapote	A
	Cichlasoma labiatum	red devil	C
	Cichlasoma salvini *	yellowbelly cichlid	A
	Cichlasoma trimaculatum *	threespot cichlid	A
	Geophagus brasiliensis	pearl eartheater	C
	Heros severum *	banded cichlid	A
	Labeotropheus sp.	(no common name)	Z
	Melanochromis auratus	gold mbuna	A
	Melanochromis johanni	blue mbuna	A
	Oreochromis niloticus	Nile tilapia	B
	Pseudotropheus zebra	zebra mbuna	A
	Pterophyllum scalare	angelfish	A/C
	Pterophyllum sp.	discus	A/C
	Tilapia sparmanni	banded tilapia	C
Anabantidae	*Anabas testudineus* *	climbing perch	C
	Betta splendens *	Siamese fightingfish	A
	Colisa fasciata	banded gourami	A
	Colisa labiosa	thicklip gourami	C
	Colisa lalia	dwarf gourami	C

Table 2. Continued.

Family	Scientific Name	Common Name	Source
	Ctenopoma nigropannosum *	twospot ctenopoma	C
	Helostoma temmincki	kissing gourami	C
	Macropodus opercularis *	paradisefish	C
	Trichogaster leeri	pearl gourami	C
	Trichogaster trichopterus	threespot gourami	C
Channidae	*Channa micropeltes*	giant snakehead	A
Pleuronectidae	*Platichthys flesus*	European flounder	U

culture ponds in Washington, D.C. (Baird 1893). Although they were subsequently distributed to several states and territories destined to become states, both species have experienced significant range reductions, unlike common carp (Courtenay et al. 1984, 1986b).

Importation of the rudd, *Scardinius erythropthalmus* (Linnaeus) into the U.S. took place in the late 1800s but the source of this translocation is unknown. Its popularity in Europe as a food and game species suggests that it may have been introduced for both purposes. It first appeared in Central Park in New York City in the late 1800s. It was reported by Bean (1897) from ponds in Central Park as a variety of the golden shiner with permanent vermillion color of fins. The unannounced appearance of this species in open waters of New Jersey and New York during the 1920s (Myers 1925) suggests movement by individuals rather than state agencies. Cahn (1927) and Greene (1935) reported introductions in 1917 by the Wisconsin Conservation Department. Although the species was established and later became extirpated in New Jersey and Wisconsin, it has persisted in Maine, apparently the result of a more recent introduction, and two places in New York (P.G. Walker, personal communication; Smith 1985).

During the last two decades, there has been renewed interest in rudd as an outdoor ornamental fish and a potential bait species. It is being cultured in Arkansas and Virginia (possibly in other states) and has been distributed to bait stores in at least 16 central and eastern states. It has been taken in open waters of at least eight states that we are aware of. The failure of rudd in New Jersey and Wisconsin should not be predictive of its potential, or lack thereof, to establish in waters of other states. In fact, its potential to establish in central, eastern, and southern states, and particularly in many states in the American west, is great. Fishermen feel they are acting humanely or feeding future catches when they dump contents of bait buckets at the end of a fishing day, a long-proven method of establishing fishes beyond their native ranges. The potential adverse impacts from releases of rudd are unknown and extreme caution is warranted. Six states, Alabama, Connecticut, Florida, Louisiana, Texas, and Virginia have recently passed regulations prohibiting importation of rudd. Arkansas has new regulations prohibiting their sale and use as a bait fish in Arkansas, but allows the production and sale of rudd outside the state. Several other states are currently considering regulations prohibiting importation of rudd.

A probability exists that rudd will hybridize, intergenerically, with native golden shiner, *Notemigonus crysoleucas* (Mitchill), with unknown consequences to wild populations of golden shiner, a primary forage species of many native game fishes. First generation hybrids (F_1) offspring should show heterosis (or hybrid vigor), but the introduced "genetic pollution" in subsequent generations could prove detrimental due to a variety of factors including spawning behavior and success toward recruitment.

It might surprise many North American sport fishermen to learn that importation of brown trout, *Salmo trutta* Linnaeus, from Europe in the early 1880s was in part to establish a new food fish (Goode 1903). This species was first introduced into Michigan in 1883, and is being cultured for stocking in most states where it can exist in the wild. Considered one of this nation's most successful fish introductions, it has had some adverse effects. For example, its introduction into the Kern River system in California resulted in serious negative impacts on populations of native golden trout, *Oncorhynchus aguabonita* (Jordan), the "state fish" of California (E.P. Pister, personal communication). Ironically, sport fishermen who praise brown trout also complain it is difficult to catch.

With perhaps the exception of common carp, it is obvious that the other cyprinids and one salmonid just discussed were not imported primarily as new food fishes, but that aspect was probably used to garner public support for these and future introductions. With the possible exception of common carp, all were popular sport species in Europe and all provide acceptable table fare. All were cultured prior to release in the United States, and two, rudd and brown trout, remain in culture today.

More recently but particularly during the 1970s, tilapias became popular in aquaculture as food fishes. The two most widely cultured species in the United States are Mozambique tilapia and blue tilapia, *Oreochromis aureus* (Steindachner). The Mozambique tilapia escaped from aquarium fish farms and both species have been intentionally introduced for biological control purposes into several states, within and beyond the so-called "sun belt" states. Blue tilapia releases have also occurred from aquaculture facilities where this species was being reared as a food resource. In Florida there have been at least two such releases, both from aquaculture facilities rearing channel catfish, *Ictalurus punctatus* (Rafinesque), that began to culture blue tilapia shortly before closing their operations, one in Dade County and another in Palm Beach County. Blue tilapia either escaped or, more likely, were released, and became established in nearby canals. Both populations have increased their ranges of distribution through southeastern Florida's extensive network of interconnected canals. The Dade County aquaculture facility was the source of blue tilapia that subsequently invaded the Taylor Slough portion of Everglades National Park, where the species has become a major management problem for the National Park Service (Courtenay 1989; Loftus 1989).

Blue tilapia has also been utilized in food-oriented aquaculture in North Carolina, Oklahoma and Pennsylvania, with escapes from these operations followed by local establishment. In North Carolina, it was introduced into Hyco Lake in 1984 as part of an aquaculture evaluation program, subsequently escaped, and is presently established. A second species, the redbelly tilapia, *Tilapia zilli* (Gervais), was introduced along with blue tilapia in Hyco Lake and is now established (McGowan 1988). Escape from an aquaculture facility, apparently operated by Oklahoma Gas and Electric Company, near Harrah, Oklahoma, resulted in establishment of blue tilapia through a portion of the North Canadian River (Pigg 1978), and that population currently persists there (J. Pigg, personal communication). The other release was from a Pennsylvania Power and Light Company aquaculture operation associated with a power plant site on the lower Susquehanna River, Pennsylvania (Skinner 1984). Blue tilapia were established in thermally heated waters of the lower Susquehanna River from 1982 through 1986, when condenser cooling water was deliberately released at lower lethal temperatures during December 1986, killing the populations in culture ponds and the river (Stauffer et al. 1989).

Blue tilapia escaped from aquaculture facilities in the San Luis Valley, Colorado, into thermal spring effluents where it persists (Zuckerman and Behnke 1986). The species has been cultured for aquaculture purposes in a few thermally heated reservoirs in Texas, where it also persists (Stickney 1979; G.P. Garrett and C. Hubbs, personal communication).

Although the Mozambique tilapia is established in open waters of several states, most introductions, originating from culture sources, were escapes from aquarium fish farms, aquarium fish releases by hobbyists, or deliberate introductions made for aquatic weed control. To date we are aware of only one instance where this species, perhaps a hybrid with the Wami tilapia, *Oreochromis urolepis hornorum* (Trewavas), escaped from a food fish culture facility. Juveniles of a red to pink coloration were collected in the Bruneau River, southern Idaho, below an effluent from an aquaculture facility, in September 1986 (Courtenay et al.

1988). The receiving waters are heated by thermal springs, and the probability of establishment by any species of tilapia there is high.

Future escapes or releases of tilapias from aquaculture sources will occur, and establishment will likely follow in waters with suitable thermal regima. Presently there is great interest among aquaculturists in California and Florida, and other states to culture the Nile tilapia, *Oreochromis niloticus* (Linnaeus). This species has been recorded from open waters of the U.S. recently, from a reservoir on the Tallapoosa River in eastern Alabama, which receives drainage from the fish culture ponds at Auburn University. Unless agencies within affected states are prepared to establish requirements for escape-proof culture facilities and/or bonding by applicants to pay for eradication procedures when an escape is first detected, permitting the use of this tilapia within their borders amounts to condoning introduction of the species into open waters when escapes occur. The same applies to requests for permits from any source to culture any exotic aquatic species, plant or animal. The aquaculture industry can avoid such requirements through some relatively simple, although perhaps not inexpensive, procedures to be discussed later.

Finally, Courtenay and Robins (1989) noted that Atlantic salmon, *Salmo salar* Linnaeus, is being cultured in net pens in inshore waters of the Pacific northwest. They stated, "It is difficult to use floating pens and other enclosures in open coastal waters without having some stock escape," and pointed out vulnerability of such pens to storms. Currently, regulations concerning mariculture of non-native species in the U.S. are essentially non-existent with the exception, in disturbingly few states, of requiring state approval for the importation of exotic marine organisms for culture purposes. Existing mariculture laws and regulations have been reviewed by the International Council for the Exploration of the Sea (ICES 1972, 1982). The risks from escapes are very similar to those from aquaculture. Some mariculture operations have been successful, but others have created problems including introductions of pests, parasites, and diseases (ICES 1982).

Species from Aquaculture for Sport Purposes

In most instances, species intended for introduction or stocking for sport purposes are cultured in "hatcheries." Such facilities are often operated by governmental agencies, but some are privately owned. Governmental facilities often culture organisms, sometimes exotic species, as forage for the fishes they utilize. On several occasions, there have been escapes of forage species from such hatcheries that resulted in temporal or permanent establishment of the forage stock in open waters (Courtenay et al. 1984). This has occurred with goldfish, *Carassius auratus* (Linnaeus), but most introductions of this exotic cyprinid have been made by aquarists or fishermen (bait release). Taylor et al. (1984) stated that introductions of exotic forage species may result in complex, unpredictable consequences to the stability of native populations.

Exotic fishes imported and cultured for sport fishing purposes were treated above, because all were also suggested for use as food fishes. The only exotic fishes destined for introduction as sport species that apparently did not undergo culture prior to introduction were bairdiella, *Bairdiella icistia* (Jordan and Gilbert), and orangemouth corvina, *Cynoscion xanthulus* Jordan and Gilbert, imported from the Gulf of California and released into the Salton Sea, southern California (Walker 1961; Courtenay and Robins 1989). These and associated introductions into the Salton Sea are probably the most outstanding examples of well-planned releases where no adverse impacts were likely (Courtenay and Robins 1989).

Recently, North Dakota has shown interest in introducing zander, *Stizostedion lucioperca* (Linnaeus), as a sport fish. The original intent may have been introduction of this European percid to a major mainstream Missouri River reservoir, Lake Sakakawea (Courtenay and Robins 1989). The purpose was to replace or supplement depleted stocks of native walleye, *S. vitreum* (Mitchill), which has experienced reduced recruitment due largely to pollution and manipulations of water levels in the reservoir and perhaps from predation by introduced rainbow smelt, *Osmerus mordax* (Mitchill). The latest proposal calls for stocking of two isolated lakes with zander to test their success in novel environments, but the species will doubtless be cultured prior to and probably during periods of introduction and evaluation. Unless the culture facilities are truly secure, escape into other waters is possible before the evaluation process has been completed. As with any culture facility, the possibility always exists that employees may disseminate the stock beyond confinement.

Species from Aquaculture for Biological Control Purposes

Within recent decades, there has been great interest in using exotic fishes for biological control of pest organisms, usually algae, rooted aquatic plants (several of which are themselves exotic), or insects. A fallacy in this intent is that fishes are not monophagous, and dietary changes typically occur through early and sometimes later life history stages. Thus, control of a pest organism by a fish can never be achieved without impacts on non-target species, in strong contrast to the record of success with many insects for biological control (Courtenay 1979).

Among the species cultured are grass carp, *Ctenopharyngodon idella* (Valenciennes), for control of rooted aquatic vegetation, silver carp, *Hypophthalmichthys molitrix* (Valenciennes), for phytoplankton control, bighead carp, *H. nobilis* (Richardson), for control of zooplankton, and several species of tilapias for aquatic weed or insect control. All of the exotic cyprinids are known to have escaped from culture facilities, as have most of the tilapias. Grass carp, for example, escaped from governmental culture facilities into waters in Arkansas in the early 1960s (Stroud 1972; Courtenay and Robins 1975), prior to intentional introductions within that state. Proponents for importation and introduction of grass carp argued through the early 1960s that conditions for establishment of this fish did not exist in U.S. waters. Stanley (1976) predicted that grass carp could spawn successfully in the U.S.

and noted that the large numbers released in the Mississippi River should reach sexual maturity in 1978 or 1979. Subsequently, successful reproduction of grass carp was documented in the lower Mississippi River basin (Conner et al. 1980). There are recent indications of spawning by this species in the lower Trinity River in Texas (Courtenay et al. 1986b), Red River in Louisiana, and the Missouri River in Missouri. Diploid stock of grass carp was recently introduced into Tennessee River reservoirs in northern Alabama by a local citizens group and subsequently by the Tennessee Valley Authority. The long-term consequences of the introduction of grass carp may not be known for another decade or two, but based on reports from a variety of sources, populations are increasing (unpublished records, U.S. Fish & Wildlife Service, Gainesville, Florida).

In the early 1970s, silver carp were collected from open waters in Arkansas, apparently resulting from escape from aquaculture sources. This species has since been collected from waters in Alabama, Illinois, and Missouri. Bighead carp, first introduced into U.S. waters in Arkansas and Alabama in 1972, have been found in open waters in Kentucky (Jennings 1988; K. Cummings, personal communication), and recently in Illinois, Kansas (F. B. Cross, personal communication), and Missouri (Pflieger 1989).

Currently there are proposals emerging from aquaculture interests requesting permits to culture silver carp and bighead carp as food fishes. A proposal submitted to Illinois in 1988 used many of the same arguments that were used for grass carp over two decades ago to rationalize why escape of bighead carp into Illinois waters would not result in establishment of the species.

Several tilapias, most of which are being reared as food fishes, have been promoted for use as biological control agents for aquatic weeds. Among these are blue tilapia and Mozambique tilapia, although the effectiveness of both species is questionable. Another is redbelly tilapia, which appears to be more effective in controlling growth of rooted aquatic plants. All have escaped from culture facilities and become established in the wild, except for some redbelly tilapia which were introduced from culture sources.

Blue tilapia has been introduced to two states (Arizona, Florida), possibly a third (Georgia), for algal or other aquatic plant control. There is some question as to what species of tilapia was introduced in Georgia, but it was probably blue tilapia. This species was imported from Israel to Auburn University in 1954, and cultured in ponds there to test its potential in aquaculture and for sport purposes. Although it was found to have no value as a sport species (Swingle 1960), Florida introduced stocks from Auburn to experimental ponds at Pleasant Grove Research Station, near Tampa, in 1961 to examine the species as a biological control for aquatic plants (Buntz and Manooch 1968). This project, conducted by the Florida Game and Fresh Water Fish Commission and sponsored under the Dingell-Johnson Federal Aid in Fish Restoration Program, was supposedly being conducted to examine the sport potential of the fish, a factor already disproven at Auburn. From the culture and research source at Pleasant Grove, this species quickly found its way into open waters of Florida. Before 1968, it was established in 12 counties and now occupies waters in 18 counties (Courtenay et al. 1984, 1986b).

Dispersal of blue tilapia from Pleasant Grove was not due to escape via effluents or flooding. It appears to have occurred via two routes — Commission personnel providing private citizens with fish to release, and sport fishermen, thinking the Commission was delaying the introduction of a new sport fish, entering the facility and removing young off nests for stocking open waters (Buntz and Manooch 1969).

A major problem with using tilapias in aquatic weed control is their reproductive potential. Tilapias spawn frequently over periods of many months in suitable habitat and temperature regimes, and provide their young with parental care, resulting in large populations of offspring. A result of such population growth is overcontrol of vegetation, resulting in loss of cover and food organisms for early life history stages of many native fishes (Nobel and Germany 1986). This is the same concern that many scientists share about introduced grass carp. Other consequences are displacement of native fish populations by crowding and/or almost continuous occupation by tilapias of spawning sites preferred by native fishes, especially centrarchids (Nobel and Germany 1986; Taylor et al. 1984). The crowding factor can also affect tilapias, resulting in populations of stunted individuals that reproduce at a small size. There also may be as yet undetermined behavioral interactions between tilapias and native fishes that result in displacement of the latter. In many open waters in the U.S. where tilapias have become established and have increased in biomass, there have been sharp declines in populations of one to several native and introduced fish species (Taylor et al. 1984).

In southern California, a hybrid between Mozambique and Wami tilapias has been used for biological control of chironomid midges (Legner et al. 1980; Legner and Pelsue 1977). Its success there, under feral conditions, is largely unknown. Since introduction of this hybrid into certain flood control channels in Los Angeles and Orange counties, it has become the dominant fish (Courtenay et al. 1984, 1986b). Stocks for these releases were cultured at facilities in southern California.

Finally, the quest for an effective biological control fish species for some particular, often man-induced or introduced, problem can be expected to continue. New species will be proposed for introduction whenever some entrepreneur finds a reference that a certain species is known to consume some particular dietary item of interest. Promotion of that certain species, as history has shown and continues to show, will ignore the fact that nearly all fishes are omnivores and, therefore, may indicate control of one or several organisms if given only those dietary choices under experimental conditions. Not even the mosquitofishes, *Gambusia* spp., deserve their common name based on their purported ability to control mosquito larvae (Courtenay and Meffe 1989).

Species from Aquaculture for Aquarium Purposes

In the U.S., the history of the aquarium fish industry may serve as a "barometer" of what could happen in the future with the utilization of exotic species in aquaculture. Of the 46 species of exotic fishes now established in the continental U.S., approximately 60% are known or presumed to have originated from the aquarium fish trade (Table 1). Many escaped or were released from aquarium fish culture facilities, particularly in Florida and California, and some were introduced by aquarists into open waters of many states. Additionally, about 50 species, none of which is presently established, can be traced to this industry, many the result of escapes from culture facilities (Table 2, Courtenay and Stauffer, 1990). To some, these numbers may be surprising; to others, they are understandable, considering the longer history of the aquarium fish industry than most other aspects of aquaculture in the U.S.

Releases of aquarium fishes and their subsequent establishment in open waters is not restricted to the U.S. In fact, this is a common phenomenon in many parts of the world (Courtenay and Stauffer 1984; Welcomme 1988; Courtenay, in press a).

Escapes or releases from aquarium fish culture facilities have been and remain an environmental problem. There have been instances where certain species of aquarium fishes, declared or listed as prohibited species by a particular state, appear to have been purposefully released to avoid detection and prosecution (Courtenay and Robins 1975; Courtenay and Miley 1975; Courtenay and Hensley 1979). Future escapes can probably be reduced, but the goal should be **no** escapes or releases.

It is more difficult to convince owners of pet fishes that releasing their unwanted fishes into local waters can create equal or, in some parts of the nation or world, worse problems. Moreover, it may be impossible to change attitudes of those who know exotic fishes can cause problems with native species and introduce the former maliciously, as may have happened in at least one instance.

The convict cichlid, *Cichlasoma nigrofasciatum* (Günther), was introduced into thermal springs in Clark and Lincoln counties, Nevada, in the early 1960s (Deacon et al. 1964; Hubbs and Deacon 1965). In Lincoln County, it was found to be established in

Ash and Crystal springs, in the Pahranagat Valley, where it has apparently contributed to the decline of an endemic, endangered fish, the White River springfish, *Crenichthys baileyi* (Gilbert). The springfish in Crystal Springs is a subspecies, *C. b. grandis* Williams and Wilde, known as the Hiko White River springfish for its type locality, Hiko Spring, located 7.4 km north of Crystal Springs. This unique sub-species was eliminated from Hiko Spring by introduced fishes between February 1966 and June 1967 (Minckley and Deacon 1968; Deacon 1979; Williams and Wilde 1981). Following an attempt to reestablish this fish in Hiko Spring in early 1984, someone introduced convict cichlids, apparently maliciously, probably from stock captured in Crystal Springs. This appears to have been an attempt to disrupt or negate reestablishment of the springfish (Baugh et al. 1985; Courtenay et al. 1985).

In addition to fishes, there have been several escapes or releases of exotic aquatic plants and invertebrates from the aquarium trade and/or hobbyists. Among the plants is hydrilla, *Hydrilla verticillata* Royle, apparently released into open waters of southeastern Florida during the early 1960s. The source of this introduction is unknown, although the species is known to have been grown by several aquaculture facilities specializing in aquarium plants, often in addition to their culture of fishes. It proliferates readily in aquaria to the point that the aquarist must dispose of excess growth. Hydrilla has been spread to many areas of the U.S., usually by small parts of the plant being carried on boats or their trailers from one drainage system to another. It has been found as far from southern Florida as California, Iowa, and the Potomac River between Maryland and Virginia. This rooted aquatic plant has demonstrated its potential to clog bodies of water to the detriment of native biota and recreational use by humans and, therefore, has become a major pest species. Annual costs of control measures for hydrilla and other exotic aquatic plants are substantial, amounting to millions of dollars, paid by taxpayers who bear no responsibility for the introductions.

Although there has been no comprehensive study of invertebrates released by the aquarium fish trade, one species that is known to have become established from this source is an Oriental thiarid snail, *Melanoides tuberculata* (Müller), present in open waters of Collier and Dade counties, Florida, certain spring systems in Clark, Lincoln, and Nye counties, Nevada, Harney County, Oregon, and Bexar and Comal counties, Texas. It was established in Maricopa County, Arizona, until floods in 1965 and 1966 destroyed the population. This snail has been suggested as a predator on eggs of fishes, and its presence in Florida waters may present a public health threat. It is known to be an intermediate host for trematode parasites, the Oriental lung fluke, *Paragonimus westermani* (Kerbert), and the liver fluke, *Opisthorchis sinensis* (Cobbold). Neither of these parasites has yet been detected in Florida, but intermediate hosts (crustaceans for the lung fluke and fishes for the liver fluke) and definitive hosts are present (Roessler et al. 1977). Thus the stage has been set, through introduction of a snail, for the proliferation and spread of two serious disease organisms if, or when, these parasites are introduced.

Impacts of Introduced Aquatic Species from Aquaculture

Taylor et al. (1984) reviewed known and potential impacts from introductions of non-native fishes in U.S. waters that apply to natural waters anywhere. Direct predation by the introduced species is the most obvious threat to the well-being of native species. With fishes, most predatory species are introduced for sport purposes. There is a wide and complex spectrum of factors often summarized as "competition" that impact native biota. In most instances competition for food resources is not a factor for omnivorous species, but competition for a preferred and necessary food item could prove to be a major negative factor. Competitive aspects including spatial and other behavioral competition between introduced and native biota are perhaps more serious, but are also more difficult to measure than competition for food resources. Some of these can be measured under experimental laboratory conditions, but laboratory conditions rarely begin to duplicate what happens in nature. With exotic species, the possibility of hybridization with native species, resulting in pollution of gene pools, is remote, but is of great concern where intercontinental congeners exist.

To some, the finding of an alien species in a novel environment may be only of curiosity. Nevertheless, whenever a species is introduced, there will be adjustments made by resident species to the alien. Adjustments may be minor, perhaps not measurable by scientists due largely to their lack of understanding or study of the ecosystem prior to an introduction. Conversely, adjustments can result in major changes in native species, sometimes including extirpation of local populations or, at worst, extinctions of one or more species. Severe negative impacts, however, may not become evident for years or several decades following an introduction, a factor that is unrecognized or perhaps conveniently ignored by proponents of purposeful introductions (Courtenay 1979). When negative impacts are noted, control measures may be instituted, in the name of eradication (impossible by then), or some additional introduction is suggested as a control, which is an almost predictable mistake. In all too many cases, introductions of feral aquatic species have been costly to native species, in several cases to purposefully introduced species, and to humans who must pay the bills for control measures when the introduction proves undesirable or destructive.

Escape of the walking catfish, *Clarias batrachus* (Linnaeus), into Florida waters in the late 1960s resulted in regulations prohibiting importation or possession of this species or its congeners in several states (Courtenay, in press b). It also led to the *only* federal listing of a prohibited fish under provisions of the Lacey Act. Early, in fact premature, fears over this introduction led to these regulations. Although this exotic has the potential to cause severe damage to native fishes, it has not created what could be considered a major environmental impact in Florida, the only state in which it has become established to date. Ironically, this exotic catfish has adversely impacted the industry responsible for its introduction by invading aquarium fish culture ponds and destroying valuable stocks of fishes. In several areas of Florida, fences have been erected around culture ponds to exclude walking catfish (Courtenay and Miley 1975). In addition, the many species of cichlids that escaped from aquarium fish culture in Florida have largely added to the overall fish biomass while not creating local extirpations of native fishes. There have been negative impacts on native fishes, such as behavioral competition from cichlids and, to some degree, declines in native fish populations (J.N. Taylor, unpublished data). The full impacts of these introductions will probably not be known for years to come.

The situation in several areas of the American southwest is considerably different. Introduced species have often quickly created major problems (Minckley and Deacon 1968; Deacon 1979; Courtenay et al. 1986a). When introduced among depauperate faunas in limited or confined habitats, alien fishes are known to contribute to and sometimes create severe population declines and extinctions of native fishes; when introduced to faunas of a similar kind and habits, aliens can lead quickly to extinctions (Myers 1965). Therefore, introductions of non-native species from any source into any receiving waters should be viewed as having potential to create negative impacts. For example, although the native fish fauna of Florida is currently strong and resilient enough to buffer severe negative impacts from introductions, this may not be so a decade or more in the future.

Establishment of alien species in novel environments is known to be enhanced by habitat alterations. There are no developed nations without habitat alterations, and most developing nations are quickly achieving this unfortunate status. Those who claim that one cannot place blame on introductions as a cause for decline of native species are wrong. When one views the sequencing of impacts, it is true in many cases that habitat alterations prior to introductions stressed or brought about population declines of native biota, enhancing the chances of establishment of alien species. It is equally true that introductions of alien species have provided the ultimate insult to the continued integrity of several native species, have the potential to do so with others, and, in fact, probably act synergistically with the earlier man-induced changes (Courtenay et al. 1986a).

Finally, importation of exotic species for aquaculture purposes poses a threat of accidentally introducing exotic parasites and disease organisms that may adversely impact native biota. Such introductions have occurred, and the potential for serious consequences exists (Deacon 1979; Hoffman and Schubert 1984; Farley, Lightner et al. and Ganzhorn, this volume). This factor should never be underestimated. The aquarium fish industry has recognized the problem of introduced parasites and diseases, and has taken steps to reduce this threat through prophylaxis of imported stock (Shotts and Gratzek 1984).

Aquaculture's Role in Environmental Responsibility

The fact that aquatic organisms can escape or be released from culture sources into open waters, and have the potential to become established and create or contribute to environmental problems is a matter of concern to biologists and resource managers. It should also be of concern to aquaculturists for several reasons.

First, it is not good culture or business practice to contribute what can be considered "biological pollution" to already stressed ecosystems. One mistaken introduction that could become an environmental disaster or one that results in serious harm to some other sector of business would probably result in regulations so stringent that only the largest corporations could afford to be in this business. Second, it is not good economics to allow a product, in this case living organisms, to escape or be carried off by employees for whatever purpose. Lost products are lost profit, or perhaps the margin between financial disaster and "breaking even." Third, business reputations among the public, the consumers, are typically built on responsibility. In an increasingly environmentally aware society, this is an important factor.

Most persons in business dislike regulations. At present, several states regulate what can be grown in aquaculture and some specify what kinds of facilities can be used for aquaculture. Many persons in aquaculture see these regulations as impeding their development or potential to compete with operations in states with less stringent or no regulations; some see these as simply obstructionist actions. This is not the case. Regulations concerning what species can be cultured and in what kinds of facilities are designed to protect the natural resources of a state, the major responsibility of the regulating agency. Regulations concerning aquaculture do not exist to justify the jobs of law enforcement personnel or the biologists and managers that help design the regulations. In our view, it is the duty of state (fish and game, environmental protection, etc.) and federal (U.S. Fish and Wildlife Service) agencies to carefully consider, evaluate, and assist in experimentation with species proposed for importation toward use in aquaculture. In fact, this is one of the major assigned missions of the U.S. Fish and Wildlife Service's National Fisheries Research Center, located in Gainesville, Florida.

That some states have more stringent regulations concerning aquaculture than others is a reflection of the attitudes of those states toward the importance of native biota and its protection. Therefore, these differences should not be viewed as one or several states being opposed to aquaculture and others in favor. We know of no state or any federal agency that is opposed to development of aquaculture. It is true, however, that a few states (e.g., Alabama, Arkansas, Mississippi) have strongly supported aquaculture through research, extension services, and open promotion of the activity. If the primary charge of an agency responsible for environmental protection is just that, then it is usually impossible, from a fiscal or manpower standpoint, for that agency to promote or assist activities that do not fall under its jurisdiction. This is, in part, why many aquaculturists favor jurisdiction over their activities by agricultural rather than conservation agencies. If a major responsibility of a conservation agency is protection and enhancement of aquatic resources of a state, as all are, it follows that they should be concerned with what is imported for culture purposes when they know that escape or release is likely, if not inevitable. Even if agriculture agencies take over major responsibilities for aquaculture, the responsibilities of the conservation agencies will not change, and there will be interagency conflicts that should be avoided. Such conflicts lead to increased costs, paid through taxes, and dilution of the missions of the agencies involved, a "no-win" situation. Furthermore, placing political pressures on conservation agencies will not result in winning new friends, but may result in setbacks to development of aquaculture.

In our view, aquaculture has a responsibility to the public and to our natural resources to confine its culture stock when what is being cultured is not native to the regional drainage system. This can be accomplished in several ways — through intensive culture in closed systems; by diking outdoor ponds to protect from effects of flooding; by assuring that all effluents are passed through sand and gravel filtration to prevent escape of eggs and/or larvae. In some areas where birds such as terns and kingfishers are a problem with their habits of moving live fishes, some method of screening outdoor ponds may be necessary. Obviously intensive culture systems are expensive, and so is diking. The costs of simple but effective sand and gravel filter systems are comparatively low, but should be requisite in all operations producing effluents that have any possibility of draining into open waters (Courtenay and Robins 1975). In comparison with what heavy industry must spend to protect the environment, all of these costs to aquaculture are minimal.

Finally, conservation agencies should view protection measures taken by developing aquaculture as responsible actions. Where protection is assured, requests for permits to culture non-native species should be considered favorably, but only when a market for the products can be justified. If no market exists, business failure is predictable and so are releases of stock. Conservation agencies in many states should, as part of their permitting procedures, require that they be notified when a culture facility is closing so that they can advise or assist in disposal of remaining, unwanted stocks of non-native culture organisms. The other option, less desirable from the standpoint of industry, is to require the posting of a bond for disposal of the stock should the facility fail to do so.

Acknowledgments

We are most grateful to C.W. Butler, J.E. Deacon, D.P. Jennings, P.C. Marsh, J. A. McCann, C.R. Robins, and those authors whose papers are cited herein as in press for providing or sharing information used in this contribution. We also thank P.M. McCoy for assistance in manuscript preparation.

The views expressed by the authors are theirs, and do not necessarily reflect opinions or official positions taken by the agencies they represent.

Literature Cited

Baird, S.F. 1879. The carp, p. 40-44. *In* Report of the Commissioner. Rept. U.S. Fish Comm. 1876-77. U.S. Govt. Printing Office, Washington, D.C.

Baird, S.F. 1893. Report of the Commissioner. Rept. U.S. Fish Comm. 1890-91. U.S. Govt. Printing Office, Washington, D.C., pp. 1-96.

Balon, E.K. 1974. Domestication of the carp *Cyprinus carpio* L. Royal Ontario Mus., Life Sci. Misc. Publ.

Bardach, J.E., J.H. Ryther and W.O. McLarney. 1972. Aquaculture, the farming and husbandry of freshwater and marine organisms. Wiley-Interscience, New York.

Baugh, T.M., J.E. Deacon and D. Withers. 1985. Conservation efforts with the Hiko White River springfish. J. Aquar. Aquat. Sci. 4:49-53.

Bean, T.H. 1897. Notes upon New York fishes received at the New York Aquarium 1895-1897. Bull. Am. Mus. Nat. Hist. 9:327-353.

Buntz, J. and C.S. Manooch. 1968. *Tilapia aurea* Steindachner, a rapidly spreading exotic in south central Florida. Proc. Annu. Conf. SE Assoc. Game Fish Comm. 22:495-501.

Cahn, A.R. 1927. An ecological study of southern Wisconsin fishes, the brook silverside *(Labidesthes sicculus)* and the cisco *(Leucichthys artedi)* in their relations to the region. Ill. Biol. Monogr. 2:1-151.

Conner, J.V., R.P. Gallagher and M. Chatry. 1980. Larval evidence for natural reproduction of the grass carp *(Ctenopharyngodon idella)* in the lower Mississippi River. Proc. 4th Larval Fish Conf. Biol. Serv. Prog., Natl. Power Plant Team, Ann Arbor, Mich. FWS/OBS/43: 1-19.

Contreras-B., S., and M.A. Escalante-C. 1984. Distribution and known impacts of exotic fishes in Mexico, p. 102-130. *In* W.R. Courtenay, Jr. and J.R. Stauffer, Jr. (eds.), Distribution, biology, and management of exotic fishes, Johns Hopkins University Press, Baltimore, Maryland.

Cooper, E. L. 1987. Carp in North America. Amer. Fish. Soc., Bethesda, Maryland.

Courtenay, W.R., Jr. 1979. The introduction of exotic organisms, p. 237-252. *In* H.P. Brokaw (ed.), Wildlife and America, Council on Environmental Quality, U.S. Govt. Printing Office, Washington, D.C.

Courtenay, W.R., Jr. 1989. Exotic fishes in the National Park system, p. 3-10. *In* L. K. Thomas, (ed.), Proceedings, 1986 Conference Science National Parks, Vol. 5. Management of exotic species in natural communities, U.S. Park Service and George Wright Society, Washington, D.C.

Courtenay, W.R., Jr. In press a. The invasive fish problem from an international perspective. Trans. R. Soc. S. Afr.

Courtenay, W.R., Jr. In press b. Regulation of aquatic invasives in the United States of America, with emphasis on fishes. Trans. R. Soc. S. Afr.

Courtenay, W.R., Jr., J.E. Deacon, D.W. Sada, R.C. Allan, and G.L. Vinyard. 1986a. Comparative status of fishes along the course of the pluvial White River, Nevada. Southwest. Nat. 30: 503-524.

Courtenay, W.R., Jr., and D.A. Hensley. 1979. Range expansion in southern Florida of the introduced spotted tilapia, with comments on its environmental impress. Environ. Conserv. 6: 149-151.

Courtenay, W.R., Jr., D.A. Hensley, J.N. Taylor, and J.A. McCann. 1984. Distribution of exotic fishes in the continental United States, p. 41-77. *In* W.R. Courtenay, Jr. and J.R. Stauffer, Jr. (eds.), Distribution, biology, and management of exotic fishes. Johns Hopkins University Press, Baltimore, Maryland.

Courtenay, W.R., Jr., D.A. Hensley, J.N. Taylor, and J.A. McCann. 1986b. Distribution of exotic fishes in North America, p. 675-698. *In* C.H. Hocutt and E.O. Wiley, (eds.), Zoogeography of North American freshwater fishes. John Wiley & Sons, New York.

Courtenay, W.R., Jr. and C.C. Kohler. 1986. Exotic fishes in North American fisheries management, p. 401-413. *In* R.H. Stroud (ed.), Fish culture in fisheries management. American Fishery Society, Bethesda, Maryland.

Courtenay, W.R., Jr. and G.K. Meffe. 1989. Small fishes in strange places: A review of poeciliid introductions, p. 319-331. *In* G.K. Meffe and F.N. Snelson (eds.), Ecology and evolution of livebearing fishes (Poeciliidae) Prentice-Hall, Englewood Cliffs, New Jersey.

Courtenay, W.R., Jr. and W.W. Miley II. 1975. Range expansion and environmental impress of the introduced walking catfish in the United States. Environ. Conserv. 2:145-148.

Courtenay, W.R., Jr. and C.R. Robins. 1975. Exotic organisms: An unsolved, complex problem. BioScience 25:306-313.

Courtenay, W.R., Jr. and C.R. Robins. 1989. Fish introductions: Good management, mismanagement, or no management? Rev. Aquat. Sci. 1:159-172.

Courtenay, W.R., Jr., C.R. Robins, R.M. Bailey and J.E. Deacon. 1988. Records of exotic fishes from Idaho and Wyoming. Great Basin Nat. 47:523-526.

Courtenay, W.R., Jr. and J.R. Stauffer, Jr. editors. 1984. Distribution, biology, and management of exotic fishes. Johns Hopkins University Press, Baltimore, Maryland.

Courtenay, W.R., Jr. and J.R. Stauffer, Jr. 1990. The introduced fish problem and the aquarium fish industry. J. World Aquacult. Soc. 21:145-159.

Crossman, E.J. 1984. Introduction of exotic fishes into Canada, pp. 78-101. *In* W.R. Courtenay, Jr. and J.R. Stauffer, Jr. (eds.), Distribution, biology, and management of exotic fishes. Johns Hopkins University Press, Baltimore, Maryland.

Deacon, J.E. 1979. Endangered and threatened fishes of the west. *In* The endangered species: A symposium. Great Basin Nat. Mem. 3:41-64.

Deacon, J.E., C. Hubbs and B.J. Zahuranec. 1964. Some effects of introduced fishes on the native fish fauna of southern Nevada. Copeia 1964:384-388.

DeKay, J.E. 1842. Zoology of New York - IV: Fishes. W. and A. White and J. Visscher, Albany, New York.

de Moor, I.J. and M.N. Bruton. 1988. Atlas of alien and translocated indigenous aquatic animals in southern Africa. S. Afr. Nat. Sci. Prog. Rept. 144:1-310.

Elton, C.S. 1958. The ecology of invasions by plants and animals. Chapman and Hall, London.

Erdman, D.S. 1984. Exotic fishes in Puerto Rico, p. 162-176. *In* W.R. Courtenay, Jr. and J.R. Stauffer, Jr. (eds.), Distribution, biology, and management of exotic fishes. Johns Hopkins University Press, Baltimore, Maryland.

Goode, G. B. 1903. American fishes. L. C. Page, Inc. Boston, Massachusetts.

Greene, C.W. 1935. The distribution of Wisconsin fishes. Wisc. Conserv. Comm., Madison, Wisconsin.

Hoffman, G.L. and G. Schubert. 1984. Some parasites of exotic fishes, p. 233-261. *In* W.R. Courtenay, Jr. and J.R. Stauffer, Jr. (eds.), Distribution, biology, and management of exotic fishes. Johns Hopkins University Press, Baltimore, Maryland.

Hubbs, C. and J.E. Deacon. 1965. Additional introductions of tropical fishes into southern Nevada. Southwest. Nat. 9:249-251.

ICES. 1972. Report of the working group on introduction of non-indigenous marine organisms. ICES Coop. Res. Rept. 32.

ICES. 1982. Status (1980) of introductions of non-indigenous marine species to North Atlantic Waters. ICES Coop. Res. Rept. 116.

Jennings, D.P. 1988. Bighead carp *(Hypophthalmichthys nobilis):* A biological synopsis. U.S. Fish & Wildl. Serv. Biol. Rept. 88.

Laycock, G. 1966. The alien animals. Nat. Hist. Press, Garden City, New York.

Legner, E.F., R.A. Medved and F. Pelsue. 1980. Changes in chironomid breeding patterns in a paved river channel following adaptation of cichlids of the *Tilapia mossambica-hornorum* complex. Ann. Entomol. Soc. Amer. 73:293-299.

Legner, E.F. and F.W. Pelsue. 1977. Adaptations of *Tilapia* to *Culex* and chironomid midge ecosystems in south California. Proc. Calif. Mosq. Contr. Assn. 45:95-97.

Ling, Shao-Wen. 1977. Aquaculture in southeast Asia, a historical overview. Washington Sea Grant Publication. University of Washington Press, Seattle, Washington.

Loftus, W.F. 1989. Distribution and ecology of exotic fishes in Everglades National Park, p. 24-34. *In* L. K. Thomas (ed.), Proceedings, 1986 Conference Science National Parks, Vol. 5. Management of exotic species in natural communities. U.S. National Park Service and George Wright Society, Washington, DC.

Maciolek, J.A. 1984. Exotic fishes in Hawaii and other islands of Oceania, p. 131-161. *In* W.R. Courtenay, Jr. and J.R. Stauffer, Jr. (eds.), Distribution, biology, and management of exotic fishes, Johns Hopkins University Press, Baltimore, Maryland.

McGowan, E.G. 1988. An illustrated guide to larval fishes from three North Carolina Piedmont impoundments. Carolina Power & Light Co., New Hill, North Carolina.

Minckley, W.L. and J.E. Deacon. 1968. Southwestern fishes and the enigma of "endangered species." Science 159:1424-1432.

Myers, G.S. 1925. Introduction of the European bitterling *(Rhodeus)* in New York and of the rudd *(Scardinius)* in New Jersey. Copeia 140:20-21.

Myers, G.S. 1965. *Gambusia,* the fish destroyer. Trop. Fish Hobb. 13:31-32, 53-54.

Nobel, R. L. and R. D. Germany. 1986. Changes in fish populations of Trinidad Lake, Texas, in response to abundance of blue tilapia, p. 455-461. *In* R. H. Stroud (ed.), Fish Culture in Fisheries Management. American Fishery Society, Bethesda, Maryland.

Pflieger, W. L. 1989. Natural reproduction of bighead carp *(Hypophthalmichthys nobilis)* in Missouri. Am. Fish. Soc. Introduced Fish Sec. Newsletter 9(4):9-10.

Pigg, J. 1978. The tilapia *Sarotherodon aurea* (Steindachner) in the North Canadian River in central Oklahoma. Proc. Oklahoma Acad. Sci. 58:111-112.

Pullin, R.S.V. 1986. The worldwide status of carp culture, p. 21-34. *In* R. Billard and J. Marcel (eds.), Aquaculture of cyprinids. Paris, INRA.

Roessler, M.A., G.L. Beardsley and D.C. Tabb. 1977. New records of the introduced snail, *Melanoides tuberculata* (Mollusca: Thiaridae) in south Florida. Fla. Sci. 40:87-94.

Shelton, W.L. and R.O. Smitherman. 1984. Exotic fishes in warmwater aquaculture, p. 262-301. *In* W.R. Courtenay, Jr. and J.R. Stauffer, Jr. (eds.), Distribution, biology, and management of freshwater fishes. Johns Hopkins University Press, Baltimore, Maryland.

Shotts, E.B., Jr., and J.B. Gratzek. 1984. Bacteria, parasites, and viruses of aquarium fish and their shipping waters, p. 215-232. *In* W.R. Courtenay, Jr. and J.R. Stauffer, Jr. (eds.), Distribution, biology, and management of exotic fishes, Johns Hopkins University Press, Baltimore, Maryland.

Skinner, W.F. 1984. *Oreochromis aureus* (Steindachner: Cichlidae), an exotic fish species, accidentally introduced to the lower Susquehanna River, Pennsylvania. Proc. Pennsylvania Acad. Sci. 58:99-100.

Smiley, C.W. 1886. Some results of carp culture in the United States. Rept. U.S. Fish Comm. 1884:657-890.

Smith, C.L. 1985. The inland fishes of New York State. N.Y. State Dept. Env. Conserv., Albany, New York.

Stanley, J. G. 1976. Reproduction of the grass carp *(Ctenopharyngodon idella)* outside its native range. Fisheries 1(3):7-10.

Stauffer, J.R., Jr., S.E. Boltz and J.M. Boltz. 1989. Thermal tolerance of the blue tilapia, *Oreochromis aureus,* in the Susquehanna River. N. Amer. J. Fish. Manag. 8:329-332.

Stickney, R.R. 1979. Principles of warmwater aquaculture. Wiley-Interscience, New York.

Stroud, R.H. 1972. Grass carp problem. SFI Bull. 240:4-5.

Swingle, H.S. 1960. Comparative evaluation of two tilapias as pondfishes in Alabama. Trans. Amer. Fish. Soc. 89:142-148.

Taylor, J.N., W.R. Courtenay and J.A. McCann. 1984. Known impacts of exotic fishes in the continental United States, p. 322-373. *In* W.R. Courtenay, Jr. and J.R. Stauffer, Jr. (eds.), Distribution, biology, and management of exotic fishes. Johns Hopkins University Press, Baltimore, Maryland.

Walker, B.W., editor. 1961. The ecology of the Salton Sea, California, in relation to the sportfishery. Calif. Dept. Fish & Game, Fish. Bull. 113.

Welcomme, R.L. 1984. International transfers of inland fish species, p. 22-40. *In* W.R. Courtenay, Jr. and J.R. Stauffer, Jr. (eds.), Distribution, biology, and management of exotic fishes. Johns Hopkins University Press, Baltimore, Maryland.

Walker, B.W., editor. 1961. The ecology of the Salton Sea, California, in relation to the sportfishery. Calif. Dept. Fish & Game, Fish. Bull. 113.

Welcomme, R.L. 1984. International transfers of inland fish species, p. 22-40. *In* W.R. Courtenay, Jr. and J.R. Stauffer, Jr. (eds.), Distribution, biology, and management of exotic fishes. Johns Hopkins University Press, Baltimore, Maryland.

Welcomme, R.L. 1988. International introductions of inland aquatic species. FAO Fish. Tech. Pap. 294.

Williams, J.E. and G.R. Wilde. 1981. Taxonomic status and morphology of isolated populations of the White River springfish, *Crenichthys baileyi* (Cyprinodontidae). Southwest Nat. 26:485-503.

Zuckerman, L. D. and R. J. Behnke. 1986. Introduced fishes in the San Luis Valley, Colorado, p. 435-453. *In* R. H. Stroud (ed.), Fish culture in fisheries management. American Fishery Society, Bethesda, Maryland.

Introduction of Exotic Species for Aquaculture Purposes

Jack R. Davidson
James A. Brock
Leonard G. L. Young

Abstract: The introduction of exotic species for economic and natural resource development is not an easy path. Weighty concerns for inadvertent spread of pests and pathogens having environmental and external costs should be considered. We review issues and concerns in the United States with emphasis on the Hawaii aquaculture development experience. Both regulatory and aquaculturist points of view are considered. We propose, herein, public management of exotic aquatic species for aquaculture development.

Introduction

Deliberate introduction of organisms dates back to the beginnings of human civilization. There is reason to believe such transplantations included aquatic as well as terrestrial organisms. Reasons for introducing aquatic species in the present and recent past include fishery management, pest control, marketing of ornamental organisms, and aquaculture. During the past two decades, increasing support for aquaculture research has been available from government and private sources. With increasing emphasis on economic diversification and improvement of trade balances for fishery products, a wide variety of promising species have been investigated. The growth of aquacultural activities has greatly accelerated interest and activity in importing exotics and interstate transfers for culture.

In many instances, aquaculturalists' interests may appear to conflict with those of environmentalists and natural resource managers. Aquaculturalists usually want to expedite the process of importing organisms, while natural resource managers may want to control the process. Aquaculturists view confinement-culture systems as offering no risks to the natural environment. To natural resource managers, introductions hold many of the same threats as attempts to introduce organisms into native systems, because of possibilities of escape or pathogen release. As a result, resource managers may resist introductions simply because they see their options narrowing rapidly once the animal is in their area. Interest of state resource managers and aquaculturalists most nearly converge on the issue of disease. Most aquaculture systems in the United States require large capital commitments for facilities and operating costs. The danger for introducing pathogens is a constant threat to the economic viability of such enterprises.

While this paper will focus on introductions of exotics for aquaculture purposes, the more serious problems associated with introduction of exotics are certainly not unique to aquaculture. Rosenthal (1976) suggested the following potential problems that may arise from species introductions:

1. Reduced growth and development of the introduced species owing to less favorable environmental conditions.
2. Population explosion of the introduced species leading to competition and possible elimination of native species.
3. Concomitant introduction of diseases, parasites, and pests harmful to introduced or resident species.
4. Destructive activities of the introduced species affecting other fields of economic interest.

This paper is not an attempt to make a statement on the status of exotic introductions. Rather, it is based on our concerns as people who share responsibility for the development and management of aquaculture. By using Hawaii's experiences in developing an aquaculture industry based on non-native species, we will reaffirm the contemporary importance of non-native introductions, discuss current concerns around the country, and highlight potential areas for improving the management of introductions.

Exotic Introductions and Modern Hawaii Aquaculture

By 1967, aquatic animal introductions into Hawaii numbered nearly 70 species, 36 of which had become established (known to be propagating outside of captivity) (Kanayama *1976)*. Eight of the established species were introduced before 1900. These include the eastern oyster *(Crassostrea virginica)*, bass *(Micropterus* sp.*)*, common carp *(Cyprinus carpio)*, and Chinese catfish *(Clarius fuscus)*. A number of established species were ornamentals, probably discarded into streams and ponds by tropical fish hobbyists. The mosquito fish *(Gambusia affinis)* and the sailfin molly *(Poecilia latipinna)* were introduced under government sponsorship for control of mosquitoes early in the twentieth century. Rainbow trout *(Oncorhynchus mykiss)* were introduced in 1920 on Maui and Kauai with marginal reproduction because of warmwater temperatures. State rivers and reservoirs were stocked with channel catfish *(Ictalurus punctatus)* in the 1950s and 1960s. This species has become established in reservoirs on Oahu. The Oscar *(Astronotus ocellatus)* and the tucunare *(Cichla ocellaris)* were introduced as candidates for freshwater fishing, which started in the 1960s.

Starting in the 1930s and extending into the 1980s, a number of attempts were made to supplement the supply of baitfish available to the skipjack tuna fleet through introductions of salt- and freshwater species. Kanayama (1967) notes the introduction of California anchovy *(Anchoa compressa)* in 1932, and threadfin shad *(Dorosoma petenense)*, four species of tilapia *(Oreochromis mossambicus, Tilapia zilli, T. melanopleura, O. macrochir)*, and Marquesan sardine *(Harengula vittata)* in the 1950s as potential baitfish. The freshwater species became established but were not accepted widely as bait. The Marquesan sardine became the first marine fish established by transplantation; however, it is not abundant. The California anchovy did not survive. Interest in an alternative to using nehu *(Stolephorus purpureus)* as bait remained high with several attempts at importing or culturing supplies extending into the 1980s. Interest has declined to some degree since the closure of the Hawaiian Tuna Packers, Inc. Cannery in 1985.

The marine grouper roi *(Cephalopholis argus)* and two snappers, Toau *(Lutjanus vaigiensis)* and taape *(L. kasmira)*, and two species of freshwater prawns, *Macrobrachium lar* in 1956 and *M. rosenbergii* in 1964, were introduced by the Department of Land and Natural Resources, State of Hawaii, for the express purpose of establishing these species in Hawaii. Roi, the two snappers, and *M. lar* have become established. *M. rosenbergii* is not established and survives through aquaculture propagation. *M. rosenbergii* culture developed in the mid-1970s and reached a peak in 1984 of 90 pond hectares, with an annual production of 144,000 kg valued at $1,706,000 (State of Hawaii 1987). There are currently seven farms in prawn production; five of these farms are small and provide a secondary income for their owners (Steven R. Lee, Economic Development Specialist, Hawaii Aquaculture Development Program, DLNR, 1988, personal communicaton).

The interest in development of commercial aquaculture of other species, private aquaculture, research and development, and consulting ventures has not abated and continues to grow. Commercial ventures produce a variety of both animal and plant species (Table 1) contributing to Hawaii's agricultural economy. The industry can be described as small but highly diversified, with many species and technologies. The aquaculture industry in Hawaii in 1988 generated $18.2 million in revenues (Figure 1). Of that amount, $5.5 million was attributed to the production sector and $12.7 million to the research, training, and technology-transfer sector. The trend is toward intensification of culture systems.

The group of highest current interest is the marine shrimp. With the exception of lobsters and freshwater prawns, no other

Table 1. Aquaculture species cultured in the State of Hawaii (January 1989).

Common Name	Scientific Name
Commercial Culture	
Freshwater prawn[3,7,10]	*Macrobrachium rosenbergii*
Marine shrimp[5,10]	*Penaeus vannamei*
(Tiger)[5,10]	*P. monodon*
Chinese carfish[3,6]	*Clarius fuscus*
Channel catfish[2,6]	*Ictalurus punctatus*
Gass carp[2,7]	*Ctenopharyngodon idella*
Silver carp[2,7]	*Hypopthalmichthys molitrix*
Bighead carp[2,5]	*Aristichtys nobilis*
Ornamental carp (Koi)[3,4]	*Cyprinus carpio*
Goldfish[6,10]	*Carassium auratus*
Tilapias[3,6]	*Oreochromis* and *Sarotherodon* sp.
Trout[2]	*Oncorhychus mykiss*
Salmon[3,5]	*Oncorhychus kisutch*
Baitfish[3,6]	*Poecilia* sp.
Brine shrimp	*Artemia* sp.
Ogo[3]	*Gracilaria* sp.
Frogs[3,6]	*Rana catesbiana*
Aquatic snails[8]	Two species
Abalone[3,5]	*Haliots* sp.
Kelp[3,5]	*Macrocystis pyrifera*
Nori[3,5]	*Porphyra tenera*
Freshwater aquarium fish[4]	Various species
Species grown in Hawaiian Fishpond Cultrue	
Mullet[9]	*Mugil cephalus*
Milkfish[9]	*Chanos chanos*
Samoan crab[6]	*Scylla serrata*
Bonefish[9]	*Albula vulpes*
Threadfin[9]	*Polydactylus sexfilis*
Species Under Investigation	
American lobster[2,5]	*Homarus americanus*
Clams[5,6]	*Mercinaria mercinaria*
Oysters[5,6,7]	*Crassostrea virginica*
	C. gigas
Microalfae[3,5]	*Spirulina, dunaliella*
Mahimahi[9]	*Coryphaena hippurus*
Freshwater mussels[5]	*Proptera alata*
Opai (bait shrimp)[9]	*Palomon debilis*

[1]Introduced exclsively for aquaculture
[2]Restocking for aquaculture farms by importation from out of state
[3]Restocking mainly from local sources
[4]Frequently interoduced in large numbers in the pet trade
[5]Large quantities of live animals imported for human consumption
[6]Species was intentionally released and is established
[7]Species was intentionally released and is not established
[8]Species was unintentionally introduced and is established
[9]Indigenous species
[10]Reproducing in Hawaii but only in captivity

crustacean species are being cultured in the State. Although a number of commercial attempts to mass culture marine shrimp — starting in the early 1980s — have failed, interest continues virtually unabated. Five marine shrimp species *(Penaeus vannamei, P. stylirostris, P. monodon, P. japonicus,* and *P. chinesis)* have been introduced for aquaculture or for aquaculture research. None of these crustaceans have become established in Hawaiian waters, although *P. vannamei, P. stylirostris, P. japonicus,* and probably *P. monodon* have escaped unintentionally from captivity. Penaeids introduced but no longer in Hawaii include *P. japonicus* and *P. chinesis.* The small introduced groups of *P. chinesis* were terminated for pathogen-control reasons. Varieties of strongest current interest for commercial production include *P. vannamei* and *P. monodon.*

Today, more than 25 aquatic species are cultured commercially in Hawaii (Table 1). Twenty of these species are introduced exotics. Land-based aquaculture operations require regular reintroduction of five of these species *(Ctenopharyngodon idella, Hypophthalmichthys molitrix, Aristichtys nobilis, Ictalurus punctatus, Oncorhynchus mykiss),* because their natural reproduction is lacking or provides only limited number of offspring for producers. Ten are exotics that were established years ago in local waters of the state. These include the fish *Clarius fuscus, Carassius auratus, Oreochromis* and *Sarotherodon* spp., *Poecilia* sp., and *Cyprinus carpio;* the molluscs *Crassostrea virginica* and *C. gigas,* and freshwater snails; and the amphibian *Rana catesbiana.* All of these species were introduced and became established in Hawaii prior to their use in aquaculture rearing.

In Hawaii, the importance of introductions to support current and future aquaculture development is obvious. Clearly, the majority of species under cultivation are exotics. Continued development of aquaculture in Hawaii will be influenced strongly by work with these and other candidate species.

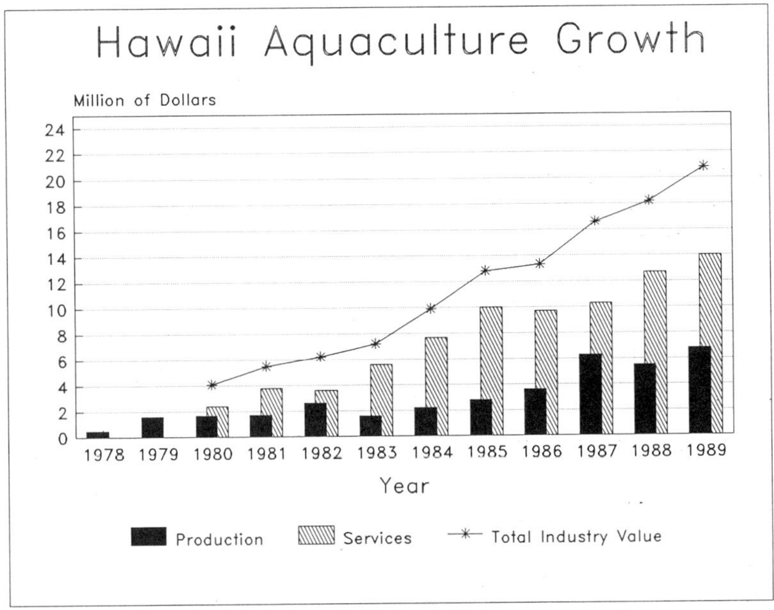

Figure 1. Growth in millions of dollars of the Hawaiian aqaculture industry.

Hawaii Species Import Regulatory Procedures

Requests for introduction of aquatic organisms into Hawaii are processed in a similar fashion to requests for introduction of other types of plants and animals (Brock 1986). Regulations pertaining to live organism imports are on a species list organized into the following categories: approved for entry, admitted under certain conditional requirements, prohibited possession by private individuals, and prohibited entry. This species list may be modified by the Board of Agriculture. For aquatic organisms on the approved species list, import permits are issued by the Plant Quarantine Branch of the Hawaii Department of Agriculture upon receipt of a written application (Figure 2). The Subcommittee on Invertebrate and Aquatic Biota, along with the Committee on Plants and Animals, the Plant Quarantine Branch, and the Board

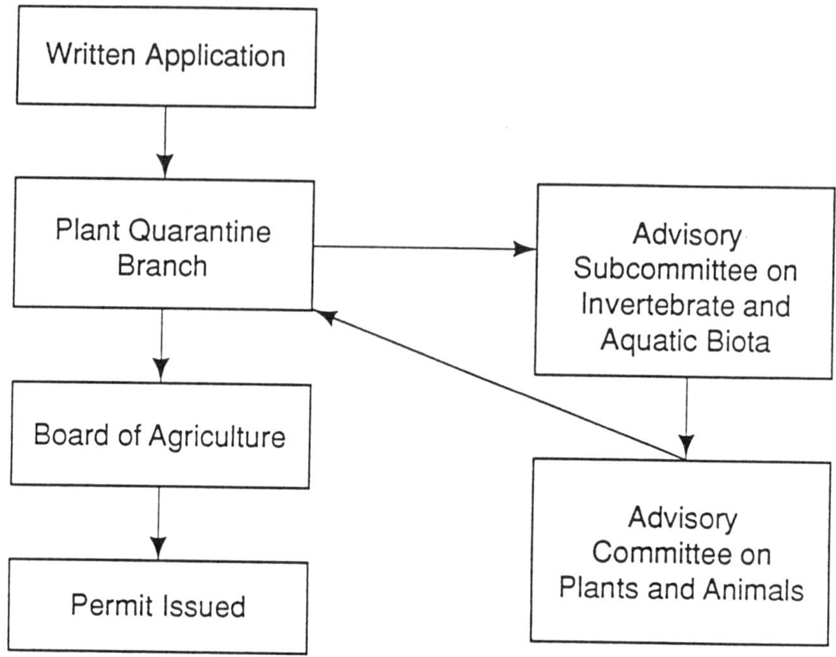

Figure 2. The aquaculture application process in Hawaii.

of Agriculture, reviews applications for new species not on the approved list. Permits are usually issued with conditions.

All imports are inspected by Plant Quarantine Branch inspectors at the port of entry. These are visual inspections to assure that the species listed on the import permit conform to those in the lot and that shipment documentation is in order. Postentry examination for parasites and pathogens is conducted for most aquatic animals imported for aquaculture. In rare cases, parasite and pathogen examination is done for aquatic animals introduced for the pet trade or for human consumption.

All imported aquatic species must be kept in captivity unless the import permit specifically states otherwise. Thus, these permits allow introduction of the species but with the expressed understanding that they are not approved for release from captivity without prior action by the Board of Agriculture. This would require written application for a variance of the import permit, a full-review process through the subcommittees, and a decision rendered by the Board.

Under current policy in Hawaii, introduced aquatic animals for aquaculture development are permitted to enter with the understanding that they will be kept isolated, usually on the premises of the importer, with the effluent disinfected prior to discharge or injected into a dispersion well. The purpose of holding the animals in isolation is predominantly for parasite and pathogen control. The duration of isolation varies depending on the species and life stage; the current knowledge of its pests, predators, and pathogens; the disease history of the species at the point of origin; and the type and result of inspections carried out prior to and after the species arrival in the state.

The Aquaculture Development Program Department of Land and Natural Resources, through its aquaculture-disease specialist, provides technical and advisory assistance to the Plant Quarantine Branch in matters pertaining to disease control for imported aquatic animals. Plant Quarantine Branch staff members inspect isolation and growout systems that hold imported aquatic species, and they inspect imported shipments upon entry into Hawaii.

Federal agencies also play an active role in some cases of exotic species introductions into Hawaii, including providing expert testimony on proposed introductions; assuring compliance to applicable federal regulations; and allocating federal funds for aquaculture research and development projects in which exotics are cultured. Federal regulations for importation and exportation of wildlife and other organisms are covered by the Lacey Act (74 Stat. 754, 18 U.S.C. 42), its amendments of 1981 (Public Law 97-79), and its implementation by Chapter 50 of the Code of Federal Regulations Parts 14, 16, 17, and 23.

Generally, actual decision-making for importation of species having economic value or as a natural resource lies with the state governments. Federal authority interacts with the shipping of organisms across national borders. States are responsible for the management of a large percent of the natural resource endowments within their boundaries. When state managers express concern with introduction of exotic species, it is with the risks of any possible new non-native transplantation. As Welcomme (1986) pointed out, the transfer of species from one country to another may be ecologically no more significant than transfers across geographical boundaries in the same country or same region.

Potential Problem Areas, Issues, and Concerns in Hawaii

The state of Hawaii system of regulations and security measures for risk management of introduced aquatic species for aquaculture development has many strong points. The system provides a timely expert review of proposed species for introduction; additional review for release into the environment; and laboratory evaluation for parasites and pathogens. This system is managed within an existing agency framework and has been implemented to cover aquaculture introductions with modest expenditure of public funds.

The system also has potential areas of weakness. These principally involve practical aspects of risk management once groups of organisms have entered the islands. Areas of concern are unauthorized escape of permitted species, and entry and establishment of nonpermitted organisms that may accompany shipments of permitted species. The nonpermitted introductions involve both pathogens, such as infectious hypodermal and hematopoietic necrosis virus of shrimp (see Lightner et al., this volume) found within the tissues of imported animals, as well as epibionts on the surface of shellfish and organisms present in the shipment water (see Farley, this volume). A polychaete worm, *Polydora nuchalis,* recently discovered in an aquaculture pond and drainage canal in Hawaii appears to have been translocated and established in Hawaii with an introduction of animals for aquaculture development (Bailey-Brock 1990). This points out the need to review the postentry security steps for introductions and, possibly, to implement increased control measures. However, if increased effort in terms of enforcement is required, this may require additional resource allocation for implementation.

Another issue is the burden of economic liability. A case in point is an introduction that does not result in viable aquaculture development. For example, a company investing in the importation of penaeid shrimp may need to destroy its shipment owing to pathogen-control reasons. Obviously, the investment is lost, but it does not end there. Loss of economic development opportunity for the company and the spin-offs such as jobs and tax revenues for the state are considerable. Further, an introduction of an exotic can pose a pathogen risk to other cultured stocks on the premises from inadvertent transfer between introduced and existing animal stocks. Should this occur by accident, these other stocks may also suffer high disease losses or be destroyed during pathogen-eradication procedures. The potential of pathogen risks to other aquafarms, the ecosystem, and fisheries resources arising from an aquatic animal importation is a further consideration. This is a risk potential that extends beyond the property of the importer. The direct consequences to the importer may be minimal, implying a greater potential for less concern by the importer for this risk. It may not be especially realistic to expect the importer to expend substantial resources to prevent this risk. This is an area of great concern for resource management. At present there is no indemnification program established in Hawaii to compensate a producer if stocks are depopulated for reasons of pathogen control.

For aquaculturists, the decision to make a species introduction or transfer for aquaculture development is influenced heavily by economic considerations. A private company will invest in administrative, rearing facilities, feeds, staff, purchase, and transportation costs if it believes that these expenditures will result in a positive business development. In most cases, an aquaculture-development activity with restricted capital would require a short-term return on this investment. A government agency or private firm with substantial capital would be in a better position to absorb a loss or defer the return on an investment. This feature points to two desirable features of having government sponsorship of selected exotic introductions for aquaculture development: potentially less economic risk to the private sector and increased control by direct government-sponsored supervision of the introduction.

Reintroduction of approved species is another concern because the potential for pathogen entry increases as more and more introductions of a species are made. The current review process in Hawaii focuses considerable attention on the initial introduction and less so when reintroductions are made. The potential problem is not the species approved for entry into Hawaii but the organisms and pathogens that may accompany these shipments. The day-to-day postentry security for imported groups is carried out by the importer. The regulatory agency cannot always check if importers are carefully following all permit conditions. Thus, rigorous enforcement of permit conditions cannot always be assured. In addition, disease monitoring on selected shipments of aquatic animals is difficult because of resource limitations, implying a need to allocate more resources for this purpose or reduce the number of permitted shipments. Alternatives need to be explored.

The issues and concerns discussed are important for maintaining the success of the aquaculture industry and integrity of the environment. Improvements can be made to the existing security system for aquatic animal translocations. Hawaii has much to lose should any of these potential problems occur. These issues and concerns are not unique to Hawaii.

Examples of Species Introductions and Concerns Elsewhere

There are a few classic examples of purposeful introductions and intentional translocations. Marine species include introduction of the Japanese oyster, *Crassostrea gigas,* and clam, *Mercenaria mercenaria,* to the west coast of the United States and Canada and attempts at establishing the Pacific salmon, *Oncorhynchus tschawytscha*, on the Atlantic coast. Aside from pests, parasites, and disease organisms introduced with spat shipments of the Japanese oyster, history seems to have judged their transplantation a success. The success of translocating the Pacific salmon to the Atlantic is pending upon establishment of a significant reproducing-return population. Other examples include the many successful translocations of trout (salmonids) worldwide, for example, salmon farming is well established in Chile and Peru, and the recent introduction of striped bass to the west coast of the United States. To the extent these translocations are judged successful in retrospect, their success may be more fortuitous than the result of careful planning. Controversial examples include common carps, tilapias, and grass carps, and accidental release of marine shrimps from experimental facilities.

Florida and southern California provide prime examples of situations in which fish species have become established without official sanction in recent years, mostly from accidental or intentional release from aquarium farms. For instance, in Florida, at least 38 species and several hybrids of exotic fishes have been found in fresh- and brackish waters. Twenty species and five hybrids are established as reproducing populations. Associated problems with accidental and intentional releases of aquarium fishes include replacement of native species of fish; threat to native aquatic plants, which could result in management problems with both native fish and waterfowl; and spread and range extension of fish to other states through linked waterways (Courtenay et al. 1974; Shafland 1979).

Current interest in Louisiana focuses on the grass carp, *Ctenopharyngodon idella*, which was first released by Arkansas to control aquatic weeds (Ronald Becker, associate director, Louisiana State University Sea Grant College Program, 1989: personal communication). It subsequently has invaded many tributaries of the lower Mississippi River including those in Louisiana. The concern is that this species will invade coastal marshes and destroy aquatic plants, especially the widgeon grass, *Ruppia maritima*, an important food and attractant for waterfowl. Aquaculturists want to raise triploid grass carp with catfish and possibly the freshwater prawn, *Macrobrachium rosenbergii.* Opponents believe this may open the door to importation of diploid fish with possibly disastrous effects on aquatic habitats.

Jack Greer (University of Maryland Sea Grant Program, 1989: personal communication) indicates that working with hybrids of traditional species is difficult in Maryland. A popular species for aquaculture in the mid-Atlantic region is the hybrid striped bass, a cross between white bass and striped bass *Morone saxatilis.* Productive hybrid striped bass aquaculture enterprises are developing. Owing to a crash in the wild populations of striped bass in Chesapeake Bay, the state of Maryland had a moratorium against the catching, possession, or sale of striped bass. Largely due to pressures from commercial fishermen, who fear competition from aquaculturists, the hybrid striped bass was included in the moratorium. As a result, it was illegal to possess a hybrid striped bass without special permission from the Maryland Department of Natural Resources. It was impossible to sell hybrid striped bass, though several demonstration projects begun by the University of Maryland and state of Maryland had already produced market-size fish. Though the hybrid bass does not involve the introduction of exotic species, there are many who fear that its escape into waterways will result in the contamination of the wild stocks. Further, commercial fishermen and law-enforcement officials have argued that hybrid striped bass would be confused with wild striped bass, complicating enforcement issues.

Granvil Treece (Mariculture specialist, Texas A & M Sea Grant College Program, 1989: personal communication) indicates a continuing, strong commercial interest for shrimp aquaculture in Texas. The interest resulted in many introductions for commercial operations and for university research. These include *Penaeus vannamei, P. monodon, P. japonicus,* and most recently *P. penicillatus* as a potential winter-crop species. There have been a few accidental releases of shrimp, but no known problems have been documented.

On the other hand, Treece cites tilapia as an example of a troublesome species. Texas Department of Parks and Wildlife brought tilapia into Texas and stocked it into Texas lakes. There have been many escapes from overflooded farm ponds and bait releases. By 1978, tilapia inhabited 14 reservoirs and three rivers in Texas. They appear to have spread to at least 13 more lakes since that time. Although this is a warmwater fish which will perish at temperatures below 50°F, many of the reservoirs are warmed by power plant discharges and some rivers by warm springs. There is speculation that physiological races of tilapia are developing that are more cold tolerant.

Management of Aquaculture Introductions

The current high level of environmental awareness together with economic lessons learned from prominent deleterious examples make contemporary state natural resource managers more wary of new introductions into natural systems, either in the form of purposeful or accidental release. This is reflected in the American Fisheries Society's Position on Introductions of Aquatic Species (Kohler and Courtenay 1986). The society's position calls for all species initially to be prohibited and considered undesirable unless they are evaluated and found to be desirable on the following basis:

1. Published information on candidate species has been reviewed.
2. Import species have more desirable qualities than the native species.
3. Preliminary assessment of candidate species for their impact on target aquatic ecosystems is benign.
4. Candidate species have been studied in their biotope for potential impacts on target aquatic systems.
5. Provision has been made for public review.

If a species passes these steps, a research program should be initiated by an agency to test the import in confined waters. The evaluation or recommendations would be circulated among interested scientists and presented for publication. Any negative decision along this pathway would result in restrictions for further study, importation, introduction, and release.

Kohler and Stanley (1984) in developing a model for evaluating exotic fish introductions suggested five levels of review. At each level, a favorable decision for introduction and recommendation was required to carry the review through the next level or to reject the application. Level 1 review required information on the purposes of introduction, abundance in native range, potential disease and parasites associated with the species to be introduced, and site of introduction (see Elston, this volume). Review level 2 permitted the proposing entity to conduct research with a limited number of specimens under confined conditions for the purpose of obtaining data. Level 3 concentrated on benefits to risk analysis of the species to the natural system and humans. Review level 4 required evaluation of a literature review based on the format for a Food and Agriculture Organization species synopsis with the possibility of requiring additional sections concerning impacts. Level 5 review required further research to complete the species synopsis and specific research to assess potential impacts on the receiving systems and native species (see Kohler, this volume).

The 1979 International Council for the Exploration of the Sea (ICES) Code of Practice to Reduce the Risks of Adverse Effects Arising from Introductions of Nonindigenous Marine Species (Sindermann 1986) further recommended that when proceeding with an introduction, a brood stock should be established in an approved quarantine situation and . . . that the first-generation progeny of the introduced species can be transplanted into the natural environment if no disease or parasites become evident, but not the original import. A continuing study should be made of the introduced species in its new environment, and progress reports submitted . . ." (see Sinderman, this volume).

We believe a model management program for exotics introduction for aquaculture is one that gives priority to and specifically addresses the unique needs of aquaculture development. Importantly, acceptable solutions would be found and implemented for problems posed by introductions or regional transfers for those species with a high potential for aquaculture development. If disease is a limitation, then specific pathogen-free or genetic-resistant stocks of the desired animals would be developed to remove this constraint (see Elston, this volume). If the targeted species is desirable for aquaculture but poses unacceptable risks to fisheries resources or ecosystems, then an alternate species would be sought or an acceptable compromise reached to allow the initial organism entry. The responsibility for providing healthy aquaculture stocks, those that would not pose a threat to fisheries or ecosystem resources, would be an acknowledged goal of the program. Public funding would be used to finance the introduction of selected exotics with high potential for aquaculture development. Control over the introductions would follow the ICES guidelines. Adequate funding would be necessary to carry the introduced group through a minimum of one generation. Once available, high-quality seed stocks would be distributed to potential farmers for cultivation. Support would be provided to stimulate development of locally produced seed for restocking farms so that reintroduction could be avoided altogether or at least minimized. In this scenario the economic risks taken for an introduction would be largely supported by expenditure of public funds.

In terms of diseases, regional inventory of diseases in major fisheries species and cultured animals would need to be carried out. And from a regulatory perspective, it does not make sense to try to control movement of a particular pathogen between regions if the organism is already present in both areas (Elston 1988). Funds would also have to be provided for education programs to keep industry and resource management informed on these matters. As Elston (1988; see this volume) points out, and our experience confirms, industry will support disease prevention when it understands the needs and benefits of such programs.

Obviously, we are aware of the limited availability of public funds. This dictates that a key issue is in the selection of candidate species for the type of development process we have suggested. Thus, only a few, high-probability-of-success-for-profitable-culture candidate species could be dealt with adequately. The highest level of quality control practical for these selected species would be practiced and, in this way, it will contribute to development of a strong basis for future economic stability and growth.

Literature Cited

Bailey-Brock, J.H. 1990. *Polydora nuchalis* (Polychaeta, Spionidae), a new Hawaiian record from aquaculture ponds. Pac. Sci. 44(1), 81-87.

Brock, J.A. 1986. An overview of Hawaii's animal/plant species importation regulations and pest, predator and pathogen control procedures for aquatic animal introductions for aquaculture development. Aquaculture 86, WMS Special Session: Certification, Transplantation and Catastrophic Diseases.

Courtenay, W.R., Jr, H.F. Sahlman, W.W. Miley, II and D.J. Herrema. 1974. Exotic fishes in fresh and brackish waters of Florida. Biol. Conserv. 6(4):292-301.

Elston, R. 1988. Comments on the regulation of aquatic animal transports. FHS/AFS Newsl. 16(4):7.

Hawaii, State of. 1987. The State of Hawaii data bank: A statistical abstract, Honolulu.

Kanayama, R.K. 1967. Hawaii's aquatic animal introductions. Forty-seventh Annual Conference of the Western Association of State Game and Fish Commissioners, July, 1967, Honolulu, Hawaii.

Kohler, C.C. and W.R. Courtenay, Jr. 1986. American Fisheries Society position on introductions of aquatic species. Fisheries 11(2): 2-3.

Kohler, C.C. and J.G. Stanley. 1984. Suggested protocol for evaluating exotic fish introductions in the United States. p. 387-406. *In* W.W. Courtenay and J.R. Stauffer, Jr. (eds.), Distribution, biology, and management of exotic fishes. The Johns Hopkins University Press, Baltimore, Maryland.

Rosenthal, R. 1976. Implications of transplantations to and ecosystems. FAO Technical Conference on Aquaculture, May 26, through June 2, 1976, Kyoto, Japan.

Shafland, P.L. 1979. Non-native fish introductions with special reference to Florida. Contribution No. 14, Non-Native Fish Research Laboratory, Florida Game and Fish Commission, Boca Raton, Florida.

Sindermann, C.J. 1986. Strategies for reducing risks from introductions of aquatic organisms: A marine perspective. Fisheries 11(2): 10-15.

Welcomme, R.L. 1986. International measures for the control of introduction of aquatic organisms. Fisheries 11 (2):4-9.

Introduction of Marine Plants for Aquacultural Purposes

MICHAEL NEUSHUL
CHARLES D. AMSLER
DANIEL C. REED
RAYMOND J. LEWIS

Abstract: The accidental introduction of the Japanese kelp, *Laminaria japonica*, to the northern coast of China, and its subsequent domestication and transplantation now provides the basis for the most extensive and productive marine farms in the world. A similar Japanese kelp, *Undaria pinnatifida*, has also been accidentally introduced to the Mediterranean Sea, and then transplanted into the Atlantic, where it is being cultivated. *Undaria* has very recently been found in Tasmania, where its effects on native vegetation are not yet known; however, it could serve as a useful food for abalone mariculture there. Many might consider these unintended introductions to be advantageous; however, the accidentally introduced brown alga, *Sargassum muticum*, is viewed as a weed. The green alga, *Codium fragile*, and the brown alga, *Colpomenia peregrina*, introduced to the North Atlantic, have become nuisances to oyster growers on both sides of the Atlantic. Intentionally introduced marine crop plants include the tropical carrageenophyte, *Kappaphycus alvarezii*, and the Japanese sea-vegetable, *Porphyra yezoensis*. The giant kelp, *Macrocystis pyrifera*, was introduced into European waters, and then removed because of concern that it might escape and spread. It has recently been introduced to China where it has become established, but so far it has not spread. It is being grown in ponds in Hawaii as well. Studies of the reproductive biology and genetics of *Macrocystis*, at the University of California, Santa Barbara, have shown that it is possible to follow the dispersal of microscopic life-history stages, and suggest that it might be possible to genetically modify this and other macroalgal crop plants, as part of the domestication process, so that like corn they would require human intervention for reproduction. Such plants would remain non-reproductive in the sea and could be safely introduced elsewhere in the world.

Introduction

Anyone contemplating the advantages that might be gained by introducing a new macroalgal crop plant to U.S. waters should be careful, since it may be unlawful to do so. The modified Lacey Act, which became law on November 16, 1981 (U.S.C. 3371-3378) is intended to provide enforcement tools for wildlife agencies seeking to control the illegal trade in fish and wildlife. This law states that, "It is unlawful for any person — (1) to import, export, transport, sell, receive, acquire, or purchase any fish or wildlife *or plant* taken or possessed in violation of any law, treaty, or regulation of the United States or in violation of any Indian tribal law. . . ." A violation of the act may be treated as a misdemeanor, punishable by fines of up to $10,000 or a year's imprisonment, or both. If the imported "fish" is worth $350 or more the violation may be treated as a felony, and criminal penalties of up to $20,000 and five years imprisonment or both can be imposed. Since each separate transaction is considered a separate violation of the act, a separate penalty can be assessed for each violation.

The severity of the Lacey Act is undoubtedly attributable, in part at least, to the lessons learned in Australia after importing rabbits and the prickly pear cactus and in Florida after introduced walking catfish, Brazilian pepper trees and water hyacinth escaped and spread. These and other similar lessons form the basis for the generally accepted view that one should not indiscriminately introduce species that are likely to escape from cultivation or captivity. This cautiousness has now been extended to include lifeforms that have been genetically transformed in the laboratory, with the general feeling that it is better to err by being overly careful. Nonetheless it is important to remind ourselves that mankind can benefit greatly from the introduction of agriculturally useful plants. When Cortez invaded Aztec Mexico in 1519, he found extensive gardens that surpassed any to be found in Europe at the time. The early American agriculturists had domesticated corn, cotton, agave, cacao, tomato, sweet potato, squash, pumpkin, beans, peanuts, pineapple, avocado, pepper, papaya, cassava, banana, potato, tobacco, prickly pears and rubber plants. One wonders what the course of history might have been if the Lacey Act had been in effect and vigorously enforced in 1519. Cortez might have spent the rest of his life in prison for having these new crop plants transported back to Europe.

The advantages and disadvantages of introducing marine crop plants from one part of the world to another have been the subject of discussion for many years. Druehl (1973), in a letter to *Science* discussing marine transplantations, points out that transplantations can cause considerable damage, citing as an example the accidental introduction of *Sargassum muticum* to the eastern North Pacific coast along with oyster spat. He predicted (accurately) that this plant would also become established in European waters along with introduced spat. It is certainly worthwhile now to reconsider Druehl's 1973 prediction that it would be many years before we will be able to predict with any degree of certainty the effects an introduced species may have on an existing ecosystem. We might now ask what new information is needed if we wish to predict the rate of spread, and other effects an introduced macroalga will have?

In response to Druehl's plea for international regulation of marine transplantation projects, North (1973) pointed out that some marine scientists at the time felt that the adverse effects of introduced species were sometimes exaggerated, and notes that there are enormous benefits to be derived from cultivating a plant like *Macrocystis*. As a world expert on the biology of this plant, he had been approached by several Japanese biologists interested in introducing *Macrocystis* to Japan, but because of the strong objections of several American scientists, this request was withdrawn.

The Japanese were not the first to be interested in introducing and growing the giant kelp. The controversy that has arisen concerning this plant illustrates some of the concerns to be considered before attempting long-range marine transplantations. The Scottish Seaweed Research Association had proposed to introduce the giant kelp to Scotland in 1950. Walker (1952) brought gametophytes from British Columbia to Prestwick, Scotland, in August 1949 and successfully cultivated juvenile kelps, but was prohibited from introducing them into the sea. This proposal was considered by a governmental advisory council and was rejected as a dangerous experiment with possibly undesirable results. The main objections were to the possibilities of interference to boat traffic, including fishermen, and to the unpredictable changes in the ecosystem that may occur (Franklin 1974).

In the early 1970s, French scientists proposed to introduce the giant kelp to the Brittany coast of France as a new source of material for their alginate industry. Perez (1972) outlined the advantages of introducing this plant with its favorable growth and alginate yields, and expressed the opinion that the risk to the "balance of nature" was negligible. Perez and colleagues obtained sporophylls from Chilean *Macrocystis* and produced young plants from spores. These plants had grown to about 5 cm in length in February 1972, when they were placed in the sea. Seven months later, they had grown to a length of 13 m and were removed before they became fertile (Braud et al. 1974; Franklin 1974). Based on this success, the French proposed a large-scale experiment, which attracted the attention of other marine scientists in Europe. Perez and colleagues (1973) studied this opportunity in Chile and California where this plant naturally occurs. One of us (M.N.) arranged for them to inspect the Santa Barbara County kelp beds by air in 1973, where the plants form a nearly continuous belt of vegetation along the coast. It was felt that the sight of these extensive kelp forests would emphasize what might happen along the coasts of France, if the plant escaped cultivation there. A special meeting of the International Counsel for the Exploration of the Sea (ICES) working group on the introduction of non-indigenous marine organisms was held in February 1974. The French proposal was rejected mainly because of the great potential of *Macrocystis* spreading to other places in Europe and the possible interference with fishing and shipping. The French scientists modified their proposal to include precautions designed to control the spread of giant kelp, but these precautions were thought to be inadequate at another meeting of the ICES working group in October 1974 (Franklin 1974). Resolutions opposing the introduction of giant kelp to Europe were passed by the British Phycological Society (Boalch 1981) and at the VIIIth International Seaweed Symposium (Anon. 1975). Further French proposals to introduce *Macrocystis* have been similarly opposed and their effort has been suspended (Boalch 1981).

The purpose of this paper is to reexamine the long-standing question of whether or not it is possible to reap the obvious benefits of cultivating macroalgal crop plants, like the giant kelp, while at the same time protecting the environment from the recognized dangers of self-propagating plants that escape from cultivation. To do this we have briefly reviewed the literature describing intentional and accidental introductions of macroalgae, and have considered our own experiences with plants introduced for experimental purposes. Recent studies here of macroalgal spores in the laboratory and sea, and of macroalgal genetics, suggest that, with more directed studies, it may one day be possible to produce cultivars that will not self-propagate when introduced.

Unintentional Introductions of Marine Plants

An examination of the estimated dates of introduction, the rates of spreading, and the effects of unintentionally introduced marine plants gives us some idea of what might happen if intentionally introduced crop plants were to escape from cultivation. These plants have been introduced unintentionally by such vectors as shipping or associated with marine organisms (such as oysters) that are intentionally imported for cultivation. The best documented examples of unintentionally introduced macroalgae that are viewed as invasive and damaging are those of the brown algae *Sargassum muticum* and *Colpomenia peregrina* and the green alga *Codium fragile*. Farnham (1980) discussed these and other taxa that were unintentionally introduced into the British Isles. For example *Colpomenia peregrina* was introduced to Britain early in this century, where it spread, causing losses from the northern French oyster-beds, because the globose thallus becomes filled with gas and floats away with oysters to which it is attached. *Codium fragile* damages oyster beds in the same way.

The brown alga *Sargassum muticum,* which is native to the western North Pacific Ocean, was introduced to the coasts of the eastern North Pacific, eastern North Atlantic, and the Mediterranean in this century. This plant, because of its large size, distinctiveness, rapid growth and rapid spread, has attracted much attention in its introduced habitats and a summary of investigations into the spread and effects of this plant provides an informative case study of an introduced marine plant that has proven to be very invasive. Interestingly, in its native habitat, it is a minor component in a community of many species of *Sargassum* that occur there. Its exact distribution in the western Pacific is not clearly known because of taxonomic confusion with entities which may or may not be conspecific (Critchley 1983).

Sargassum muticum was probably introduced from Japan to the eastern North Pacific coast as early as 1930 as a contaminant from oyster spat (Deysher 1984). It was first recognized as a new entity for this coast in 1947 at Coos Bay, Oregon, and first identified by E.Y. Dawson from Willapa Bay, Washington, in 1953. Scagel (1956) provides evidence that, because of its similarity to the native fucoid *Cystoseira,* it had been collected earlier and misidentified. The earliest collected specimens date to 1944 to 1947 from several areas around the Strait of Georgia, British Columbia. Scagel's careful examination of specimens from collectors who filed specimens prior to 1944 showed that it was probably not present prior to 1937.

The first *Sargassum* collections were made long after the first oyster spat imports, which began in 1902 in Washington and 1912 in British Columbia, with the greatest volume imported from 1920 to 1931 (Deysher 1984). *Sargassum* must have been introduced repeatedly with the many shipments of oyster spat that were made.

The time gap between this mode of introduction and identification of *Sargassum* in established populations may be due to the early confusion with *Cystoseira* and to the time necessary to develop populations which are large enough to be noticed. Deysher (1984) pointed out that there are other possible modes of introduction, such as via shipping or drifting of natural plants, but that the coincidence of *Sargassum* with oyster spat imports supports this as the mode of introduction.

Sargassum muticum grows in warm temperate waters, and has spread to the south from the first populations found in British Columbia, Washington, and Oregon, as summarized in Table 1. It was first found in northern California in 1963 at Crescent City. In 1969, it was found in San Diego, and it has spread rapidly in several areas of southern California. It was not found in central California until 1974, when it was found at the Berkeley Marina. It was also found in Baja California, Mexico, in 1971. The southernmost distribution noted by Deysher (1984) is San Quintin Bay in Baja California. Recently, it has also been reported from Alaska (S.C. Lindstrom, personal communication), indicating that it is able to grow in some localities to the north.

Table 1. Spread of *Sargassum muticum* in the eastern North Pacific.

Locality	**Year**	**Reference**
Washington/British Columbia	1930s	Scagel 1956, Deysher 1984
Oregon	1947	Scagel 1956
California (northern)	1963	Norton 1981
California (southern)	1969	Deysher 1984
California (Catalina I.)	1971	Setzer and Link 1971
Baja California, Mexico	1971	Setzer and Link 1971
California (central)	1974	Deysher 1984
Alaska	1988	Lindstrom (pers. comm.)

Sargassum muticum is largely limited to warmer and more protected localities in its eastern North Pacific distribution. Scagel (1956) hypothesized that it occupies a niche in the low intertidal and subtidal zones that is not occupied by other large algae. He pointed out, however, that it may outcompete eel grass, *Zostera marina,* where the two may co-occur. In the Strait of Georgia, British Columbia, *S. muticum* competes with two red algae in the low intertidal zone, *Neorhodomela larix* and *Lithothrix aspergillus;* however, none of these algae are outcompeted to extinction by any of the others (DeWreede 1980). After *Sargassum* was found at Santa Catalina Island in 1971, a die back of *Macrocystis* occurred in 1976. *Sargassum* invaded some of the vacated space, preventing the reestablishment of the kelp (Ambrose and Nelson 1982). Thus the local distribution of *Sargassum* may be restricted by competition with established *Macrocystis* but once it gains a foothold it is capable of excluding this kelp.

The conclusion that *Sargassum* was introduced to the eastern North Pacific via oyster spat led to the prediction by Druehl (1973) that this plant would be introduced into the North Atlantic, particularly in France, since oyster spat was being imported to French waters from Japan and British Columbia in the late 1960s and early 1970s. This prediction was fulfilled in the discovery of a population of attached *Sargassum* near Portsmouth, England in February 1973 (Farnham et al. 1973), with fertile drift specimens, which were initially misidentified, being found in the English Channel by 1971 (Farnham 1980). It appears possible that mature plants were introduced to France with oyster spat, and that these plants drifted across the English Channel to initiate the populations discovered in England. Critchley et al. (1983) point out that the attached plants first found in England were at least 2 years old, pointing to an introduction in the late 1960s.

From this initial introduction, this plant has spread rapidly, as summarized in Table 2. Critchley et al. (1983) and Belsher and Pommellec (1988) provide more detailed chronologies. It has spread predominantly to the east, following the prevailing currents in the English Channel. Drift plants were often observed before attached plants were found. Attached populations were first noted across the English Channel in France in 1976. Attached populations both to the west and east along the coasts of France and England, including the Channel Islands, were found in subsequent years. Attached populations were first found in the Netherlands in 1981. It also spread further south along the Brittany coast of France. More significantly, it was found on the Mediterranean coast of France in 1981, also associated with oyster beds. Apparently, the transplantation of oysters from northern French oyster beds to the Mediterranean was responsible for this introduction. By 1987, it had spread down the Brittany coast to La

Table 2. Spread of *Sargassum muticum* in Europe.

Locality	Year	Reference
England	1973	Farnham et al. 1973
France (English Channel)	1976	Gruet 1976 *
France (Atlantic Coast)	1981	Critchley et al. 1983
Netherlands	1981	Nienhuis 1982
France (Mediterranean)	1981	Critchley et al. 1983

*cited by Critchley et al. 1983

Rochelle, but only spread to a few new areas in the Mediterranean (Belsher and Pommellec 1988).

After the initial discovery of *Sargassum* in England, attempts were made to eradicate it. Voluntary efforts at hand-picking were carried out initially, with "*Sargassum* wanted" posters being used to publicize the effort, since this would not be ecologically damaging. This may have slowed the spread of *Sargassum*, but it did not eradicate it, and it was abandoned in 1976. Other methods were examined, including the use of herbicides and biological control, but these methods were not selective enough and were not used (Critchley et al. 1986).

In England, *S. muticum* occupies a zone from the mid intertidal zone to the sublittoral fringe. In this habitat, it co-occurs with many marine algae, and appears to outcompete other fucalean brown algae. A particularly thick population was noted in Portsmouth Harbor, which had only a few low-lying algae present before (Fletcher and Fletcher 1975).

The green alga, *Codium fragile* was first discovered along the coast of the northeastern United States in 1957, and was probably introduced shortly before then. Loosanoff (1975) observed Codium plants on the hulls of ships from Europe in Long Island Sound during World War II, which suggests shipping as the probable agent of dispersal. This species has been identified as *Codium fragile* subspecies *tomentosoides,* which is native to the western Pacific (Silva 1955).

Codium has an adverse effect on shellfisheries, since it is a perennial that can grow to 20 cm or more in length and breadth and will attach to shellfish. It is called "oyster thief" by fishermen because the plants can add both drag and floatation, since photosynthetically-produced oxygen can accumulate in the spongy thallus. Consequently shellfish with attached *Codium* are swept onto beaches by storms, or simply float away (Wassman and Ramus 1973). *Codium* plants can also kill scallops by becoming so large that the animals cannot swim and have been implicated as an important factor in the destruction of scallop fisheries (Wassman and Ramus 1973).

The dispersal of *C. fragile* throughout the North Atlantic ocean has been well documented. It first appeared in Holland in 1900, although its transportation vector from the western Pacific is unknown (Silva 1955). It spread throughout Scandinavian waters by the 1920s and 1930s and later to the British Isles and France (Table 3).

Codium fragile was presumably introduced to the western North Atlantic from Europe and several possible vectors have been suggested including transportation on ships or on oyster shells (Carlton and Scanlon 1985). Since there are no native species of *Codium* or morphologically similar genera in the western Atlantic

Table 3. Spread of *Codium fragile* ssp. *tomentosoides* in the eastern North Atlantic and adjacent waters.

Locality	Year	Reference
Holland	1900	van Goor 1923 *
Denmark	1920	Rosenvinge 1922
Helgoland	1930	Schmidt 1935 **
Sweden	1938	Silva 1955
England	1939	Silva 1955
Ireland	1941	Parkes 1941
France	1946	Silva 1955
Norway	1952	Silva 1955

*cited by Silva 1955
** cited by Silva 1957

north of Cape Hatteras (North Carolina), it is unlikely that the introduction of *Codium* went unnoticed for very long, and records of its spread are probably accurate. It first appeared on the east side of Long Island Sound on Long Island in 1957 and by 1961 had spread to the west side of the Sound and north to southern Massachusetts, as well as to the outer coast of Long Island (Table 4). Its dispersal around Cape Cod took 10 years and has been documented in detail by Carlton and Scanlon (1985), although an isolated population appeared in Maine in 1964 (Table 4). *Codium* spread south from Long Island to New Jersey by 1966 and to Virginia by 1975. *Codium fragile* was first observed south of Cape Hatteras in 1979, which is an important event since the long sandy coast in this area is considered to be a major biogeographical barrier for marine macroalgae and the northern limit of the western North Atlantic warm temperate biogeographic zone (Hoek 1975; Searles et al. 1984). It is possible that *C. fragile* may not have the adverse effects on shellfisheries observed further north since two native *Codium* species co-occur with it in North Carolina (Searles et al. 1984); however, *C. fragile* now appears to be displacing these native species in many parts of the state (D.F. Kapraun, J. Ramus, personal communication). It has also been found as drift in South Carolina but attached plants have yet to be collected (R. Zingmark, R. Wiseman, personal communication).

Table 4. Spread of *Codium fragile* ssp. *tomentosoides* in the western North Atlantic.

Locality	**Year**	**Reference**
New York (inner Long Island)	1957	Bouck & Morgan 1957
New York (outer Long Island)	1961	Taylor 1967
Connecticut	1961	Carlton & Scanlon 1985
Massachusetts (south shore Cape Cod)	1961	Carlton & Scanlon 1985
Maine	1964	Coffin & Stickney 1967
New Jersey	1966	Taylor 1967
Massachusetts (north of Cape Cod)	1971	Carlton & Scanlon 1985
Virginia	1975	Hillson 1976
North Carolina	1979	Searles et al. 1984
New Hampshire (sexually reproducing)	1982	Prince 1988

Since *C. fragile* appears to have warm temperate floristic affinities, it is likely that it will continue to spread further south.

Most western North Atlantic *C. fragile* populations reproduce only asexually by either motile cells, plant fragmentation, or whole-plant buoyancy from gas entrapment (reviewed by Carlton and Scanlon 1985; Prince 1988). These mechanisms probably account for the rapid spread of the plants over short distances and, probably, for much of the observed long distance coastal dispersal as well (particularly drifting plants or fragments releasing motile cells). Carlton and Scanlon (1985) reviewed four possible vectors for the artificial dispersal of *C. fragile:* (1) as a fouling organism on ships moving along the coast or through the inland waterway; (2) on transplanted commercial oysters; (3) by fisherman after moving fouled nets to new locations, and (4) as discarded packing material for fisheries products such as lobsters or bait worms.

The Japanese brown alga *Undaria pinnatifida* was introduced accidentally to France. It was first found in 1971 in association with oyster parks in the Mediterranean (Perez et al. 1981). Since the oyster spat was imported from Japan, it is fairly certain that

Undaria was introduced by this vector. Since *Undaria* is widely cultivated as a food in Asia, attempts were made to exploit this accidental introduction. Perez et al. (1984, 1988) report on recent attempts made to cultivate this alga along the Brittany coast of France, where it is reported to grow to a larger size than plants grown in Japan. Hay and Lucken (1987) reported the finding of *Undaria pinnatifida* in New Zealand, and Sanderson (1988) has recently found this plant in Tasmania, Australia, as well, where it may have been introduced with ballast water from cargo vessels transporting wood chips to Japan. Sanderson (personal communication) has estimated that the first introduction occurred, probably with ballast water, in 1982 and has found that since that time the plant has spread along 10 km of coastline. The role of shipping, and particularly the discharge of ballast water, in accidental transplantations has been recently examined by Williams et al. (1988).

An example of an accidentally introduced alga which has provided the basis for a new seaweed industry is the case of *Laminaria japonica,* which was introduced from Japan to Dalian, northern China, in 1927. From 1927 to 1949, this plant was harvested from wild and semi-wild populations, but the production did not meet the demand in China. Mariculture started in 1952, and this industry has grown remarkably, with over 250,000 tons of dry *Laminaria* produced in 1980 (Tseng 1987). *L. japonica* was also accidentally introduced to the Mediterranean coast of France, also associated with oysters (Perez et al. 1984), but no further reports of the persistence or spread of this plant are known.

There are many other examples of accidentally introduced macroalgae, including the red algae *Acanthophora* (Russell 1981) and *Polysiphonia*. Doty (1961) has carefully documented the introduction of *Acanthophora* to Oahu in the Hawaiian islands, its encirclement of this island, and its spread to Kauai. Kapraun and Searles (1990), discovered the filamentous eastern Atlantic and Mediterranean species, *Polysiphonia breviarticulata,* in North Carolina in December 1982. It remained as a small and inconspicuous population until the Spring of 1988, when sterile free-living plants appeared in the plankton in great quantities along 180 km of coastline. These drifting plants were a serious problem for fishermen, whose nets were fouled, and to those visiting the beaches where they formed thick mats several meters wide, prior to decomposing. Wilce et al. (1982), reported similar dramatic problems produced by an unusual free-living form of the native filamentous brown alga *Pilayella littoralis*. These examples show that even small filamentous macroalgae can cause serious problems. *Ishige isiforme*, a brown alga from Japan, was introduced to the Gulf of California in the 1920s by shrimp boats dumping ballast water there (Dawson 1966). Norris (1975) found *Spyridia filamentosa* as an introduced species in the Gulf of California, where the interesting floristic similarity to Japan has stimulated considerable speculation about many possible unintentional introductions.

Intentional Introductions of Marine Plants

Intentionally introduced marine plants generally have not escaped from cultivation and damaged natural ecosystems. A possible exception to this is *Undaria pinnatifida*, which was transplanted from the Mediterranean coast of France to Brittany, where natural populations are now found (Floc'h and Pajot 1989). In a few cases those that have been intentionally introduced have survived without spreading. Most of the successful introductions have been from Japan to Korea or China (e.g.,*Undaria pinnatifida*), or from Northern to Southern China, where because of extreme seasonal changes in temperature, seedstock has to be produced in refrigerated tanks on land and then planted out in the sea.

The tropical red-algal carrageenophytes, *Eucheuma* and *Kappaphycus*, were domesticated by M. S. Doty and his coworkers (Doty 1973) and have since become the basis for a major marine farming industry in the Philippines and elsewhere. *Kappaphycus alvarezii* (=*Eucheuma alvarezii*, kappa-carrageenan, Doty 1988) and *Eucheuma denticulatum* (iota-carrageenan) are two species that are particularly important sources of the phycocolloid carrageenan and are cultivated in many areas, typically starting from introduced seedstock. One particular strain, *K. alvarezii* var. *tambalang,* has been particularly productive in cultivation. Doty (1985) transplanted this plant from the Philippines to Hawaii in 1971 (under a permit from the Hawaii State Department of Agriculture) where it prospered but did not spread (Russell 1983). From Hawaii, it has been introduced to Ponape, Guam, Kiribati, Tonga, Fiji and French Oceania (Russell 1982; Doty 1985), and reintroduced to Fiji from Tonga after the plants were lost in a cyclone in 1980 (Luxton et al. 1987). It has recently been seen as an escapee from cultivation in the Bora Bora lagoon (M. Littler and J. Norris, personal communication), where the vegetative fragments tangle in the coral heads (E. Zablackis, personal communication).

Both species mentioned above were introduced to Kiribati in 1977 (Why 1985). Braud and Perez (1979) report on the introduction of *E. denticulatum* to Djibouti from Malaysia in 1977 and its success in pilot cultivation trials. Farming trials in Bali, Indonesia, were initially tried with indigenous *Eucheuma* plants, but growth and carrageenan yields were low. Therefore, both commercial species were introduced from the Philippines in 1984, but growth and carrageenan yields remained low. It was concluded that the local people needed better training in farming practices in order for *Eucheuma* cultivation to be successful in that area (Adnan and Porse 1987).

The tambalang strain has been a particularly important strain in cultivation. It reproduces almost entirely vegetatively and does not form attachments to other objects or its own branches (Doty 1985). Fragments of plants that were transplanted to Hawaii did escape enclosures on the reef flat where they were grown at Coconut Island and colonized Kaneohe Bay; however, when the source of the fragments was removed, the population in Kaneohe Bay eventually disappeared (Russell 1983). *Eucheuma* is cultivated from vegetative cuttings, so growth by vegetative means is favorable both for farming of this plant and for the prevention of its spread into the natural environment.

The Japanese marine crop plant *Porphyra* has recently been introduced to Washington and British Columbia for commercial purposes. Nets are seeded in the sea with shells bearing the conchocelis life-history stage (pers. obs. R. Lewis). The introduced cultures are reported to be axenic, thereby preventing the chance introduction of any associated organisms. Indeed nets are still being seeded in Japanese waters, removed, partially dried and frozen prior to exportation. There is no evidence to-date that this technique is introducing either *P. yezoensis* or other plants or microbes to the natural ecosystem (S.C. Lindstrom, personal communication).

Macrocystis pyrifera was introduced to China in 1978. Many sporophylls and 48 young plants were collected from Santo Tomas, near Ensenada, Baja California, Mexico in August 1978. Spores were released and gametophyte cultures were started. The gametophytes and young sporophytes were then transported to Qingdao, China, a journey which took 9 days. The young plants were raised in the sea and gametophytes were raised in the laboratory. Approximately one hundred thousand sporophytes were grown from the gametophyte cultures (Liu et al. 1981). In China, *Macrocystis* grows well during much of the year, but is limited seasonally by low temperatures in the winter and high temperatures in the summer. As a result, these plants are not fertile during all the year, as they are in southern California, but are fertile during two periods each year between the extreme seasons (Liu et al. 1984). In this environment of extremes, *Macrocystis* seems to survive in China only because of artificial cultivation efforts. North et al. (1988) observed plants from this introduction grown in Bohai Bay, near Yantai on the north side of the Shandong Peninsula in 1986. According to North and his coworkers, the plants have not spread because of predation by extensive sea urchin populations, although small sporophytes have been seen on holdfasts and rocks used as weights on the culture lines.

The introduction of *Macrocystis* to China used seedstock collected in Mexico, where the seawater temperatures are warmer, so that these plants would hopefully be able to survive the warm summer water temperatures in China. Similarly, *Macrocystis integrifolia* from Monterey in northern California was grown in the warmer waters near Santa Barbara, along with *M. pyrifera* from Santa Barbara and Catalina in southern California (Lewis et al. 1986). In this study, it was found that *Macrocystis* plants from the three localities were interfertile. Both intraspecific and interspecific progeny from the parental sporophytes were grown in the ocean near Santa Barbara. It was found that the growth rates of the progeny were significantly different among the various combinations. If the parents of these progeny were from Catalina or Santa Barbara, southern California, the plants grew significantly faster than plants that were hybrids between southern California and northern California plants. Intraspecific progeny from *M. integrifolia* from Monterey grew the least well in the Santa Barbara area. In addition, reproductive plants were not obtained from the intraspecific crosses of *M. integrifolia* plants or from the *M. pyrifera/M. integrifolia* cross. The other seven combinations produced fertile progeny in these trials. Therefore, it appears that *M. integrifolia* is not well suited for propagation in the Santa Barbara area; however, with the kind of artificial propagation system used by the Chinese, it is possible that this plant may be successfully cultivated in this area without it escaping to form wild populations.

Our studies of marine plant reproduction in the laboratory, greenhouse and sea, and the few introductions we have tried for experimental purposes, are worth mentioning here. Given the speed of modern air travel we have been able to successfully ship or hand-carry living macroalgae from Tasmania, New Zealand, southern Argentina, and Korea to our seawater-supplied greenhouse. We have not introduced these into the sea with only one exception. The exception was *Eucheuma uncinatum* (thought by J. Norris to belong to a new genus) which was brought from Guaymas, in the Gulf of California, Mexico. Permission was sought and granted by the California Department of Fish and Game to introduce male plants only. Only one attempt was made to grow these plants in the sea, and this was not successful, although they grew surprisingly well in our colder waters under greenhouse conditions.

Our one attempt to grow intergeneric hybrid kelps in the sea failed (Neushul 1981). Hybrids were produced in the laboratory using standard crossing methods. These morphologically distinctive plants, bearing two elongated pneumatocysts, were apparently not well-adapted for survival in the sea, and died soon after being outplanted. The only plant that lived to form sori (after 1 year in the greenhouse) did not produce functional sporangia, although a natural hybrid collected from the sea (J. Coyer, personal communication) did produce viable spores, and gametophytes, which produced sporophytes in our cultures. The latter did not survive, but at least initially appeared to be normal.

Characteristics of Invasive Species

The global concern that invading species will alter the structure and function of natural communities has led to the establishment of international programs to investigate the ecology of biological invasions. Certainly a central question to be asked by the mariculturalist when considering introducing an alga to a new area is "how invasive might this species be should it escape cultivation?" and if it does escape, "how will it alter the invaded habitat?" Some maintain that species with certain demographic and physiological attributes are more likely to invade than others and that these attributes can form a useful guide in developing a risk analysis (Ehrlich 1986; Bazzaz 1986); however, there has been considerable debate as to whether accurate predictions can be made as to which particular species will be successful invaders (Simberloff 1986; Roughgarden 1986).

Predicting the invasive ability of different algae is complicated by the fact that, as a group, algae have an extremely diverse array of life-history strategies causing a wide assortment of demographic and physiological attributes. In their work on parasites, Dobson and May (1986) suggest that, other things being equal, species with direct life-histories are much more likely to be successful invaders than those with indirect life-histories. Based on past introductions, this appears to be true for algae as well. *Laminaria* and *Macrocystis* are two species that have been successfully introduced without escaping cultivation. Like all kelps, both have a complex heteromorphic life-history in which the large visible stage (sporophyte) alternates with a microscopic, sexual reproductive stage that is dioecious (gametophyte). The physiological requirements of the two stages can differ considerably. Conditions suitable for growth and reproduction of the gametophyte often occur sporadically in time and space and may not always coincide with those suitable for the sporophyte (Deysher and Dean 1986). Propagation of the sporophyte (the potentially invasive stage) is totally dependent on the microscopic gametophyte stage; there is no evidence for vegetative or asexual reproduction of the sporophyte in nature. Thus the relatively complex, indirect life-history of kelps may severely limit their ability to successfully invade.

In contrast, *Codium fragile,* a rather successful invader, has a much more direct life-history. Although one population is capable of sexual reproduction (gametes are produced from meiosis in diploid plants), most western North Atlantic *C. fragile* populations reproduce only asexually by either motile cells, plant fragmentation, or buoyancy of whole plants or fragments from gas entrapment (reviewed by Carlton and Scanlon 1985; Prince 1988). These mechanisms undoubtedly aided in the rapid spread of this species over short distances in the western north Atlantic and likely contributed to its dispersal over longer distances on this coast as well (particularly drifting plants or fragments releasing motile cells) . In addition, a population of *C. fragile* in North Carolina was found to consist of only haploid female plants that had two to four times more nuclear DNA than other, diploid, *Codium* species (Kapraun and Martin 1987; Kapraun et al. 1988). The authors suggest these plants are opportunistic autopolyploids that are functionally similar to a wide variety of polyploid "weedy" vascular plants. If this is also true of other western North Atlantic populations, it would help explain the markedly invasive and competitive nature of this species.

The invasive character of *Sargassum muticum* has also been well-studied. First, it is both monoecious and self-fertile, so a single plant is able to reproduce. Gametes are formed by meiosis, and both male and female gametes are formed simultaneously. The gametangia are borne on small branches, receptacles, which occur along the upper portions of the plant. These portions also possess small gas-filled floats which bring the plant to the surface of the water. Late in the reproductive season of this plant, the fronds typically degrade or break off. A detached frond, bearing both floats and reproductive structures, would be able to float to another locality and disperse zygotes locally. This may be the reason for the rapid but erratic spread seen along the Northeast Pacific coast (Deysher and Norton 1982). In addition, *S. muticum* has a wide temperature tolerance for growth, with both germlings and laterals of adult plants showing increasing growth with increasing temperature from 5 to 25° C (Norton 1977).

A crucial life-history or demographic feature of an invasive plant is an efficient means of dispersal, especially since it is highly unlikely that a colonizing species will establish following a single introduction of a small number of propagules (Bazzaz 1986). Although *Macrocystis* and other large kelps produce an enormous number of spores (Neushul 1959; Chapman 1984; Reed 1987), spore dispersal in kelps is generally limited to a few meters (Sundene 1962; Anderson and North 1966; Dayton 1973; Paine 1979, 1988) and only sporadically occurs over distances approaching kilometers (Reed et al. 1988). In contrast, small filamentous brown algae in the family Ectocapaceae produce relatively few spores that are morphologically similar to those of kelps yet they disperse over relatively long distances (Reed et al. 1988). This difference in dispersal ability between these two algal groups may be related to differences in the behavior of their spores. Like spores of most brown algae, ectocarpoid spores are positively phototactic. Such behavior may enable them to remain in the water column longer where they can be transported greater distances by prevailing currents (Amsler and Searles 1980; Reed et al. 1988). There is no evidence that kelp spores display phototaxis: they lack eyespots and other morphological features which are correlated with phototactic behavior (Henry and Cole 1982; Kawai 1988). Thus, in addition to a complex life-history, inefficient means of dispersal may also be reason why kelp has yet to escape cultivation. In contrast, efficient dispersal in ectocarpoid algae coupled with a life-history which is often direct or in which both the gametophyte and sporophyte stage are potentially invasive (the two stages are isomorphic and appear to have similar growth requirements) are traits that would seem to enhance invasiveness. This may in part explain why many ectocarpoid species appear to have cosmopolitan distributions (cf. Clayton 1974; Amsler 1985).

In addition to the demographic and physiological attributes of a species that make it a good invader, the invaded community or habitat must permit invasion if it is to occur. As appropriately stated by Bazzaz (1986), "the colonizer and the colonized are partners in the process." Elton (1958) in an extensive review of invasions by plants and animals concluded that invaders were most likely to establish viable populations in agricultural or otherwise disturbed and unusually simplified communities. Subsequent research has not changed this view (Orians 1986). Indeed, one of the most easily invaded habitats may be the farms themselves where the aquacultural crop is produced. This seems to be the case so far for *Macrocystis* in China.

Discussion and Conclusions

It seems likely that the accidental introduction of undesirable plants, like those discussed here, will continue to occur. It is also likely that increasing numbers of requests will be made to introduce marine crop plants like *Porphyra, Eucheuma, Kappaphycus,* and the giant kelp, *Macrocystis*. Consequently we need a rational, rather than a "finger in the dike," approach to the problem of marine plant introductions. We must carefully consider what has been learned from accidental and intentional introductions. It is also important that during the domestication process we select strains that require human intervention to reproduce. The best example to date of a strain of a commercially important marine plant which requires human intervention to be propagated is the tambalang strain of *Kappaphycus*. This plant grows nearly entirely vegetatively, and is cultivated using vegetative cuttings. This strain was selected during the early stages of *Eucheuma* farm operations in the Philippines, when a genetic improvement strategy was incorporated into the farming practices (Doty 1973); however, this example is exceptional, as shown by Van der Meer's (1988) recent review of efforts made to genetically improve marine crop plants, which shows that comparatively little is known about marine plant genetics generally, and that very few efforts have been made to genetically modify and select desirable cultivars. Here we must look at algal reproduction, as quantitative life-histories with "turnstiles" between each successive stage where one or more of these must be controlled by the mariculturalist.

Using *Macrocystis* as an example, there are several strategies possible for developing reproductively controllable cultivars which could be safely introduced. Although a perennial, *Macrocystis* could be cultivated as an annual in areas where adverse summer or winter conditions kill spores, gametophytes or young sporophytes thereby preventing dispersal and spreading (this is one reason that introduced *Laminaria japonica* has not escaped in parts of China). Selection of isolates with restricted temperature (or light or nutrient) tolerances would facilitate such a strategy. The filamentous gametophytic phase of kelps is particularly amenable to this type of manipulation and can be grown in light-temperature or other cross-gradient culture much like other small brown algae (Amsler 1985; Amsler and Lewis unpublished). Ideally, one would like to select for strains which are unable to reproduce at all in nature. Ongoing studies of kelp genetics in our laboratory (Lewis et al. 1986; Neushul 1987) have identified a number of *Macrocystis* gametophytes which are unable to produce gametangia under normal conditions, perhaps because they require substances (or concentrations of them) not normally present in seawater. Such isolates that could reproduce in the laboratory when provided with an organic or other substance not found in nature would be ideal candidates for producing sporophytic material for introduction.

We have recently discovered that kelp spores swim toward nutrients which stimulate gametogenesis in gametophytes (the next step in the life history) and swim away from some nutrients at concentrations which inhibit gametogenesis (Amsler and Neushul 1989). Using a screening technique based on this observation, one might be able to select gametophytes with unusual nutrient requirements by collecting spores with unusual chemotactic responses. It would also be possible to select for spores without chemotactic ability and, presumably, lower fitness under natural conditions since nutrient chemotaxis by spores is probably an adaptation to facilitate settlement in optimal microhabitats. Some such strains might also produce sperm without functional chemoreceptors. Chemotactic ability of kelp sperm is very important for successful fertilization and may determine the minimum gametophyte density necessary for sporophyte production in nature (Reed 1990). Consequently, *Macrocystis* strains without chemotactic ability in either spores or sperm would be less likely to escape cultivation.

Some thought should be given to the impact that an introduced plant of the same species might have. The exemplary studies of Müller (e.g., 1979) on interfertility of *Ectocarpus*, and West et al. (e.g., 1983) on *Mastocarpus*, have shown that strains collected from different seas are often interfertile. We have shown that three species of *Macrocystis* from California are interfertile (Lewis et al. 1986) and have recently discovered that strains from the eastern North Pacific and Tasmania, Australia, are also interfertile (Lewis 1989). What would be the effect on natural populations if introduced plants interbred with those of the indigenous populations? Also, would it be desirable to cultivate large numbers of genetically identical plants, which it is now possible to do using cultivated gametophytes, and risk reducing the genetic diversity in nearby native populations?

We know of no studies that have focused on determining the likelihood that a specific strain would escape from cultivation and become self reproductive. Studies that have been made of accidentally introduced *Sargassum, Codium,* and *Undaria* and on invasive terrestrial plants and animals give us some idea of the types of macroalgae that would be likely to spread once introduced. As mentioned above, before a species can be introduced it is crucial to have a detailed understanding of how it is dispersed in its native habitat (e.g., Reed et al. 1988) and of its dispersal potential under ideal conditions in the laboratory. It is also particularly important to consider the potential for vegetative reproduction and dispersal. *Codium*, for instance, may successfully spread via vegetative fragments (cf. Carlton and Scanlon 1985) and populations of some species can be maintained solely by vegetative fragmentation (e.g., Amsler 1984). A related concern is that life-history variability among the red algae could greatly complicate attempts to introduce plants thought to reproduce only vegetatively (Van der Meer and Todd 1977; Van der Meer 1981; Maggs 1988).

The task of determining how an introduced species might impact other organisms in an ecosystem, which seemed to Druehl (1973) to be years away from any solution, may not be as difficult as he imagined. Defining the environmental conditions needed for macroalgal crop plant reproduction, using light and temperature gradient tables, would be a first step in determining the likelihood that a given strain would escape cultivation and become self reproducing. Obviously both *in vitro* and extensive greenhouse cultivation studies of large numbers of plants are needed. Although the facilities for large-scale greenhouse cultivation of marine macrophytes are expensive to construct and labor-intensive to operate, they are available at the University of California Santa Barbara and in a few other places. The seawater from our greenhouse empties into a salt pond which acts as a "trap" for propagules that might wash out of the tanks. With such greenhouses, where water motion is produced in easy-to-clean glass tanks, even the largest of kelps can be raised to reproductive maturity (Sanbonsuga and Neushul 1979).

Finally, it seems obvious that tissue culture methods and the techniques of biotechnology have much to offer. Perhaps it would be possible to select for sterile plants that can be propagated only by using tissue culture methods (Polne-Fuller 1988). Another approach would be to select for heterotrophy, particularly in those species with microscopic gametophytic, or "*Conchocelis* life-history phases. The recent discovery of algal plasmids by Goff and Coleman (1988), and molecular genetic studies with marine macroalgae (e.g., Shivji and Catollico 1987; Fain et al. 1988, Olsen et al. 1988) are interesting first examples of how modern molecular genetic methods can be used effectively with macroalgae.

In summary, it does now seem possible to safely reap the benefits of cultivating marine macroalgae. The tools seem to be at hand to begin working on domesticating new marine crop plants with the goal of producing strains that can be safely introduced for maricultural purposes anywhere in the world.

Acknowledgments

We would like to acknowledge the advice and encouragement of our colleagues, who took a great deal of time to answer our questions, offer suggestions and even send unpublished manuscripts and observations. We would particularly like to thank Bill Bushing, Rene Delepine, Donald Kapraun, Sandra Lindstrom, James Norris., Joseph Ramus, Craig Sanderson, Reid Wiseman, Earl Zablackis, and Richard Zingmark.

Literature Cited

Adnan, H. and H. Porse, 1987. Culture of *Eucheuma cottonii* and *Eucheuma spinosum* in Indonesia. Hydrobiologia 151/152:355-358.

Ambrose R. F. and B. V. Nelson, 1982. Inhibition of giant kelp recruitment by an introduced brown alga. Botanica Marina 25: 265-267.

Amsler, C.D. 1984. Culture and field studies of *Acetinospora crinita* (Carmichael) Sauvageau (Ectocarpaceae, Phaeophyceae) in North Carolina (USA). Phycologia 23:377-382.

Amsler, C.D. 1985. Field and laboratory studies of *Giffordia mitchelliae* (Phaeophyceae) in North Carolina. Bot. Mar. 28:295-301.

Amsler, C.D. and M. Neushul. 1989. Chemotactic effects of nutrients on spores of the kelps *Macrocystis pyrifera* and *Pterygophora californica*. Mar. Biol. 102:557-564.

Anderson, E.K. and W.J. North. 1966. *In situ* studies of spore production and dispersal in the giant kelp *Macrocystis pyrifera*. Proc. Int. Seaweed Symp. 5:73-86.

Anon., 1975. Resolution passed at the VIIIth Int. seaweed Symp.. Phycologia 14:123.

Bazzaz, F.A., 1986. Life history of colonizing plants: some demographic, genetic and physiological features, p. 96-110. *In* H. A. Mooney and J. A. Drake (eds.), Ecology of biological invasions of North America and Hawaii. Springer Verlag, New York.

Belsher, T. and S. Pommellec. 1988. Expansion de l'algue d'origine japonaise *Sargassum muticum* (Yendo) Fensholt, sur les cotes francaises, de 1983 a 1987. Cah. Biol. Mar. 29:221-231.

Boalch, G. T. 1981. Do we really need to grow *Macrocystis* in Europe? Proc. Xth Int. Seaweed Symp., p. 657-667, W. de Gruyter & Co., New York.

Bouck, G. B. and E. Morgan. 1957. The occurrence of *Codium* in Long Island Waters. Bull. Torrey Bot. Club 84:384-387.

Braud, J.P. and R. Perez. 1979. Farming on pilot scale of *Eucheuma spinosum* (Florideophyceae) in Djibouti waters. Proc. Int. Seaweed Symp. 9:533-539.

Braud, J.P., H. Etcheverry and R. Perez. 1974. Développement de l'algue *M. pyrifera* (L.) Ag. sur les côtes Bretonnes (Development of the alga *M. pyrifera* (L.) Ag. along the Britanny coasts) Sci. Peche 233:1-15.

Carlton, J. T. and J. A. Scanlon. 1985. Progression and dispersal of an introduced alga: *Codium fragile* ssp. *tomentosoides* (Chlorophyta) on the Atlantic coast of North America. Bot. Mar. 28:155-165.

Chapman, A.R.O. 1984. Reproduction, recruitment and mortality in two species of *Laminaria* in southwest Nova Scotia. J. Experimental Mar. Biol. Ecology 78:99-109.

Clayton, M.N. 1974. Studies on the development, life history, and taxonomy of the Ectocarpales (Phaeophyta) in southern Australia. Aust. J. Bot. 22:743-813.

Coffin, G. W. and A. P. Stickney. 1967. *Codium* enters Maine waters. Fish. Bull. 66:159-161.

Critchley, A.T. 1983. *Sargassum muticum:* a taxonomic history including world-wide and western Pacific distributions. J. Mar. Biol. Assoc. U.K. 63:617-625.

Critchley, A.T., W.F. Farnham and S.L. Morrell. 1983. A chronology of new European sites of attachment for the invasive brown alga, *Sargassum muticum,* 1973-1981. J. Mar. Biol. Assoc. U.K. 63:799-811.

Critchley, A.T., W.F. Farnham and S.L. Morrell. 1986. An account of the attempted control of an introduced marine alga, *Sargassum muticum,* in Southern England. Biol. Conserv. 35:313-332.

Dawson, E.Y. 1966. New records of marine algae from the Gulf of California. J. Ariz. Acad. Sci. 4:55-66.

Dayton, P.K. 1973. Dispersion, dispersal and persistence of the intertidal alga *Postelsia palmaeformis.* Ecology 54:433-438.

DeWreede, R.E. 1980. The effect of some physical and biological factors on a *Sargassum muticum* community, and their implication for commercial utilization. *In* I.A. Abbott, M.S. Foster and L.F. Eklund (eds.), p. 32-44, Pacific Seaweed Aquaculture. California Sea Grant College Program, La Jolla, California.

Deysher, L., 1984. Recruitment processes in benthic marine algae. Ph.D. Dissertation, University of California, San Diego.

Deysher, L.E. and T.A. Dean. 1986. Interactive effects of light and temperature on sporophyte production in the giant kelp *Macrocystis pyrifera.* Mar. Biol. 93:17-20.

Deysher, L. and T.A. Norton. 1982. Dispersal and colonization in *Sargassum muticum* (Yendo) Fensholt. J. Exp Mar. Biol. Ecol. 56:179-195.

Dobson, A. P. and R. M. May, 1986. Patterns of invasions by pathogens and parasites. p. 58-78. *In* H. A. Mooney and J. A. Drake (eds.), Ecology of biological invasions of North America and Hawaii, Springer Verlag, New York.

Doty, M.S. 1961. *Acanthophora,* a possible invader of the marine flora of Hawaii. Pac. Sci. 15: 547-552.

Doty, M.S. 1973. Farming the red seaweed, *Eucheuma,* for carrageenans. Micronesica 9:59-73.

Doty, M.S. 1985. *Eucheuma alvarezii* sp. nov. (Gigartinales, Rhodophyta) from Malaysia. p. 37-45. *In* I.A. Abbott and J.N. Norris (eds.), Taxonomy of economic seaweeds: with reference to some pacific and caribbean Species. california sea grant College Program, La Jolla, California. Report No. T-CSGCP-011.

Doty, M.S. 1988. *Prodromus ad Systematica Eucheumatoideorum:* A tribe of commercial seaweeds related to *Eucheuma* (Solieriaceae, Gigartinales). p. 159-207. *In* I.A. Abbott (ed.), Taxonomy of economic seaweeds: with reference to some pacific and caribbean species, Vol. II. California Sea Grant College Program, La Jolla, California. Report No. T-CSGCP-018.

Druehl, L. D. 1973. Marine transplantations. Science 179:12.

Ehrlich, P. R. 1986. Which animal will invade? pp. 79-95. *In* H. A. Mooney and J. A. Drake (eds.), Ecology of biological invasions of North America and Hawaii, Springer Verlag, New York.

Elton, C.S. 1958. The ecology of invasions by animals and plants. Methuen, London.

Fain, S.R., L.D. Druehl, and D.L. Baillie. 1988. Repeat and single copy sequences are differentially conserved in the evolution of kelp chloroplast DNA. J. Phycol. 24:292-302.

Farnham, W. F., R. L. Fletcher and L. M. Irvine. 1973. Attached *Sargassum* found in Britain. Nature 243: 231-232.

Farnham, W. F. 1980. Studies on aliens in the marine flora of Southern England, p. 875-914. *In* J.H. Price, D.E.G. Irvine and W.F. Farnham (eds.), The shore environment, Vol. 2: Ecosystems. Academic Press, New York.

Fletcher, R.L. and S.M. Fletcher. 1975. Studies on the recently introduced brown alga *Sargassum muticum* (Yendo) Fensholt. I. Ecology and reproduction. Bot. Mar. 18:149-156.

Floc'h, J.Y. and R. Pajot. 1989. The Asian brown alga *Undaria pinnatifida* in its new biotope on the Atlantic coast of France. XIIIth Int. Seaweed Symp., Vancouver, Canada, August 13-18, 1989, p. A-82. (abstract only)

Franklin, A. 1974. Giant kelp for Europe? New Scientist, Dec. 12, 1974, pp. 812-813.

Goff, L.J. and A.W. Coleman. 1988. The use of plastid DNA restriction endonuclease patterns in delineating red algal species and populations. J. Phycology 24:357-368.

Hay, C. H. and P. A. Lucken, 1987. The Asian kelp *Undaria pinnatifida* (Phaeophyta: Laminariales) found in a New Zealand harbour. N Z. J. Bot. 25: 364-366.

Henry, E.C. and K.M. Cole 1982. Ultrastructure of swarmers in the Laminariales (Phaeophyceae). I. Zoospores. J. Phycol. 18:550-569.

Hillson, C.J. 1976. *Codium* invades Virginia waters. Bull. Torrey Bot. Club 103:266-267.

Hoek, C. van den. 1975. Phytogeographic provinces along the coasts of the northern Atlantic Ocean. Phycologia 14:317-330.

Kapraun, D.F., M.G. Gargiulo, and G. Tripodi. 1988. Nuclear DNA and karyotype variation in species of *Codium* (Codiales, Chlorophyta) from the North Atlantic. Phycologia 27:273-282.

Kapraun, D.F. and D.J. Martin. 1987. Karyological studies of three species of Codium (Codiales, Chlorophyta) from coastal North Carolina. Phycologia 26:228-234.

Kapraun, D.F. and R.B. Searles. 1990. Planktonic bloom of an introduced species of *Polysiphonia* (Ceramiales, Rhodophyta) along the coast of North Carolina, USA. Hydrobiologia 204/205:269-274.

Kawai, H. 1988. A flavin-like autofluorescent substance in the posterior flagellum of golden and brown algae. J. Phycol. 24:114-117.

Lewis, R.J. 1989. Interhemispheric hybridization in *Macrocystis.* XIIIth Int. Seaweed Symp., Vancouver, Canada, August 13-18, 1989. p. A-43. (abstract only)

Lewis, R.J., M. Neushul, and B.W.W. Harger. 1986. Interspecific hybridization of the species of *Macrocystis* in California. Aquaculture 57:203-210.

Liu, T., R. Suo, Z. Liu, D. Hu, S. Cao and G. Liu. 1981. Introduction of giant kelp *(Macrocystis pyrifera)* from Mexico to China and artificial cultivation of its juvenile sporophytes. Mar. Fish. Res. 3:68-79. (in Chinese, with English abstract)

Liu, T., R. Suo, X. Liu, D. Hu, Z. Shi, G. Liu, Q. Zhou, S. Cao, S. Zhang, J. Chen and F. Wang. 1984. Studies on the artificial cultivation and propagation of giant kelp *(Macrocystis pyrifera)* . Hydrobiologia 116/117:259-262.

Loosanoff, V.J. 1975. Introduction of *Codium* in New England. Fish. Bull. 73:215-218.

Luxton, D.M., M. Robertson and M.J. Kindley. 1987. Farming of *Eucheuma* in the south Pacific islands of Fiji. Hydrobiologia 151/152:359-362.

Maggs, C.A. 1988. Intraspecific life history variability in the Florideophycideae (Rhodophyta). Bot. Mar. 31:465-490.

Müller, D.G. 1979. Genetic affinity of *Ectocarpus siliculosus* (Dillw.) Lyngb. from the Mediterranean, North Atlantic, and Australia. Phycologia 18:312-318.

Neushul, M. 1959. Studies on the growth and reproduction of the giant kelp *Macrocystis*. Dissertation. University of California, Los Angeles, California.

Neushul, M. 1981. The domestication and cultivation of Californian macroalgae. Proc. Int. Seaweed Symp. 10:71-96.

Neushul, M. 1987. Biomass production by marine crops: Genetic manipulation of kelps. p. 395-408. *In* D.L. Klass (ed.), Energy from biomass and wastes. X. Inst. of Gas Technology, Chicago.

Nienhuis, P.H. 1982. Attached *Sargassum muticum* in the S.W. Netherlands. Aquat. Bot. 12:189-195.

Norris, J.N. 1975. Marine algae of the northern gulf of California. Ph.D. Dissertation. University of California, Santa Barbara.

North, W.J. 1973. Regulating marine transplantation. Science 179:1181.

North, W. J., T. Liu, J. Chen and R. Suo, 1988. Cultivation of *Laminaria* and *Macrocystis* (Laminariales, Phaeophyta) in the People's Republic of China. Phycologia 27:298-299.

Norton, T.A. 1977. Ecological experiments with *Sargassum muticum.* J. Mar. Biol. Assoc. U.K. 57:33-43.

Norton, T.A. 1981. *Sargassum muticum* on the Pacific Coast of North America. Proc. 8th Int. Seaweed Symp., Bangor Wales. pp. 449-455.

Olsen, J.L., W.T. Stam, P.V.M. Bot, and C. van den Hoek. 1988. scDNA-DNA hybridization studies in Pacific and Caribbean isolates of *Dictyosphaeria cavernosa* (Chlorophyta) indicate a long divergence. J. Phycol. 24(suppl.):24.

Orians, G. H. 1986. Site characteristics favoring invasions, pp. 133-148. *In* H. A. Mooney and J. A. Drake (eds.), Ecology of biological invasions of North America and Hawaii. Springer Verlag, New York.

Paine, R.T. 1979. Disaster, catastrophe, and local persistence of the sea palm, *Postelsia palmaeformis.* Science 205:685-687.

Paine, R.T. 1988. Habitat suitability and local population persistence of the sea palm *Postelsia palmaeformis.* Ecology. 69:1787-1794.

Parkes, H.M. 1941. Records of *Codium* species in Ireland. Proc. R. Irish Acad. (B). 75:125-134.

Perez, R. 1972. Opportunité de l'implantation de l'algue *Macrocystis pyrifera* sur les côtes Bretonnes (Advisibility of implanting seaweed *Macrocystis pyrifera* on the Brittany coasts). Sci. Pêche 216:1-9.

Perez, R. , J. Couespel du Mesnil, Y. Colin, L. Le Fur and H. Didou. 1973. Etudes sur l'opportunité d'introduire l'algue *Macrocystis* sur le littoral Français (Study on the opportunity of introducing the alga *Macrocystis* to the French littoral). Rev. Trav. Inst. Pêches Marit., Nantes. 37(3):307-361.

Perez, R., J. Y. Lee and C. Juge. 1981. Observations sur la biologie de l'algue japonaise *Undaria pinnatifida* (Harvey) Suringar introduite accidentellement dans l'étang de Thau. Sci. Pêche, 315: 1-12.

Perez, R., R. Kaas, and O. Barbaroux, 1984. Culture expérimenale de l'algue Undaria *pinnatifida* sur les côtes de France. Sci. Peche, 343:3-15.

Perez, R., P. Durand, R. Kaas, O. Barbarouz, V. Barbier, C. Vinot, M. Bourgeay-Causse, M. Leclercq, and J.Y. Moigne, 1988. *Undaria pinnatifida* on the French coasts, cultivation method, biochemical composition of the sporophyte and the gametophyte, p. 315-327. *In* Stadler, Mollion, Verdus, Karamanos, Morvan and Christiaen (eds.), Algal biotechnology. Elsevier, London and New York.

Polne-Fuller, M. 1988. The past, present and future of tissue culture and biotechnology of seaweeds. p. 17-31, *In* Stadler, Mollion, Verdus, Karamanos, Morvan and Christiaen (eds.), Algal Biotechnology. Elsevier, London and New York.

Prince, J.S. 1988. Sexual reproduction in *Codium fragile* ssp. *tomentosoides* (Chlorophyta) from the northeast coast of North America. J. Phycol. 24:112-114.

Reed, D.C. 1987. Factors affecting the production of sporophylls in the giant kelp *Macrocystis pyrifera* (L.) C. Ag. J. Exper. Mar. Biol. Ecol. 113:61-69.

Reed, D. C. 1990. The effects of variable settlement and early competition on patterns of kelp recruitment. Ecology 71:776-787.

Reed, D. C., D. R. Laur and A. W. Ebeling. 1988. Variation in algal dispersal and recruitment: the importance of episodic events. Ecol. Monogr., 58:321-335.

Rosenvinge, L.K. 1922. Om nogle i nyere Tid invandrede Havalger i de danske Farvande. Bot. Tidsskr. 37:125-135.

Roughgarden, J., 1986. Predicting invasions and rates of spread. p. 179-190. *In* H. A. Mooney and J. A. Drake (eds.), Ecology of biological invasions of North America and Hawaii. Springer Verlag, New York.

Russell, D.J. 1981. The introduction and establishment of *Acanthophora spicifera* (Vahl.) Boerg. and *Eucheuma striatum* Schmitz in Hawaii. Ph.D. Dissertation, Department of Botany. University of Hawaii.

Russell, D. J. 1982. Introduction of *Eucheuma* to Fanning Atoll, Kiribati, for the purpose of mariculture. Micronesica 18:35-44.

Russell, D.J. 1983. Ecology of the imported red seaweed *Eucheuma striatum* Schmitz on Coconut Island, Oahu, Hawaii. Pac. Sci. 37 (2): 87-107.

Sanbonsuga, Y. and M. Neushul. 1979. Cultivation and hybridization of giant kelps (phaeophyceae). Proc. Int. Seaweed Symp. 9:91-96.

Sanderson, J.C. 1988. *In* Letters to the Editor, Australian Mar. Sci. Bull. 102:13.

Sanderson, J.C. (Msc.) A preliminary survey of the distribution of the introduced macroalga, *Undaria pinnatifida* (Harvey) Suringar on the East Coast of Tasmania, Australia (Submitted to Botanica Marina).

Scagel, R.F., 1956. Introduction of a Japanese alga, *Sargassum muticum,* into the Northeast Pacific. Fisheries Research Papers, State of Washington Department of Fisheries, 1:49-59.

Searles, R.B., M.H. Hommersand, and C.D. Amsler. 1984. The occurrence of *Codium fragile* subsp. *tomentosoides* and *C. taylorii* (Chlorophyta) in North Carolina. Bot. Mar. 27:185-187.

Setzer, R. and C. Link. 1971. The wanderings of *Sargassum muticum* and other relations. Stomatopod 2:5-6.

Shivji, M.S. and R.A. Cattolico. 1987. Structure and organization of chloroplast DNA from the red alga *Porphyra yezoensis.* J. Phycol. 23(suppl.):4.

Silva, P.C. 1955. The dichotomous species of *Codium* in Britain. J. Mar. Biol. Assoc. U.K., 34:565-577.

Silva, P.C. 1957. *Codium* in Scandinavian waters. Svensk. Bot. Tidskr. 51:117-134.

Simberloff, D. 1986. Introduced insects: A biogeographic and systematic perspective. p. 3-26. *In* H. A. Mooney and J. A. Drake (eds.), Ecology of biological invasions of North America and Hawaii. Springer Verlag, New York.

Sundene, O. 1962. The implications of transplant and culture experiments on the growth and distribution of *Alaria esculenta.* Nytt Mag. Bot. (Oslo) 9:155-174.

Taylor, J.E. 1967. *Codium* reported from a New Jersey estuary. Bull. Torrey Bot. Club 94:57-59.

Tseng, C.K. 1987. *Laminaria* mariculture in China. p. 239-263. *In* M.S. Doty, J.F. Caddy and B. Santelices (eds.), Case studies of seven commercial seaweed resources. FAO Fish. Tech. Pap. 281.

Van der Meer, J. P. and E. R. Todd, 1977. Genetics of *Gracilaria* sp. (Rhodophyceae, Gigartinales). IV. Mitotic recombination and its relationship to mixed phases in the life history. Can. J. Bot. 55:2810-2817.

Van der Meer, J. P. 1981. Genetics of *Graciliaria tikvahiae* (Rhodophyceae) VII. Further observations on mitotic recombination and the construction of polyploids. Canadian J. Botany 59:787-792.

Van der Meer, J.P. 1988. The development and genetic improvement of marine crops. p. 1-15, *In* Stadler, Mollion, Verdus, Karamanos, Morvan and Christiaen (eds.), Algal biotechnology. Elsevier, London and New York.

Walker, F.T. 1952. Chromosome number of *Macrocystis integrifolia* Bory. Ann. Bot. 16:23-26.

Wassman, E.R. and J. Ramus. 1973. Seaweed Invasion. Nat. Hist. 8:24-36.

West, J., M. Guiry, and M. Masuda. 1983. Further investigations on the genetic affinities and life history patterns of the red alga *Gigartina*. Proc. China-US Phycol Symp. Sci. Press pp. 137-165.

Why, S. 1985. *Eucheuma* Seaweed Farming in Kiribati, Central Pacific (1983). Report: Overseas Development Administration, London.

Wilce, R.T., C. W. Schneider, A. V. Quinlan and T. van den Bosch. 1982. The life history and morphology of free-living *Pilayella littoralis* (L.) Kjellm. (Ectocarpaceae, Ectocarpales) in Nahant Bay, Massachusetts. Phycologia 21:336-354.

Williams, R. J., F. B. Griffiths, E.J. Van der Wal and J. Kelly. 1988. Cargo vessel ballast water as a vector for the transport of non-indigenous marine species. Est. Coast Shelf Sci. 4:409-420.

CHAPTER 2

Dispersal of Pathogens, Parasites, Pests, Predators and Competitors

Mass Mortalities and Infectious Lethal Diseases in Bivalve Molluscs and Associations with Geographic Transfers of Populations

C. AUSTIN FARLEY

Abstract: Disease in many species of commercial oysters has been documented since the early 1920s when mass mortalities struck populations of oysters in the United Kingdom. In 1926, oyster mortalities were seen in Australia and in the 1930s in Malpeque Bay, Canada. Epizootiological evidence implied that an infectious process was operating.

Microbial pathogens have been implicated in most of the more recent mortalities. *Perkinsus marinus* was described in the 1940s as a causative pathogen in Gulf of Mexico mortalities. Epizootics of disease were discovered later from the middle and southern Atlantic regions. *Haplosporidium nelsoni* (known as MSX, or multinucleate spere X) was associated with mortalities in Delaware Bay that began in 1957 and continue to the present. This disease quickly spread to Chesapeake Bay and eventually to other Atlantic Coast sites. Haplosporidan disease caused by *Haplosporidium costale* (known as SSO, or seaside organism) was associated with mortalities in oysters from Chincoteague Bay and other sites characterized by high salinity in the northeast. This disease was introduced to California waters, but did not become endemic in local oyster populations. A fatal oyster disease associated with a protistan pathogen, *Mikrocytos mackini*, was discovered in British Columbia, Canada, in 1960. A second, similar organism (*Bonamia ostreae*) was found in the early 1960s, originating from the hatchery at Milford, Connecticut. This disease was transferred to Elkhorn Slough in California, and later to France where the protistan agent caused severe mortality in the native flat oysters. Two other diseases, caused by an iridovirus and *Marteilia refringens*, preceded this disease in France.

Fatal sarcoma epizootics have been seen in several species of bivalve molluscs. Etiology is not known for these neoplastic diseases; however, some have been shown to be transmissible. It is the premise of this paper that most of these mass mortalities have been caused by the transfer of infectious oyster stocks.

Introduction

Numerous species of oysters have been cultured in many parts of the world since ancient times. The Japanese have cultured *Crassostrea gigas* for more than 1000 years and developed many techniques still in use today. Spat (recently settled "baby" oysters) are collected on hard-surfaced cultch material such as clean oyster shells, and moved to "hardening areas" from growth areas, where they are eventually harvested. Since setting areas, hardening areas, and growth areas are usually located apart from one another, mass movement of oyster crops is necessary and has become a traditional part of the culturing of oysters.

The nature of the oyster is such that it survives in extreme situations — hot sunshine at low tide, varying salinities, extremes in temperature, etc. Because of its hardiness, the oyster can survive movements over great distances and time spans measured in weeks.

Oyster culture in Europe probably began in the Middle Ages. It continues to the present time, although its prime was at the end of the 19th century, when modern transportation permitted an expansion of transfer activities in oyster culture. One of the most significant events was the probable introduction of *Crassostrea gigas* (now recognized as the Portuguese oyster, *Crassostrea angulata*) about 500 years ago via either (1) the Romans, (2) Marco Polo, or (3) the Crusaders. Oyster production in North America was primarily in the hunter-mode until early in the 20th century when seed collection and leased ground approaches began. However, natural production was the primary method in the middle Atlantic region until quite recently. These activities set the stage for the mass mortalities that occurred periodically in commercially exploited oyster populations worldwide (Table 1).

Historical Background

The earliest well documented case of mass mortality in oyster populations occurred in *Ostrea edulis* in the United Kingdom in the 1920s. This episode reported by Orton (1924) was charac-

Table 1. Mass mortalities in oysters

Date	Location	Disease Conditions	Etiology
		Crassostrea virginica	
1930s	Malpeque Bay, Canada	Inflammatory lesions	Unknown
1940s	Louisiana, Texas	Perkinsosis	*Perkinsus marinus*
1950s	Chesapeake Bay, Va.	Perkinsosis	*Perkinsus marinus*
1957	Delaware Bay, N.J.	Haplosporidiosis	*Haplosporidium nelsoni*
1959	Chesapeake Bay, Va.	Haplosporidiosis	*Haplosporidium nelsoni*
1960	Chesapeake Bay, Md.	Haplosporidiosis	*Haplosporidium nelsoni*
1960	Chincoteague Bay, Va.	Haplosporidiosis	*Haplosporidium costale*
1969	Wellfleet Harbor, Mass.	Haplosporidiosis	*Haplosporidium nelsoni*
1970	Great Bay, Maine	Herpesvirosis	Herpesvirus
		Crassostrea gigas	
1950s	Matsushima Bay, Japan	Inflammatory lesions — bacteremia	Bacteria
1960s	British Columbia, Canada	Microcytosis	*Mikrocytos mackini*
1960s	Japan-Washington USA	Focal necrosis	Bacteria
1970s	France	Gill disease	Iridovirus
1970s	France	Aber disease	*Marteilia refringens*
1970s	Korea	Egg disease	*Marteilia* sp.?
		Saccostrea commercialis	
1924	Australia	Winter disease	*Mikrocytos roughleyi*
1969,70	Australia	QX disease	*Marteilia sydneyi*
		Crassostrea angulata	
1960s	France, Portugal	Gill disease	Iridovirus
		Ostrea edulis	
1920	United Kingdom	Fatal infectious disease	Unknown
1940s	Holland	Shell disease	Fungus
1962	Chincoteague Bay, Va	Bonamiosis	*Bonamia ostreae*
1966	California	Bonamiosis	*Bonamia ostreae*
1970s	France	Aber disease	*Marteilia refringens*
1980s	France, Europe	Bonamiosis	*Bonamia ostreae*
1980s	United Kingdom	Herpesvirosis	Herpesvirus

terized by (1) mass mortality, (2) poor condition, (3) pale digestive gland, (4) pustules, (5) shell deposits, and (6) inflammatory lesions. No pathogen was identified, but Orton came to the tentative conclusion that the disease was infectious in nature. The second documentation of mass mortality was in *Saccostrea commercialis* in Australia and was described as "winter disease" by Roughley (1926), who saw (1) mortality, (2) mantle recession, (3) pale digestive gland, (4) shell and tissue pustules, and (5) general inflammatory response. A microcell type organism (*Mikrocytos roughleyi*) (Farley et al. 1988) was recently described in association with this disease. After these two well studied (for the state of the art at the time) episodes of mass mortalities occurred, severe mortalities were seen in *Crassostrea virginica* in Malpeque Bay, Canada, in the 1930s. While many of the same type of lesions were noted (mantle recession, poor condition, pustules, inflammatory response), no pathogen has been identified. The disease was thought to be infectious based on epizootic studies of introductions of oysters from other areas (Needler and Logie 1947; Logie 1958; Logie et al. 1960).

Crassostrea virginica

A disease struck *Crassostrea virginica* in the Louisiana region of the Gulf of Mexico in the 1940s which was attributed to *Dermocystidium* = *Labyrinthomyxa* = *Perkinsus marinus* and was characterized by mortality, phagocytosis, inflammatory response, and ceroid cell increase (Mackin 1951). This parasite was originally thought to be a fungus, but was later identified as a coccidian on the basis of apical complexes in biflagellated zoospores (Perkins 1976). This disease was found in association with mortality in the 1950s in lower Chesapeake Bay (Andrews and Hewatt 1957).

The modern era of oyster pathology was initiated when mass mortalities began in Delaware Bay in 1957, with more than 90% of the oysters killed by 1960. A parasite "MSX" (multinucleate sphere unknown) was found in association and later named

Minchinia = *Haplosporidium nelsoni* by Haskin et al. (1966) on the basis of spores (Couch et al. 1966). The disease was established in Wellfleet, Massachusetts, in the late 1960s from an introduction of seed oysters from the James River in Virginia (Krantz et al. 1971), and to the south shore of Cape Cod with an introduction from Connecticut waters (Kern 1986). Intensive studies at this time led to the discovery of several lethal infectious agents in *C. virginica*: (1) *Haplosporidium costale* (Wood and Andrews 1962), which was introduced to the Pacific Coast later via a transfer of *C. virginica* from Connecticut (Katkansky and Warner 1970a, b), and (2) herpesvirus (Farley et al. 1972).

Crassostrea gigas

Even though *Crassostrea gigas* had been cultured for probably more than 1000 years, few fatal diseases were evident in this species. Takeuchi et al. (1960) studied a bacterium associated with mortalities in Matsushima Bay. This is probably the same bacterium associated with "focal necrosis" seen in Japan and Washington as described by Sindermann (1970). This is now believed to be a *Nocardia* which continues to be associated with mortalities in Washington (Friedman and Hedrick 1990). Mackin found a small intracellular protistan organism in association with mortalities in Denman Island, B.C., Canada in the 1960s (Quayle 1961) which was described as *Mikrocytos mackini* (Farley et al. 1988). In recent years, a complex of non-fatal, rare diseases have been seen in Japanese oysters with possible relationships to other diseases. These include gill disease in France, caused by an iridovirus (Comps and Bonami 1977; Comps and Duthoit 1979), another iridovirus in larvae from the west coast of the United States (Elston 1979), *Marteilia refringens* in France (Balouet et al. 1979), and a protozoan cytozoic egg parasite in Korea (Chun 1970; Matsusato et al. 1977), and a haplosporidan also in Korea (Kern 1976).

Ostrea edulis

Ostrea edulis has a long, rich history of fatal epizootics, starting with Orton's study in the 1930s. Cultivation was in a steady state until the 1950s when Korringa (1951) attributed mass mortalities in Holland to a fungus that invaded the shell. The organism was described as *Ostracoblabe* (Alderman 1976). In the early 1960s, oysters (*O. edulis*) were transferred to Chincoteague Bay, Virginia, from the Milford National Marine Fisheries Service (Connecticut) hatchery. All died within one year and a small intracellular parasite was discovered in association with the mortality (Farley et al. 1988).

A second transfer of Milford *O. edulis* to Elkhorn Slough, California, in the early 1960s resulted in mass mortalities and rediscovery of the "microcell" organism (Katkansky et al. 1969) seen in the Chincoteague introduction. A third episode of this disease was seen in laboratory-held *O. edulis* in a study at the Oxford, Maryland, Laboratory (Farley et al. 1988). Mass mortality again struck *O. edulis,* this time in France in the late 1960s. An organism, *Marteilia refringens,* was described by Grizel et al. (1974) in association with these mortalities. The disease affected populations most intensively that were in the upper parts of estuaries (Abers), hence the name "Aber disease." Infections were confined to the digestive gland.

Some recovery of the industry was in effect when the second disease struck this species in France in 1979. The cause of the most recent epizootic was found to be "microcell disease," previously seen in the U. S. epizootics mentioned earlier. The parasite was described as *Bonamia ostreae* (Pichot et al. 1980). Elston et al. (1986) uncovered *O. edulis* mortalities in Puget Sound and showed that the disease was indeed associated with *B. ostreae.* They documented the transfer of this disease via movement of oysters from Elkhorn Slough to (1) Puget Sound, and (2) France! Farley et al. (1988) demonstrated initially by electron microscopy that the Elkhorn Slough epizootic of the early 1960s was clearly that species later described as *B. ostreae.*

Other Pertinent Oyster Mortalities

Crassostrea angulata, the Portuguese oyster, began dying *en mass* in 1964-65 in the French South Atlantic region and was virtually eliminated as a commercial species. The disease was characterized by lysis of gill epithelia and acute inflammatory response. It was initially diagnosed as "*Thanatostrea*" (Franc and Arvy 1970) but later found to be caused by an iridovirus (Comps and Duthoit 1976). Mortalities in Australia of *Saccostrea commercialis* in the middle 1960s were found to be associated with a second species of *Marteilia* called "QX" by Wolf (1972) and later described as *Marteilia sydneyi* (Perkins and Wolf 1976).

Epizootic Neoplasia

In 1969, a symposium was held at the Smithsonian Institution that addressed occurrences of neoplastic disease in poikilothermic animals which resulted in publication of a monograph (Dawe and Harshbarger 1969). This collection of papers described the first presumably malignant lesions in oysters (Couch 1969; Wolf 1969; Farley 1969). Soon after this information was published, the first epizootic of presumably malignant neoplasms was described in *Mytilus edulis* and *Ostrea lurida* from Yaquina Bay, Oregon (Farley 1969; Farley and Sparks 1970). Since these initial discoveries, numerous epizootics have been described in a variety of species of bivalve molluscs (Christensen et al. 1974; Farley 1975; Brown et al. 1977; Farley et al. 1986; Twomey and Mulcahy 1988). The characteristics of this complex of diseases are that (1) most of them show cells with enlarged hyperchromatic nuclei, (2) cellular proliferation as evidenced by mitosis and apparent increase in concentrations of abnormal cells, (3) diffuse invasion of connective tissue spaces and sinuses, (4) neoplasm prevalences of 10-90%, (5) seasonal peaks of activity, and (6) where documentation was possible, association with mortality.

Epizootics have been studied intensively enough using field and laboratory approaches to document progression of disease to a fatal outcome, and even transmission ability from animal to animal. Christensen et al. (1974) showed in *Macoma balthica* that animals with this disease died, while control animals did not. Cooper et al. (1982) and Farley et al. (1986) demonstrated progression and mortality in *Mya arenaria* sarcomas, and Brown (1980) and Farley (1989) showed the transmissibility of this disease by proximity and injection of sarcoma cells. A virus was proposed as the etiologic agent by Oprandy et al. (1981), but no convincing evidence was presented to confirm this hypothesis. Sunila and Farley (1989) demonstrated that sarcoma cells remained viable in seawater under rather extreme conditions of salinity, temperature, and pH for at least 6 hours, suggesting strongly that transmission from clam to clam via sarcoma cell transplantation was possible and that a viral agent was not necessary for natural transmission of the disease. More recent studies by Twomey and Mulcahy (1988) on cockle neoplasia in Ireland and by Elston et al. (1986) studying *M. edulis* in Puget Sound have demonstrated similar results of progression, lethality, and transmissibility. Elston et al. (1988) claim cell-free transmission, implying a viral involvement, but no one has yet characterized a virus in any of these diseases; Elston and his associates also found neoplasms in control animals, weakening this interpretation. It is likely that many of these epizootic neoplastic diseases are caused by transmissible agents that conceivably could be transferred geographically to spread this type of disease into new areas. It is also possible that some of the earlier fatal epizootics could have been neoplastic in nature. For example, the histopathology and cytologic characteristics of the Malpeque Bay disease resemble neoplastic manifestations.

Speculative Scenario of Disease Transfers in Bivalve Molluscs

Little evidence is available regarding initiation of epizootics caused by geographic transfers before the 1960s. However, earlier investigators (Orton 1924; Roughley 1926) had suggested this as a possibility. Canadian scientists suggested that the Malpeque

Bay epizootic originated from a transfer of *C. virginica* from U.S. waters. Anecdotal information suggested that small clandestine introductions of *C. gigas* preceded the epizootic. Gill disease virus is responsible for the demise of *C. angulata* in the middle 1960s on the South Atlantic French coast. Indeed, Comps and Bonami (1977) found a similar virus in Japanese oysters imported to this region from Japan in 1968. *C. gigas* was also thought to have been transferred from the west coast of the United States in the late 1960s. Subsequently, Elston (1979) found an iridovirus disease of *C. gigas* larvae from a hatchery on the Pacific Coast of Washington which was associated with larval mortality. It is my opinion that gill disease in *C. angulata* was introduced from *C. gigas* transferred from the west coast of the United States. However, it is possible that the disease could have originated in the Orient (early, specific evidence of transfers is lacking). The destruction of the *C. angulata* industry in France led to massive introductions of *C. gigas* from numerous places to sites in the Brittany and South Atlantic regions of France in the late 1960s. Mass mortalities began in *O. edulis* in the Brittany region of France in 1968 (Comps 1970). These were caused by a protistan parasite described as *Marteilia refringens* (Grizel et al. 1974). This parasite was later found in *C. gigas* in France (low prevalences and intensities) by Balouet et al. (1979), and similar organisms were also found in French mussels (Comps et al. 1975) and copepods (Desportes and Ginsburger-Vogel 1977). A closely related species of *Marteilia* was subsequently identified in *S. commercialis* in Australia (Perkins and Wolf 1976). A parasite of similar morphology and development was described as a cytozoic parasite of *C. gigas* ova from Korean (Chun 1970) and Japanese oysters (Matsusato et al. 1977). Until recently, no parasites of this group have been found on either coast of the United States or Canada. The initial occurrence of this unique group of protistan pathogens in Far East situations tempts the hypothesis that the recent outbreaks in Europe came from introductions from the Far East. These recently documented transfers of *C. gigas* to France support this assumption. The documentation for the initiation of *B. ostreae* epizootics in three sites in the United States (Farley et al. 1988), all associated with transfers from the Milford, Connecticut, hatchery stocks of *O. edulis*, and the final transfer from California to France, is the strongest evidence that diseases can and have been introduced by the geographic transfer of oysters from ambient infection sites to sites that were initially free of the disease, and that devastating mortalities can and have resulted from these practices. Other well documented instances are: (1) the transfer and introduction of *Mytilicola orientalis* to native species of bivalves on the Pacific Coast of America (Odlaug 1946) and in France by *C. gigas* transfers, and (2) the establishment of epizootic sarcoma in Chesapeake Bay soft-shelled clams by the presumed transfer of infected clams from New England after Hurricane Agnes in 1972 (Farley et al. 1986).

Recommendations

Agricultural policies to prevent the spread of infectious diseases have long been established at the national and international levels. The U.S. Department of Agriculture has had regulations and laws in effect for many years to prevent the importation of infective animals and materials. The United Kingdom also has had regulations that quarantine suspect animals and prevent introduction of diseases. In human situations, vaccination and certification of disease-free status have been required routinely for those wishing to immigrate, but policy in transfers of invertebrate animals has been and, for the most part still is, non-existent. The International Council for the Exploration of the Sea (ICES) Working Group for Marine Pathology established a set of guidelines in the 1970s for geographically transferring molluscs (see Sindermann, this volume). These guidelines require: (1) evaluation of the need, (2) ecological evaluation, (3) disease diagnosis of a statistically valid number of the candidate animals, and (4) if all conditions are acceptable, then a brood stock may be introduced under quarantine conditions and an F1 generation produced. If the brood stocks and the progeny remain "disease free," then a trial introduction of the F1 stocks may be made in an isolated location. A successful outcome then warrants mass establishment of the species. While risks still exist even with this procedure, it is by far a much safer approach than introducing massive numbers of wild animals. Modifications of these guidelines have been adopted by several states in the United States, but a strong national and/or international policy is still lacking in the United States and in many other countries, as well as consistent individual state disease management policies. While progress has been made with interstate agreements along this line, much remains to be done regarding within-state, interstate, national, and international laws and regulations. For a detailed treatment of this subject, see Sindermann (this volume).

In summary, (1) diagnostic regional maps of disease occurrence by species should be developed and updated on a regular basis; (2) mass movements from one ecological zone to another, and movement of disease-infected populations to non-diseased areas should be prohibited by laws and regulations; (3) these laws and regulations should be developed through interstate and national approaches of industry, management agencies, and disease research-orientated federal and state agencies and academic institutions dedicated to these types of studies; (4) limitations must include the concept that diseases may be transferred by organisms on a passive level, such as a carrier, in which the disease or diseases may not even be diagnosable in the species being transferred, i.e., human hepatitis from oysters carrying the virus in their gastrointestinal (GI) tract or hard clams carrying oyster infective *Perkinsus* stages in their GI tract, while not actually having the disease.

It is obvious that a massive problem exists, and while some progress has been made, much more is needed to establish a safe approach to stabilizing the exacerbations caused by unwise disease management policies.

Acknowledgments

I thank Dr. Rebecca Ellison and Ms. Sara V. Otto for critical review of the manuscript and Ms. Jane Keller for technical editing.

Literature Cited

Alderman, D.J. 1976. Fungal diseases of marine animals, p. 223-260. *In* E.B.G. Jones (ed.), Recent advances in aquatic mycology. Elek Science, London.

Andrews, J.D. and W.G Hewatt. 1957. Oyster mortality studies in Virginia. II. The fungus disease caused by *Dermocystidium marinum* in oysters of Chesapeake Bay. Ecol. Monogr. 27: 1-25.

Balouet, G., A. Cahour and C. Chastel. 1979. Epidemiologie de la maladie de la glande digestive de l'huitre plate: hypotheses sur le cycle de *Marteilia refringens*. Haliotis 8: 323-326.

Brown, R.S. 1980. The value of the multidisciplinary approach to research on marine pollution effects as evidenced in a three-year study to determine the etiology and pathogenesis of neoplasia in the soft-shell clam, *Mya arenaria*. Rapp. P.V. Reun. Cons. Int. Explor. Mer 179: 125-128.

Brown, R.S., R.E. Wolke, S.B. Saila and C.W. Brown. 1977. Prevalence of neoplasia in 10 New England populations of the soft-shell clam (*Mya arenaria*). Ann. N.Y. Acad. Sci. 298: 522-534.

Christensen, D.J., C.A. Farley and F.G. Kern. 1974. Epizootic neoplasms in the clam *Macoma balthica* (L.) from Chesapeake Bay. J. Nat. Cancer Inst. 52: 1739-1749.

Chun, S. K. 1970. Studies on the oyster diseases. (1) Pathologic investigation. Bull. Korean Fish. Soc. 3: 7-18.

Comps, M. 1970. Observations sur les causes d'une mortalite abnormale des huitres plates dans le basin de Marennes. Rev. Trav. Inst. Peches Marit. 34: 317-326.

Comps, M. and J.R. Bonami. 1977. Infection virale associee a des mortalites chez l'huitre *Crassostrea gigas* Thunberg. Compt. Ren. Hebdomad. Seances Acad. Sci., Paris (Serie D) 285: 1139-1140.

Comps, M. and J.-L. Duthoit. 1976. Infection virale associee a la maladie des branchies de l'huitre portugaise *Crassostrea angulata* Lmk. Compt. Ren. Hebdomad. Seances Acad. Sci., Paris (Serie D) 283: 1595-1597.

Comps, M. and J.-L. Duthoit. 1979. Infections virales chez les huitres *Crassostrea angulata* LMK et *Crassostrea gigas* Th. Haliotis 8: 301-307.

Comps, M., H. Grizel, G. Tige and J.L. Duthoit. 1975. Parasites nouveaux de la glande digestive des mollusques marins *Mytilus edulis* L. et *Cardium edule* L. Compt. Ren. Hebdomad. Seances Acad. Sci., Paris (Serie D) 281: 179-181.

Cooper, K.R., R.S. Brown and P.W. Chang. 1982. The course and mortality of a hematopoietic neoplasm in the soft-shell clam, *Mya arenaria*. J. Invertebr. Pathol. 39: 149-157.

Couch, J.A. 1969. An unusual lesion in the mantle of the American oyster, *Crassostrea virginica*. Nat. Cancer Inst. Monogr. 31: 557-562.

Couch, J.A., C.A. Farley and A. Rosenfield. 1966. Sporulation of *Minchinia nelsoni* (Haplosporida, Haplosporidiidae) in *Crassostrea virginica* (Gmelin). Science 153: 1529-1531.

Dawe, C.J. and J.C. Harshbarger, editors. 1969. Neoplasms and Related Disorders of Invertebrate and Lower Vertebrate Animals. Nat. Cancer Inst. Monogr. 31.

Desportes, I. and T. Ginsburger-Vogel. 1977. Affinites du genre *Marteilia* parasite d'huitres (maladie des Abers) et du crustace *Orchestia gammarellus* (Pallas), avec les myxosporidies, actinomyxidies et paramyxidies. Compt. Ren. Hebdomad. Seances Acad. Sci., Paris (Serie D) 285: 1111-1114.

Elston, R.A. 1979. Viruslike particles associated with lesions in larval Pacific oysters (*Crassostrea gigas*). J. Invertebr. Pathol. 33: 71-74

Elston, R.A., C.A. Farley and M.L. Kent. 1986. Occurrence and significance of bonamiasis in European flat oysters *Ostrea edulis*, in North America. Dis. Aquat. Organ. 2: 49-54.

Elston, R.A., M.L. Kent and A.S. Drum. 1988. Transmission of hemic neoplasm in the bay mussel, *Mytilus edulis*, using whole cells and cell homogenate. Develop. Comp. Immunol. 12: 719-729.

Farley, C.A. 1969. Probable neoplastic disease of the hematopoietic system in oysters (*Crassostrea virginica* and *Crassostrea gigas*). Nat. Cancer Inst. Monogr. 31: 541-555.

Farley, C.A. 1975. Epizootic and enzootic aspects of *Minchinia nelsoni* (Haplosporida) in the American oyster *Crassostrea virginica*. J. Protozool. 15: 585-599.

Farley, C.A.. 1989. Selected aspects of neoplastic progression in molluscs, p. 24-31. *In* H.E. Kaiser (ed.), Comparative aspects of tumor development. Cancer growth and progression. Kluwer Academic Publishers, The Netherlands.

Farley, C.A. and A.K. Sparks. 1970. Proliferative diseases of hemocytes, endothelial cells, and connective tissue cells in molluscs. Bibl. Haematol. 36: 610-617.

Farley, C.A., W.G. Banfield, G. Kasnic Jr. and W.S. Foster. 1972. Oyster herpes-type virus. Science 178: 759-760.

Farley, C.A., S.V. Otto and C.L. Reinisch. 1986. New occurrence of epizootic sarcoma in Chesapeake Bay soft- shell clams (*Mya arenaria*). Fish. Bull. 84: 851-857.

Farley, C.A., P.H. Wolf and R.A. Elston. 1988. A long-term study of "microcell" disease in oysters with a description of a new genus, *Mikrocytos* (g.n.) and two new species, *Mikrocytos mackini* (sp. n.) and *Mikrocytos roughleyi* (sp. n.). Fish. Bull. 86: 581-593.

Franc, A. and L. Arvy. 1969. Sur *Thanatostrea polymorpha* n.g., n. sp., agent de destruction des branchies et des palps de l'huitre portugaise. Compt. Ren. Hebdomad. Seances Acad. Sci., Paris (Serie D) 268: 3189-3190.

Friedman, C.S. and R.P. Hedrick. 1991. Pacific oyster nocardiosis: isolation of the bacterium and induction of laboratory infections. J. Invertebr. Pathol. 57:109-120.

Grizel, H., M. Comps, J.R. Bonami, F. Cousserans, J.L. Duthoit and M.A. LaPennec. 1974. Recherche sur l'agent de la maladie de la glande digestive de *Ostrea edulis* Linne. Bull. Inst. Peches Marit. Maroc No. 240: 7-30.

Haskin, H.H., L.A. Stauber and J.G. Mackin. 1966. *Minchinia nelsoni* n. sp. (Haplosporida, Haplosporidiidae): causative agent of the Delaware Bay oyster epizootic. Science 153: 273-275.

Katkansky, S.C. and R.W. Warner. 1970a. Sporulation of a haplosporidan in a Pacific oyster (*Crassostrea gigas*) in Humboldt Bay, California. J. Fish. Res. Board Can. 27: 1320-1321.

Katkansky, S.C. and R.W. Warner. 1970b. The occurrence of a haplosporidan in Tomales Bay, California. J. Invertebr. Pathol. 16: 144.

Katkansky, S.C., W.A. Dahlstrom and R. W. Warner. 1969. Observations on survival and growth of the European flat oyster, *Ostrea edulis*, in California. Calif. Fish Game 55: 69-74.

Kern, F.G. 1976. Sporulation of *Minchinia* sp. (Haplosporida, Haplosporidiidae) in the Pacific oyster *Crassostrea gigas* (Thunberg) from the Republic of Korea. J. Protozool. 23: 498-500.

Kern, F.G. 1986. Observations on the resurgence of MSX (*Haplosporidium nelsoni*) in New England and mid-Atlantic oyster stocks. Sixth Annual Shellfish Biology Workshop, Milford, Connecticut. (Abstract)

Korringa, P. 1951. Investigations on shell disease in the oyster, *Ostrea edulis* L. Rapp. P.V. Reun. Cons. Int. Explor. Mer 128: 50-54.

Krantz, G.E., L.R. Buchanan, C.A. Farley and H.A. Carr. 1971. *Minchinia nelsoni* in oysters from Massachusetts waters. Proc. Nat. Shellfish. Assoc. 62: 83-85.

Logie, R.R. 1958. Epidemic disease in Canadian Atlantic oysters (*Crassostrea virginica*). Ph.D. Thesis, Rutgers University, New Jersey.

Logie, R.R., R.E. Drinnan and E.B. Henderson. 1960. Rehabilitation of disease-depleted oyster populations in eastern Canada. Proc. Gulf Carib. Fish. Inst. 13: 109-113.

Mackin, J.G. 1951. Histopathology of infection of *Crassostrea virginica* (Gmelin) by *Dermocystidium marinum* Mackin, Owen and Collier. Bull. Mar. Sci. Gulf Carib. 1: 72-87.

Matsusato, T., T. Hoshina, K.Y. Arakawa and K. Masumura. 1977. Studies on the so-called abnormal egg-mass of the Japanese oyster, *Crassostrea gigas* (Thunberg). - I. Distribution of the oysters collected on the coast of Hiroshima Pref. Bull. Hiroshima Fish. Exper. Stat. No. 8: 1-21.

Needler, A.W.H. and R.R. Logie. 1947. Serious mortalities in Prince Edward Island oysters caused by a contagious disease. Trans. R. Soc. Can. 41 (Ser. 3, Sect. 5): 73-89.

Odlaug, T.O. 1946. The effect of the copepod, *Mytilicola orientalis* upon the Olympia oyster, *Ostrea lurida*. Trans. Amer. Microsc. Soc. 65: 311-317.

Oprandy, J.J., P.W. Chang, A.D. Provonost, K.R. Cooper, R.S. Brown and V.Y. Yates. 1981. Isolation of a viral agent causing hematopoietic neoplasia in the soft-shell clam, *Mya arenaria*. J. Invertebr. Pathol. 38: 45-51.

Orton, J.H. 1924. An account of investigations into the cause or causes of the unusual mortalities among oysters in the English oyster beds during 1920 and 1921. Part 1. Fish. Invest., London (Ser. 2) 6(3): 1-199.

Pichot, Y., M. Comps, G. Tige, H. Grizel and M.A. Rabouin. 1980. Recherches sur *Bonamia ostreae* gen. n., sp. n., parasite nouveau de l'huitre plate *Ostrea edulis* L. Rev. Trav. Inst. Peches Marit. 43: 131-140.

Perkins, F.O. 1976. Zoospores of the oyster pathogen *Dermocystidium marinum*. I. Fine structure of the conoid and other sporozoan-like organelles. J. Parasitol. 62: 959-974.

Perkins, F.O. and P.H. Wolf. 1976. Fine structure of *Marteilia sydneyi* sp. n.—haplosporidan pathogen of Australian oysters. J. Parasitol. 62: 528-538.

Quayle, D.B. 1961. Denman Island oyster disease and mortality, 1960. Fish. Bd. Can., Manuscript Report Series No. 713, pp. 1-9.

Roughley, T.C. 1926. An investigation of the cause of an oyster mortality on the Georges River, New South Wales, 1924-25. Proc. Linn. Soc. N. S. W. 51: 446-491 (+ plates).

Sindermann, C.J. 1970. Principal diseases of marine fish and shellfish. Academic Press, New York.

Sunila, I. and C.A. Farley. 1989. Environmental limits for survival of sarcoma cells from the soft-shell clam *Mya arenaria*. Dis. Aquat. Organ. 7: 111-115.

Takeuchi, T., Y. Takemoto, and T. Matsubara. 1960. Haematological study of bacteria affected oysters. Hiroshima Sulsan Shikenjo Hokuku 22(1): 1-7. (Report of Hiroshima Prefectural Fisheries Experimental Station, Hiroshima City, Japan.)

Twomey, E. and M.F. Mulcahy. 1988. Epizootiological aspects of a sarcoma in the cockle *Cerastoderma edule*. Dis. Aquat. Organ. 5: 225-238.

Wolf, P.H. 1969. Neoplastic growth in two Sydney rock oysters, *Crassostrea commercialis* (Iredale and Roughley). Nat. Cancer Inst. Monogr. 31: 563-573.

Wolf, P.H. 1972. Occurrence of a haplosporidan in Sydney rock oysters (*Crassostrea commercialis*) from Moreton Bay, Queensland, Australia. J. Invertebr. Pathol. 19: 416-417.

Wood, J.L. and J.D. Andrews. 1962. *Haplosporidium costale* (Sporozoa) associated with a disease of Virginia oysters. Science 136: 710-711.

Geographic Dispersion of the Viruses IHHN, MBV and HPV as a Consequence of Transfers and Introductions of Penaeid Shrimp to New Regions for Aquaculture Purposes

DONALD V. LIGHTNER
RITA M. REDMAN
THOMAS A. BELL
ROBERT B. THURMAN

Abstract: Introductions of penaeid shrimp for aquaculture purposes from remote areas to Hawaii, Mexico, Ecuador, Brazil, and North America have shown that certain shrimp pathogens, most notably the viruses IHHN, MBV, and HPV, can readily be transported with live shipments of shrimp. In some cases, these introductions have resulted in catastrophic disease losses to facilities which had imported the contaminated stocks. In others, the effects were moderate or insignificant.

Some important examples include the introduction of two virus-caused diseases, IHHN and MBV. These were discovered initially in populations of penaeid shrimp imported for aquaculture purposes into University of Arizona-operated shrimp culture facilities in Hawaii and Mexico. MBV was first recognized and described in a population of *Penaeus monodon* that was imported from Taiwan into Puerto Penasco, Mexico, in 1976. Later, MBV was found in populations of *P. monodon* imported into Hawaii from Taiwan, Tahiti, and the Philippines. Likewise, IHHN virus was discovered in a number of populations of *P. stylirostris* and *P. vannamei* imported into Hawaii in 1980 through 1982 from shrimp hatcheries in Florida, Panama, Costa Rica, Ecuador, and Tahiti. IHHN was found in Hawaii to be a highly lethal disease of juvenile *P. stylirostris*, frequently resulting in mortality rates approaching 90% in populations reared in high-density systems. Two populations of imported *P. vannamei* (from hatcheries in Costa Rica and Ecuador) were shown to carry IHHNV asymptomatically, and to readily transmit the disease to previously unexposed *P. stylirostris* populations.

HPV and MBV have been found recently at several North and South America shrimp culture facilities, in shrimp imported from various Asian locations. *P. vannamei* reared at the same facilities were found to have become infected by both viruses. While in neither case were the infections accompanied by significant disease, careless introductions such as these, and other as yet unrecognized pathogens, may result ultimately in the introduction of other devastating diseases, like IHHN has been to *P. stylirostris*.

Introduction

There are nearly as many viruses known in marine invertebrate and vertebrate animal species of aquaculture interest as there are species cultured. With only a few exceptions (lobsters, for example), at least one virus-caused disease is recognized in each significant marine species now being cultured. A perusal of some recent reference books on the subject of disease and pathology of these animals shows this to be true (Table 1). *The Fish Health Blue Book* (published by the Fish Health Section of the American Fisheries Society; Amos 1985) lists eight types of virus-caused diseases, six of which are of concern to the marine aquaculture industry. Another reference, *Disease Diagnosis and Control in North American Marine Aquaculture* (Sindermann and Lightner 1988), lists 15 or 16 virus diseases in cultured marine crustacea, molluscs, fish, and turtles. Sparks (1985) in his *Synopsis of Invertebrate Pathology Exclusive of Insects* lists 24 virus-caused diseases of cultured and/or commercially important invertebrate animals. In the current literature we find that several more new viruses, or new hosts or geographic records for previously known viruses, are reported each year.

Many of these virus diseases are of only minor apparent significance to the aquaculture industry. However, others are of considerable significance, causing catastrophic losses whenever host or environmental conditions favor their development. Of the viruses that we now recognize, several have significantly affected the development of commercial marine aquaculture. Some important examples include VHS, IPN, and IHN of salmonids; and

IHHNV, BP, BMN, and MBV of penaeid shrimp. The 1988 crash of Taiwan's crop of *Penaeus monodon* may cost that country as much as 500 million dollars in lost export revenue. One or more viruses (MBV and HPV) are believed to have caused and/or contributed to the epizootic (Rosenberry 1988; S.N. Chen and G.H. Kou, personal communication, July 21, 1988, National Taiwan University, Taipei).

Table 1. A partial list of the viruses of cultured marine animals from three recent reviews.

	AFS Blue Book (1985)	*Sindermann and Lightner (1988)*	*Sparks (1985)*
Finfish			
Salmonids	7	1	-
Others	1	2	-
Crustacea			
Penaeid shrimp	-	6	4
Macrobrachium	-	0	0
Crabs	-	3	13
Lobsters	-	0	0
Molluscs			
Oysters	-	2-3	5
Others	-	0	2
Turtles	-	1	-
Totals	8	15-16	24

The Viruses of Penaeid Shrimp

Six virus diseases are presently recognized in the penaeid shrimp (Table 2). These six viruses are: BP = *Baculovirus penaei* (Couch 1974a, 1974b); MBV = *P. monodon*-type baculovirus (Lightner and Redman 1981); BMN = baculoviral midgut gland necrosis (Sano et al. 1981, 1984, 1985; Momoyama 1983); IHHNV = infectious hypodermal and hematopoietic necrosis virus

Table 2. The penaeid viruses and their known natural and experimentally infected hosts.

Host Subgenus			VIRUS*			
and Species**	*BP*	*MBV*	*BMN*	*IHHNV*	*HPV*	*REO*
Litopenaeus:						
P. vannamei	+++	+		+	+	
P. stylirostris	++			+++		
P. setiferus	+			+(e)		
P. schmitti	++					
Penaeus:						
P. monodon	+	++		++	++	++
P. esculentus		+			++	
P. semisulcatus		+		+	+++	
Fenneropenaeus:						
P. merguiensis		++			+++	
P. indicus					++	
P. chinensis (=*orientalis*)					++	
P. penicillatus	++	++			++	
Marsupenaeus:						
P. japonicus			+++	++(e)		+++
P. plebejus		++				
Farfantepenaeus:						
P. aztecus	+++			+(e)		
P. duorarum	+++			+(e)		
P. brasiliensis	++					
P. paulensis	++					
P. subtilis	++					
Melicertus:						
P. kerathurus		+				
P. marginatus	+++					
P. plebejus		++				

*Abbreviations:

BP = *Baculovirus penaei*
MBV = *P. monodon*-type baculovirus
BMN = Baculoviral midgut gland necrosis
IHHNV = Infectious hypodermal and hematopoietic necrosis virus
HPV = Hepatopancreatic parvo-like virus
REO = Reo-like virus
+ = Infection observed, but without signs of disease
++ = Infection may result in moderate disease and mortality
+++ = Infection usually results in serious epizootic
e = Experimentally infected; natural infections not yet observed

**Classification according to Holthuis, 1980, FAO Species Catalog.

(Lightner et al. 1983a); HPV = hepatopancreatic parvo-like virus (Lightner and Redman 1985); and REO = reo-like virus (also known as RLV) of the hepatopancreas (Tsing and Bonami 1987). Each of these six virus "species" may actually be comprised by a multitude of individual strains, some of which are highly pathogenic to some penaeids, while being of little consequence to other penaeids (Table 2).

Three basic diagnostic procedures are used in screening penaeid shrimp for virus infections: (1) direct samples for microscopic (wet-mount) examination and/or histopathology or electron microscopy; (2) enhancement of infection followed by sampling and histopathology and electron microscopy; and (3) bioassay of a suspect shrimp population with a sensitive indicator species combined with direct sampling and examination of the indicator shrimp for signs of infection using wet-mounts or histopathology. Details of the current diagnostic procedures for the virus diseases of penaeids have been recently published elsewhere (Lightner 1988; Lightner and Redman 1990).

The Exotic Shrimp Dilemma

To an industry composed of scientists and businessmen, a major concern has been the need for testing a variety of local and exotic shrimp before they are transferred. The reasons for these transfers are sometimes justifiable, sometimes not, but shrimp growers import shrimp, seeking species that will grow well and provide more profit from their culture systems. Some desirable characteristics used to justify importing exotic species have included: a larger harvest size, faster growth rates, disease resistance, a higher market price, easier reproduction and larval rearing, and growth at colder water temperatures. Hence, larvae, postlarvae, and broodstock from shrimp farms, from experimental facilities, or from wild stocks collected by commercial fisherman have been transferred countless times from one geographic location to another for aquaculture purposes without testing, particularly for pathogens.

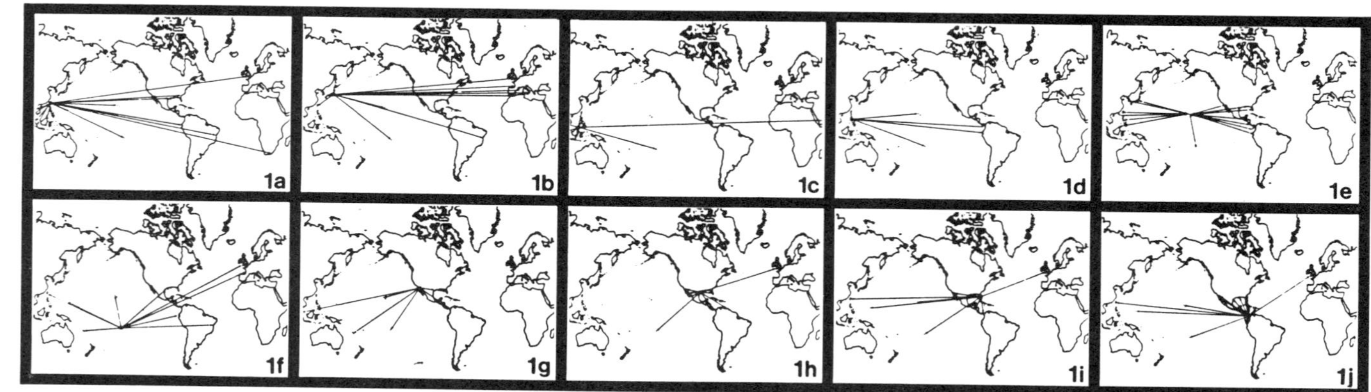

Figure 1. Examples of published and unpublished records of live shrimp transfers. Such transfers are typical of those made beginning nearly two decades ago and continuing in today's shrimp-culture industry. Specific examples shown illustrate transfers of penaeids to or from:

1a. Taiwan	*1b. Japan*	*1c. Maylasia*	*1d. Philippines*	*1e. Hawaii*
1f. French Polynesia	*1g. Puerto Penasco, Mexico*	*1h. Texas*	*1i. Florida*	*1j. Panama*

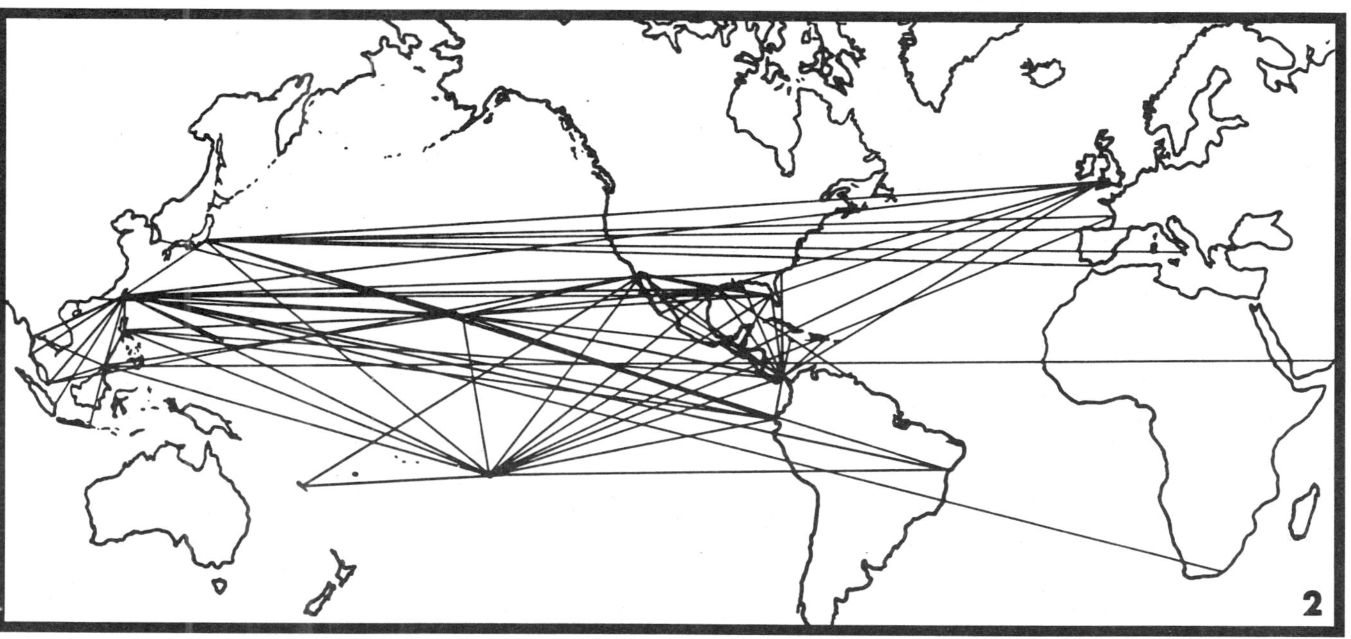

Figure 2. A hypothetical "exotic shrimp transfer network" that results from the combination of Figure 1 shrimp-transfer maps into a single map. It is apparent that if an unrecognized pathogen entered any facility in the transfer network, the mechanism exists for it to be rapidly transferred to several other facilities. Further, if the pathogen remained undetected, it could be easily introduced into all of the facilities in the transfer network. This model may explain the nearly ubiquitous distribution of IHHNV in shrimp culture facilities.

Added to this concern was our paucity of knowledge of shrimp pathogens. Less than a decade ago the shrimp culture industry knew of only one virus in the penaeids (Couch 1974a, 1974b). We now recognize at least six different virus diseases in these animals, and we suspect several more (Lightner 1988; Lightner and Redman 1990).

Figure 1 shows examples of published and unpublished penaeid shrimp transfers from one culture facility to another over a two- or three-year period bracketing 1980. When these are combined into a single map (Figure 2), they illustrate "an exotic shrimp transfer network." From the combined map (Figure 2), it is apparent that if an unrecognized pathogen entered any one facility, the mechanism exists for it to be rapidly transferred to several other facilities, and if it remains undetected, to all of the facilities in the "transfer network." Several of the shrimp viruses have entered this transfer network, as a direct result of the host being introduced to farms in regions far from their original geographic range. IHHN virus of penaeid shrimp was first recognized in imported shrimp in Hawaii that had been imported from at least five different sources (Lightner et al. 1983a, 1983b). None of the suppliers knew of the existence of IHHNV, nor did the shrimp they supplied *(P. vannamei)* show any outward signs. However, when the imported *P. vannamei* stocks were grown adjacent to stocks of *P. stylirostris,* most of the latter species were soon lost to IHHN. An interesting coincidence is that all of the sources that provided IHHNV-contaminated *P. vannamei* had previously abandoned most of their attempts at culturing the faster growing *P. stylirostris* because of its poor survival in their culture systems.

Host and Geographic Distribution of the Penaeid Viruses

The host geographic range of the known penaeid viruses has been updated several times recently (Couch 1981; Johnson 1983;

Lightner 1983, 1988; Lighter et al. 1985). In recent years, surveys and investigations of mortality problems undertaken by this laboratory and by other researchers in various shrimp-growing areas have provided new data on several of the virus diseases that affect cultured penaeid shrimp. This review summarizes the current knowledge of the natural hosts and the natural and introduced geographic distribution of the penaeid virus diseases.

Baculovirus penaei (BP)

BP is widely distributed in cultured and wild penaeids in the Americas, ranging from the Northern Gulf of Mexico south through the Caribbean and reaching at least as far as the State of Bahia in central Brazil. On the Pacific Coast, BP ranges from Peru to Mexico, and it has been observed in wild penaeid shrimp in Hawaii. BP has not yet been observed in wild, cultured, or imported (from the Americas) penaeids outside of the Americas. Recent new information on the host and geographic distribution of BP has come from Brazil, Ecuador, and Mexico. In South America (Brazil and Ecuador), BP was found to infect larvae and postlarvae of six penaeid species. In Ecuador BP was found to infect the larvae of imported *P. monodon* in a hatchery in which BP was enzootic in its stocks of *P. vannamei* (Philippe Danigo, personal communication, December 19, 1984, SEMACUA, Ecuador). BP has been found in at least two hatcheries in Brazil in native *P. schmitti, P. paulensis,* and *P. subtilis,* and in introduced *P. vannamei* and *P. penicillatus.* Five of these species (all but *P. vannamei*) represent new host species for the virus (Table 2). It is significant that the imported Asian species *P. monodon* and *P. penicillatus* were found to be infected by BP.

BP in cultured shrimp was found for the first time in Mexico in larval and postlarval *P. stylirostris* at a facility near Guaymas, Sonora, on the west coast of Mexico (Lightner et al. 1988). Because the affected facility had no history of stock importations, BP must be assumed to be enzootic in wild penaeids in the region.

Penaeus monodon-type *Baculovirus (MBV)*

MBV-type baculoviruses are similar to BP in their diverse host range and in their wide distribution on the Indo-Pacific coasts of Asia, Australia, and Africa, and in Southern Europe. However, unlike BP, MBV has been observed in the Americas in imported stocks and in an American penaeid exposed to the virus. Although MBV was first discovered in a quarantined population of *P. monodon* that had originated in Taiwan (Lightner and Redman 1981; Lightner et al. 1983c), it had not actually been demonstrated in Taiwan until it was found to be widely distributed in Taiwanese shrimp farms in a 1986 survey of the country (Lightner et al. 1987). Additional studies in 1987 and 1988 linked MBV to serious disease losses in many Taiwanese farms (S.N. Chen and G.H. Kou, personal communication, National Taiwan Univ., Taiwan).

Since the information on MBV was last summarized, MBV has been found in Texas, Ecuador, and Brazil in imported stocks of *P. monodon*. Of possible significance was the presence of MBV-like (spherical) occlusion bodies found along with a heavy BP infection of juvenile *P. vannamei* being cultured at the same farm in Ecuador with MBV-infected *P. monodon*.

A similar agent, found first in *P. plebejus*, and thus called Plebejus Baculovirus (PBV), was found in cultured penaeids in Australia (Lester et al. 1987). Other than its presence in a new host species, the agent of PBV differs little from MBV in host cell cytopathology and in the morphology of the virus. It probably represents a strain of the MBV-type viruses rather than a distinct species.

Baculoviral Midgut Gland Necrosis (BMN)

BMN has been reported only in *P. japonicus* cultured in Japan, where it is considered a major problem in the larval and early postlarval stages of that species (Sano et al. 1984, 1985; Momoyama 1983; Sano and Fukuda 1987). Despite numerous introductions of *P. japonicus* stocks (larvae, postlarvae, and broodstock) to Hawaii, France, Brazil, and other locations during the past two decades, BMN has not been detected in that species or in other penaeids cultured with introduced stocks of *P. japonicus.*

Hepatopancreatic Parvo-Like Virus (HPV)

HPV has a geographic range in Asia and Australia similar to that of MBV, and like MBV it has been introduced to the Americas with imported penaeids. More recently, HPV was found for the first time in dual infections with MBV. It was found in postlarval and juvenile *P. monodon* sampled from farms in the Pingtung area of Southern Taiwan. This region in 1987 had experienced serious disease losses in its farms due, at least in part, to MBV. The severity of HPV infections in some of the shrimp sampled suggests that HPV, while unrecognized, may have contributed to the 1987 epizootic.

Reports of HPV in captive-wild *P. esculentus* in Australia (Paynter et al. 1985), in *P. monodon* imported to Israel from Kenya (Colorni et al. 1987), and in captive-wild and hatchery-reared *P. indicus* and *P. merguiensis* in Singapore (Chong and Loh 1984) have expanded the known host and geographic distribution of this virus (Tables 2, 3 and 4). In the Singapore study of four shrimp farms surveyed, HPV incidence was highest (>50%) in the two farms that reared hatchery-derived postlarvae, and lower (<15%) in the two farms which cultured only feral shrimp collected by tidal entrapment (Chong and Loh 1984). This suggests that HPV is transmitted either vertically from parent broodstock, or horizontally from shrimp to shrimp with efficiency only during the larval stages.

HPV has been observed in the Americas. In Brazil in 1987, HPV was found in stocks of *P. penicillatus* imported from Taiwan. At the same culture facility, HPV was found in light infections in juvenile *P. vannamei,* which had been exposed to infected *P. penicillatus* indirectly as a result of normal farming practices. The discovery of HPV in cultured shrimp in Brazil represents the

Table 3. Observed and reported occurrences of the penaeid viruses in wild and cultured penaeids indicating their probable natural and introduced geographic distributions.

Virus	Region/site where found	Host Status*	Virus Status
IHHNV	Atlantic side: SE U.S., Caribbean, & Brazil	Cul	introduced
	Pacific side: Ecuador, Peru, & Central America	Cul	introduced?
	Pacific: Hawaii, Guam, Tahiti	Cul	introduced
	Asia: Taiwan	Cul	introduced
	Singapore, Malaysia, & Philippines	CW	enzootic?
	Middle East: Israel	Cul	introduced
HPV	Indo-Pacific: P.R. China, Taiwan, Philippines, Malaysia, Singapore, & Australia	Cul, CW, W	enzootic
	Africa: Kenya	W	enzootic
	Middle East: Israel & Kuwait	Cul, CW	enzootic
	Americas: Brazil, Ecuador	Cul	introduced
BP	Atlantic side: SE U.S., Caribbean, & Brazil	Cul, CW	enzootic
	Pacific side: Ecuador, Peru, & Central America	Cul, CW, W	enzootic
	Mexico	Cul, CW	enzootic
	Hawaii	W	enzootic
MBV	Indo-Pacific: P.R. China, Taiwan, Philippines, Malaysia, Singapore, & Australia	Cul, CW, W	enzootic
	Africa: S. Africa	W	enzootic
	Middle East: Israel & Kuwait	Cul, CW, W	enzootic
	Mediterranean: Italy	Cul, CW, W	enzootic
	Pacific: Tahiti, Hawaii	Cul	introduced
	Americas: Mexico, Ecuador, Texas, & Brazil	Cul	introduced
BMN	Japan	Cul, CW, W	enzootic
REO	Japan, Malaysia	Cul	enzootic
	Hawaii & France	Cul	introduced

*Abbreviations:

Cul = "cultured"; from cultured or captive-wild broodstock
CW = "captive-wild"; from wild-caught seed or from single-spawn wild broodstock
W = "wild"; from natural sources or commercial fishery

Table 4. Penaeid viruses in the Americas and their status.

Virus	Status in the Americas and Hawaii
IHHNV	Widely distributed in cultured *P. vannamei* and *P. stylirostris;* not recognized in wild penaeids; enzootic in Southeast Asian wild penaeids; recently introduced into Western Mexico from Texas and Panama with postlarval *P. vannamei.*
HPV	Enzootic in Asia, Australia and Africa; introduced to one or more sites in South America from Taiwan; can infect the American penaeid *P. vannamei.*
BP	Widely distributed in American penaeids; enzootic in wild penaeids on both Atlantic and Pacific sides of tropical and subtropical America.
MBV	Enzootic in Asia, Australia, Africa, and Mediterranean; introduced to several sites in Hawaii, North, Central, and South America; can infect *P. vannamei;* contaminated stocks eradicated from Hawaii, Mexico, and Texas.
BMN	Enzootic in Japan; not reported outside of Japan.
REO	Enzootic in Japan; introduced to Hawaii from Japan; contaminated stocks eradicated in Hawaii.

first time this pathogen has been documented in the Americas and in an American penaeid (S. Bueno, R. Meyer, and D. Lightner, unpublished observations). More recently, HPV lesions have been found in *P. vannamei* cultured in Ecuador (Bell and Lightner, unpublished data). The numerous introductions of *P. monodon* from Southeast Asia may eventually be found to have resulted in the introduction of HPV into Ecuador.

Infectious Hypodermal and Hematopoietic Necrosis Virus (IHHNV)

IHHNV has a worldwide distribution in cultured penaeid shrimp, but its distribution in wild penaeids remains unknown.

Infection by the virus causes serious disease in *P. stylirostris* and acute catastrophic epizootics in intensively cultured juveniles of that species. In other penaeids, IHHNV has been reported to cause infection and disease (Brock et al. 1983; Lightner et al. 1985; Lightner 1988), but disease severity does not approach that observed in *P. stylirostris.* The natural host(s) and natural geographic distribution of IHHNV are unknown. However, the occurrence in Southeast Asia (Singapore, Malaysia, and the Philippines) of IHHNV (or a similar agent) in shrimp-culture facilities using only captive-wild *P. monodon* broodstock suggests that Southeast Asia is within the virus' geographic range, and that *P. monodon* may be among its natural host species.

Since 1985, no new hosts for IHHNV have been demonstrated. However, the geographic distribution of the virus in culture facilities has continued to expand. In Mexico in 1987, IHHN was found in an imported population of postlarval *P. vannamei* at a facility in Baja California (Lightner, unpublished data). IHHN was also found in imported quarantined stocks of *P. vannamei* in a 1986 survey of Taiwanese shrimp-culture facilities, but not in cultured stocks of other penaeid species, including *P. monodon,* at the farms surveyed (Light-ner et al. 1987).

Reo-Like Virus (REO)

REO is the newest of the penaeid viruses. It was discovered by Tsing and Bonami (1987) in juvenile *P. japonicus* in France using electron microscopy, and subsequently in the same species in Hawaii using the same technique (Lightner et al. 1985). Most recently REO, or a closely related form, has been found associated with a serious disease syndrome in pond-cultured *P. monodon* in Southeast Asia (Nash et al, 1988). In both species, other lesions were more apparent by light microscopy, and signs of REO infection were overlooked until found by electron microscopy. The virus was located in the cytoplasm of F-cells and R-cells of the hepatopancreatic tubule epithelium, where it formed large cytoplasmic viral inclusions. The nonenveloped, icosahedral virions of REO measured about 60 nm and 50 to 70 nm in diameter, respectively, in purified preparations and in tissue sections.

While the significance of REO infections in cultured penaeids is virtually unknown, some studies have suggested that REO may be a potential serious pathogen of penaeid shrimp. Tsing and Bonami (1987) experimentally transferred the disease in juvenile *P. japonicus* by inoculation of new host shrimp with purified virus or by feeding new host shrimp pieces of REO-infected hepatopancreas. Development of the infection was slow, requiring about 45 days to develop. Secondary infections by *Fusarium solani* were common in REO-infected shrimp. A related study (Tsing et al. 1985) suggested a possible link between infection by REO and "gut and nerve syndrome" (GNS), an idiopathic condition of chronically ill populations of *P. japonicus* cultured in Hawaii (Lightner et al. 1984).

Discussion and Conclusions

The practice of transporting penaeid stocks between facilities and/or different geographic regions has resulted in the introduction of five of the six known penaeid shrimp viruses to regions where they may not have previously existed. Five of the six known types of penaeid viruses are apparently not native to the Americas, but of these five, four (IHHNV, MBV, HPV, and REO) have been introduced with shrimp intended for aquaculture (Tables 3 and 4). Whether or not these introduced viruses have escaped the culture facilities to which they have been introduced and have become established in local wild penaeid stocks is not known.

While evaluation of nonnative penaeids by the emerging shrimp culture industry is an essential component to the growth and development of that industry, introduction of pathogens like IHHNV to regions where it previously did not occur can have catastrophic consequences to the industry (Lightner et al. 1983a,1983b; Lightner 1988). Prevention of such exotic pathogen introductions is dependent upon the use of quarantine, certification, and inspection policies, and procedures that are supported by reliable diagnostic tests. Mechanisms have been proposed by a number of international groups to reduce the risks of importation of exotic pathogens and pests with transfers of aquatic species. Two examples are the FAO Guidelines (1977) and the ICES Code of Practice (Sindermann 1988). Both provide a workable mechanism to reduce this risk. However, for these guidelines to work, adequate quarantinable facilities and qualified diagnosticians must be available. The shrimp-culture industry to this day is still short of both, but the situation is improving.

The virus diseases of cultured marine fish and shellfish are indeed important factors that affect the profitability and development of the aquaculture industry. Successful management of these pathogens translates directly into jobs, commerce, foreign exchange revenues, and most importantly into new sources of high quality foodstuffs.

Acknowledgments

Grant support for this work was from the U.S. Marine Shrimp Farming Consortium, Grant No. 88-38808-3320, U.S.D.A. Cooperative State Research Service; from Sea Grant, U.S. Department of Commerce; and from the Hawaii Aquaculture Development Program, Honolulu, Hawaii.

Literature Cited

Amos, K.H. 1985. Procedures for the detection and identification of certain fish pathogens (The Fish Health Blue Book). Third Edition. Spec. Publ. Amer. Fish. Soc. 114 pp.

Brock, J.A., D.V. Lightner and T.A. Bell. 1983. A review of four virus (BP, MBV, BMN and IHHNV) diseases of penaeid shrimp with particular references to clinical significance, diagnosis and control in shrimp aquaculture. International Council for the Exploration of the Sea. C.M. 1983/Gen:10/Mini-Symposium.

Chong, Y.C. and H. Loh. 1984. Hepatopancreas chlamydial and parvoviral infections of farmed marine prawns in Singapore. Singapore Vet. J. 9: 51-56.

Colorni, A., T. Samocha and B. Colorni. 1987. Pathogenic viruses introduced into Israeli mariculture systems by imported penaeid shrimp. Bamidgeh 39: 21-28.

Couch, J.A. 1974a. Free and occluded virus, similar to *Baculovirus*, in hepatopancreas of pink shrimp. Nature 247: 229-231.

Couch, J.A. 1974b. An enzootic nuclear polyhedrosis virus of pink shrimp: ultrastructure, prevalence, and enhancement. J. Inverteb. Pathol. 24: 311-331.

Couch, J.A. 1981. Viral diseases of invertebrates other than insects, p. 127-160. *In* E. W. Davidson (ed.), Pathogenesis of invertebrate microbial diseases. Allanheld, Osmun Publ., Totowa, New Jersey.

FAO. 1977. Control of the spread of major communicable fish diseases. Report of the FAO/OIE Government Consultation on an International Convention for the Control of the Spread of Major Communicable Fish Diseases. FAO Fisheries Reports, No. 192. FID/R192 (EN). 36 p.

Holthuis, L.B. 1980. FAO Species Catalog. Vol. 1 — Shrimps and Prawns of the World. An Annotated Catalogue of Species of Interest to Fisheries. FAO Fisheries Synopsis No. 125 (FIR/S125), Vol. 1. Food and Agriculture Organization of the United Nations, Rome. 271 pp.

Johnson, P.T. 1983. Diseases caused by viruses, rickettsiae, bacteria, and fungi, p. 1-78. *In* A.J. Provenzano, Jr. (ed.), The biology of crustacea, Vol. 6. Academic Press, New York.

Lester, R.J.G., A. Doubrovsky, J.L. Paynter, S.K. Sambhi and J.G. Atherton. 1987. Light and electron microscope evidence of baculovirus infection in the prawn *Penaeus plebejus.* Dis. Aquat. Org. 3: 159-165.

Lightner, D.V. and R.M. Redman. 1981. A baculovirus-caused disease of the penaeid shrimp, *Penaeus monodon*. J. Inverteb. Pathol. 38: 299-302.

Lightner, D.V. 1983. Diseases of cultured penaeid shrimp, pp. 289-320. *In* J.P. McVey (ed.), CRC Handbook of Mariculture. Vol. 1. Crustacean Aquaculture. CRC Press, Boca Raton, Florida.

Lightner, D.V., R.M. Redman and T.A. Bell. 1983a. Infectious hypodermal and hematopoietic necrosis (IHHN), a newly recognized virus disease of penaeid shrimp. J. Inverteb. Pathol. 42: 62-70.

Lightner, D.V., R.M. Redman, T.A. Bell and J.A. Brock. 1983b. Detection of IHHN virus in *Penaeus stylirostris* and *P. vannamei* imported into Hawaii. J. World Maricult. Soc. 14: 212-225.

Lightner, D.V., R.M. Redman and T.A. Bell. 1983c. Observations on the geographic distribution, pathogenesis and morphology of the baculovirus from *Penaeus monodon* Fabricius. Aquaculture 32:209-233.

Lightner, D.V., R.M. Redman, T.A. Bell and J.A. Brock. 1984. An idiopathic proliferative disease syndrome of the midgut and ventral nerve in the Kuruma prawn, *Penaeus japonicus* Bate, cultured in Hawaii. J. Fish Dis. 7: 183-191.

Lightner, D.V. and R.M. Redman. 1985. A parvo-like virus disease of penaeid shrimp. J. Invertebr. Pathol. 45: 47-53.

Lightner, D.V., R.M. Redman, R.R. Williams, L.L. Mohney, J.P.M. Clerx, T.A. Bell and J.A. Brock. 1985. Recent advances in penaeid virus disease investigations. J. World Maricult. Soc. 16: 267-274.

Lightner, D.V., R.P. Hedrick, J.L. Fryer, S.N. Chen, I.C. Liao and G.H. Kou. 1987. A survey of cultured penaeid shrimp in Taiwan for viral and other important diseases. Fish Pathol. 22: 127-140.

Lightner, D.V. 1988. Diseases of cultured penaeid shrimp and prawns, p. 8-127. *In* C.J. Sindermann and D.V. Lightner (eds.), Disease Diagnosis and Control in North American Marine Aquaculture. Elsevier, Amsterdam.

Lightner, D.V., R.M. Redman and E.A. Almada Ruiz. 1988. *Baculovirus penaei* in *Penaeus stylirostris* (Crustacea: Decapoda) cultured in Mexico: unique cytopathology and a new geographic record. J. Invertebr. Pathol. 51: 137-139.

Lightner, D.V. and R.M. Redman. 1990. Host, geographic range and diagnostic procedures for the penaeid virus diseases of concern to shrimp culturists in the Americas, pp. 173-196. *In* W. Dougherty (ed.), Proceedings of a Symposium on the Frontiers of Shrimp Research. Elsevier, New York.

Momoyama, K. 1983. Studies on baculoviral mid-gut gland necrosis of Kuruma shrimp *(Penaeus japonicus)* — III. Presumptive diagnostic techniques. Fish Pathol. 17: 263-268.

Nash, M., G. Nash, I.A. Anderson and M. Shareff. 1988. A reo-like virus observed in the tiger prawn, *Penaeus monodon,* from Malaysia. J. Fish Dis. 11:531-535.

Paynter, J.L., D.V. Lightner and R.J.G. Lester. 1985. Prawn virus from juvenile *Penaeus esculentus,* p. 61-64. *In* P.C. Rothlisberg, B.J. Hill and D.J. Staples (eds.), Second Australian National Prawn Seminar, NPS2, Cleveland, Australia.

Sano, T. and H. Fukuda 1987. Principal microbial diseases of mariculture in Japan. Aquaculture 67: 59-69.

Sano, T., T. Nishimura, K. Oguma, K. Momoyama and N. Takeno. 1981. Baculovirus infection of cultured Kuruma shrimp *Penaeus japonicus* in Japan. Fish Pathol. 15: 185-191.

Sano, T., T. Nishimura, H. Fukuda, T. Hayashida and K. Momoyama. 1984. Baculoviral mid-gut gland necrosis (BMN) of kuruma shrimp *(Penaeus japonicus)* larvae in Japanese intensive culture systems. Helgol. Meeresunters. 37: 255-264.

Sano, T., T. Nishimura, H. Fukuda, T. Hayashida and K. Momoyama. 1985. Baculoviral infectivity trials on kuruma shrimp larvae, *Penaeus japonicus*, of different ages, p. 397-403. *In* Fish and shellfish pathology, Academic Press, New York.

Rosenberry, B. 1988. Crash in Taiwan. Aquacult. Dig. 13.9.1.

Sindermann, C.J. 1988. Disease problems caused by introduced species, p. 394-398. *In* C.J. Sindermann and D.V. Lightner (eds.), Disease diagnosis and control in North American marine aquaculture. Elsevier, Amsterdam.

Sindermann, C.J. and D.V. Lightner, editors. 1988. Disease diagnosis and control in North American marine aquaculture. Elsevier, Amsterdam.

Sparks, A.K. 1985. Synopsis of invertebrate pathology exclusive of insects. Elsevier, Amsterdam.

Tsing, A. and J.R. Bonami. 1987. A new viral disease of the shrimp, *Penaeus japonicus* Bate. J. Fish Dis. 10: 139-141.

Tsing, A., D. Lightner, J.R. Bonami and R. Redman. 1985. Is "gut and nerve syndrome" (GNS) of viral origin in the tiger shrimp *Penaeus japonicus* Bate?, p. 91. *In* Programme and abstracts of the 2nd international conference of the European assoc. of fish pathol. Montpellier, France.

Dissemination of Microbial Pathogens through Introductions and Transfers of Finfish

JACK GANZHORN
J. S. ROHOVEC
J. L. FRYER

Abstract: The introduction and transfer of fish is associated with numerous risks. One of the most important is the potential for disease dissemination. Several species of fish have been so extensively introduced that they now have nearly worldwide distribution. Several important fish diseases are thought to have been spread by means of these historical movements of fish as well as by more recent introductions and transfers. Examples include parasites such as *Bothriocephalus opsarichthydis* and *Myxobolus (Myxosoma) cerebralis*. Examples of bacterial pathogens that are thought to have been spread include *Renibacterium salmoninarum* and *Yersinia ruckeri*. Viral diseases, such as infectious pancreatic necrosis, infectious hematopoietic necrosis, and channel catfish virus disease have also been thought to have been disseminated by these means. In response to the biological and economic losses that these diseases may cause, various agencies and organizations have developed policies which address the importation of fish; however, there remains a need for continuing efforts to address this complex problem adequately.

Introduction

During this century, there have been unprecedented numbers of introductions and transfers of fish species and other aquatic organisms from one geographic location to another. Significant risks are associated with these activities and include the alteration of habitats and native community trophic structures, negative genetic impacts, and the dispersal of microbial pathogens. The introduction of disease agents is considered by the American Fisheries Society as one of the most significant threats that an introduced species may pose to a native community (Kohler and Courtenay 1986). Disease agents are easily disseminated when fish are shipped without adequate precautions. Not only may the fish or eggs be infected, but the water and the containers may also be contaminated and therefore serve as vehicles by which pathogens can be introduced into new areas. Whether a pathogen will be transferred and successfully established in a new area depends on the fate of the shipment, the biology of the pathogen, and the presence or absence of appropriate hosts. A review of the history of major fish introductions and transfers is an important background for examining the dispersal of fish diseases. In view of this background information, we will discuss examples of parasitic, bacterial, and viral agents that are thought to have been disseminated with fish. In addition, we will summarize some of the efforts that have been made on the national and international levels to minimize the risks of fish introductions and transfers.

Examples of Fish Introductions

There are important reasons why fish have been moved from one place to another. These include the establishment of a recreational fishery, utilization for aquaculture, introduction of a species to improve productivity of natural waters or to control undesirable aquatic organisms, and for the ornamental aquarium trade (Welcomme 1986). Unintentional movement of fish also frequently occurs; for example, the escapement of bait species. As a result of these intentional and unintentional introductions with subsequent natural migration, there are some species such as the common carp *(Cyprinus carpio)*, the rainbow trout *(Oncorhynchus mykiss)*, and the brown trout *(Salmo trutta)* that now have nearly worldwide distribution.

The rainbow trout serves as a good example of a species that has been widely introduced without consideration for disease dissemination. Originally restricted to western North America and the Kamchatka Penninsula, the rainbow trout has now been successfully established throughout North America and has been exported to every continent except Antarctica (MacCrimmon 1971). Not only is it widely used as an aquacultural species, but with the brown trout (MacCrimmon and Marshall 1968) and brook trout *(Salvelinus fontinalis)* (MacCrimmon and Campbell 1969), it has been introduced extensively to establish recreational fisheries. Most of these introductions were accomplished before fish health surveillance was routinely done and certainly before the infectious nature of many important diseases was understood. Many of these exports were from areas other than the native range of the species. Consequently, not only were the imported fish potentially infected with disease agents from their origin, but they also could have been infected with more recently acquired diseases. For example, the first imports of rainbow trout into Chile were eggs sent from Germany (MacCrimmon 1971). These shipments could have been infected with both European and North American fish pathogens.

Examples of Disseminated Fish Pathogens

There are many examples of fish pathogens believed to have been disseminated by fish introductions and transfers; however, documenting the specific sources of these introductions often proves difficult. Detection of a pathogen for the first time in a specific geographic location does not necessarily indicate a recent introduction, but rather may simply be a result of a new level of disease surveillance. Also, illegal or undocumented fish movement and natural migration may confuse attempts to establish the source of a pathogen in a new geographical area.

Parasites

There are a large number of parasites that infect fish, many of which have caused significant losses in cultured and wild fish populations. Many parasites have life stages that persist outside the host. These resistant stages may survive on shipment containers or in the water and ice used. Some species exhibit complex life cycles that may require specific and multiple hosts. These factors and possible environmental conditions determine whether a parasite will become established in a new location. This may explain why the geographic range of the myxosporean parasite, *Ceratomyxa shasta*, has not increased even though the spore stage of the parasite has undoubtedly occurred in other areas through introductions and natural migration of infected fish (Sanders et al. 1970, Johnson et al. 1979). On the other hand, *Ichthyophthirius multifiliis*, believed to have originated in Asia (Hoffman 1970a, 1981), has become widely dispersed perhaps because of its simple life cycle and ability to infect numerous fish species.

Hoffman (1970a) reported that nearly 50 parasites are known to have been spread intercontinentally by transfer of fish. Two examples which have significant impact on fish populations and which are thought to have been disseminated by introductions of fish are *Bothriocephalus opsarichthydis* (synonymous with *B. acheilognathi* and *B. gowkongesis*), the Asian tapeworm (Hoffman and Schubert 1984) and *Myxobolus (Myxosoma) cerebralis*, the causative agent of whirling disease (Hoffman 1970a).

The Asian tapeworm may have originated in China and the Amur River drainage, but with the extensive exportation of the Chinese carps, particularly the grass carp *(Ctenopharyngodon idella)*, the parasite has become established in numerous locations. The life cycle of *B. opsarichthydis* involves a secondary host, a planktonic crustacean, which is eaten by the fish. The presence of appropriate secondary and primary hosts has allowed this parasite to become established in the Soviet Union, Europe, Malaysia, and North America (Fernando and Furtado 1963; Molner 1970, 1982; Korting 1974; Hoffman 1976; Hoffman and Schubert 1984). This parasite can infect numerous cyprinid hosts including the mosquitofish *(Gambusia affinis)* and bait fish species *(Notemigonus crysoleucas* and *Pimephales promelas)* (Hoffman 1980; Hoffman and Schubert 1984) which are commonly shipped without strict disease-control inspections. Not only has the parasite had significant impact on cultured carp in Europe (Korting 1974), but also has infected indigenous wild fish populations in North America (Heckmann et al. 1986).

The myxosporean parasite, *M. cerebralis*, infects the cartilaginous tissue of the head and spinal column of salmonids. The typical signs of the disease in the rainbow trout are erratic swimming or whirling, a darkened caudal region, and head deformation in surviving fish. Even though the disease has been recognized since 1903 when it was observed among rainbow trout in Germany (Hofer 1903), the life cycle of the parasite has essentially remained undescribed until recently, when Wolf and Markiw (1984) proposed a life cycle involving tubificid oligochaete worms as secondary hosts. It is apparent that suitable hosts and environmental conditions are present on several continents in order to allow this protozoan to become established. Hoffman (1970a) has suggested that *M. cerebralis* originated in Europe among salmonids such as the brown trout, in which the parasite rarely causes mortality, and that the disease as observed in rainbow trout reflects a nonadapted host-parasite relationship. If *M. cerebralis* is enzootic in European brown trout, it is conceivable it was disseminated as viable spores contaminating the European trout eggs that were routinely transported in the late 19th century (Halliday 1976). It is also possible the parasite was disseminated in shipments of frozen trout destined for markets in other countries (Hoffman 1970a). These suggestions cannot be proven because many fish introductions were conducted well before the disease was recognized and surveillance instituted. However, several observations would indicate that the present distribution of *M. cerebralis* is at least in part a result of fish introductions. One of these observations is that a long time elapsed from when the disease was first recognized in Europe to the time when it was seen on the North American continent where the susceptible rainbow trout was extensively cultured. Also, it is important to note that *M. cerebralis* is present in South Africa and New Zealand (Halliday 1976) where salmonids did not occur prior to their introduction. Therefore, natural hosts for this parasite did not exist.

The present distribution of *M. cerebralis* within the United States includes California, Connecticut, Massachusetts, Michigan, Nevada, New Hampshire, New Jersey, Ohio, Pennsylvania, West

Virginia, Virginia (Hnath 1983); and more recently, Colorado, Oregon, Idaho, and Washington (Lorz et al.1989). *M. cerebralis* was first recognized in this country among brook trout in Pennsylvania in 1956 (Hoffman 1970b). The parasite's geographical range then increased rapidly throughout the Northeastern and Great Lakes regions of the country. Shipment of live fish and eggs as well as contaminated frozen fish may have contributed to its spread. The method by which the parasite was introduced to the western United States remains uncertain. Whirling disease was first detected in California in 1966 at a trout farm near Monterey (Hoffman 1970b). At about the same time, *M. cerebralis* was reported from fish at several sites in Nevada. Efforts were made to eradicate the disease at these hatcheries; however, it is known that some potentially infected fish were released in certain watersheds in Nevada. Whirling disease is now enzootic in numerous streams and lakes in northern California and Nevada (Horsch 1987). Natural migration of infected fish within and between watersheds may have contributed to the spread of the parasite. The appearance of the parasite in the Columbia River Basin may have resulted from the natural migration of infected fish, either a resident inland stock from waters of nearby Nevada or an infected anadromous stock. It could also have been a result of the transfer of infected fish in either undocumented shipments or in fish which were examined but in which the pathogen escaped detection.

Bacterial Pathogens

Many of the infectious diseases of fish are caused by bacterial pathogens. Certain of these pathogens exist in a carrier state and therefore are particularly prone to being transferred to new geographic areas despite fish-disease surveillance. In the carrier state, the bacterium exists in the host without any detectable pathology and at concentrations which are below the detection level of routine fish health examinations. Also, certain pathogenic bacteria can infect numerous fish hosts, certain of which may not be routinely inspected for fish diseases when they are imported.

There are also pathogenic bacteria which can externally contaminate eggs or actually exist inside the egg and consequently be vertically transmitted.

Enteric redmouth disease (ERM), caused by *Yersinia ruckeri,* was first recognized among rainbow trout in Idaho in the mid 1960s. It was thought to have had a restricted range and subsequent isolations of the bacterium in some western states and in Canada were attributed to fish transfers from Idaho (Rucker 1966; Wobeser 1973). It was later discovered, however, that *Y. ruckeri* had been isolated from fish in the eastern United States and Australia before the original isolations in Idaho (Bullock et al. 1978). Therefore, the original range of *Y. ruckeri* at the time of the first isolation was more extensive than originally recognized. There have been numerous reports of initial isolations of *Y. ruckeri* from fish in European countries and South Africa (Lesel et al. 1983; Roberts 1983; Dalsgaard et al. 1984; Sparboe et al. 1986; De La Cruz et al. 1986). Whether these cases indicate actual recent importation of the pathogen is not clear. They could be a result of a heightened level of fish-disease surveillance, as was the case in North America. Even though the original geographical range of *Y. ruckeri* may never be accurately known, its ability to be spread is certainly enhanced because the bacterium can exist in a carrier state (Busch and Lingg 1975; Hunter et al. 1980) and infects numerous fish hosts (McArdle and Dooley-Martyn 1985; Michel et al. 1986). Although *Y. ruckeri* is widely spread throughout the world, it has never been recognized in fish cultured in Japan.

A clearer example of the importation of a bacterial pathogen is *Renibacterium salmoninarum,* the causative agent of bacterial kidney disease (BKD). Bacterial kidney disease was first recognized among wild Atlantic salmon in Scotland in 1930 (Smith 1964) and often causes significant mortality among cultured salmonids. *Renibacterium salmoninarum* infects salmonid fish and can be vertically transmitted within the eggs of infected adults (Sanders and Fryer 1981; Evelyn et al. 1984). Its original range will probably never be known because of the extensive introductions of various salmonids throughout the world prior to the organism's discovery; however, the presence of BKD in Chile, South America, where salmonids are not indigenous is a clear example of the spread of *R. salmoninarum* (Sanders and Barros 1986).

Viral Pathogens

The dissemination of viruses is particularly significant because of the severe impact that they often can have on intensively cultured fish and because fish viral diseases are untreatable. Certain of these viruses possess characteristics that favor their transfer with shipments of fish and fish eggs. These include being able to survive for extended periods of time outside of the host, existing in a carrier state or as latent infections not easily detected, and being vertically transmitted to progeny via eggs from infected parents. Unlike parasites and bacteria, the ability to detect and identify viruses is a relatively recent achievement which occurred after significant fish introductions had already been made around the world. This makes it difficult to determine what the original ranges were for the important fish viruses; however, some inferences can be made based on what is known about the ecology of the virus and from historical observations. There are several fish viruses which have apparently been disseminated via fish introductions and transfers and have had significant impact on cultured fish populations. These include infectious pancreatic necrosis virus (IPNV), infectious hematopoietic necrosis virus (IHNV), and channel catfish virus (CCV).

Mortality among fry was common in the early years of trout culture in the eastern part of North America. M'Gonigle (1941) attributed the mortality among brook trout in Canada to a nutritional problem that he termed "acute catarrhal enteritis" and recommended that these fry be planted into streams. It is now generally accepted that this condition in young salmonids was infectious pancreatic necrosis (IPN). Wood et al. (1955) were the first to postulate a viral etiology based on histological findings. The virus was finally isolated in the late 1950s (Wolf et al. 1959) after the advent of fish tissue and cell culture techniques; however, by that time, trout eggs from the eastern United States had been extensively distributed across North America and throughout most of the world. It is possible IPNV was spread to its present range by these introductions of potentially infected eggs, because IPNV can be vertically transmitted to fry via the eggs (Bullock et al. 1976). In Europe, IPNV was first isolated in 1965 (Besse and de Kinkelin 1965a, 1965b, de Kinkelin and Besse 1966) and attributed to the import of eggs (Wolf 1966). Sano (1966) suspected that the virus had been in Japan since the mid 1960s and it was first isolated in 1971 (Sano 1971); however, no inferences concerning its source have been made. On the other hand, the presence of IPNV in Chile has been attributed to specific egg shipments from North America (McAllister and Reyes 1984). Also, the recent isolations of IPNV in Taiwan and Korea are thought to reflect introduction via shipments of fish and eggs from Japan (Chen et al. 1985; Hah et al. 1984; Hedrick et al. 1985).

Similar to IPNV, mortality caused by IHNV was observed well before the agent was isolated and identified as a virus (Guenter et al. 1959; Rucker et al. 1953; Watson et al. 1954). Burrows et al. (1951) describes mortality that occurred among sockeye salmon in the 1940s that resembles typical characteristics of infectious hematopoietic necrosis (IHN) epizootics. Further study of these epizootics indicated that an infectious agent was present and that it was filterable (Rucker et al. 1953; Watson 1954; Watson et al. 1954). The virus was isolated from sockeye fry in Oregon in 1958 (Parisot et al. 1965). The original range of IHNV is thought to have been among coastal salmonid stocks from Alaska to northern California and to have been introduced into the Snake River Valley in Idaho (Wolf 1988). Sano (1976) reported that IHNV was first introduced into Japan through the importation of sockeye salmon eggs from Alaska and that by 1975 the virus had spread through all the salmonid rearing areas on Hokkaido and Honshu. Recently, IHNV has been isolated among rainbow trout in France (de Kinkelin et al. 1987) and Italy (Bovo et al. 1987). Serological studies indicate that these new isolants are similar to a strain that infects rainbow trout in the Pacific Northwest of the United States

(Arkush et al. 1989).

Mortality due to CCV was noticed among very young channel catfish *(Ictalurus punctatus)* during the 1960s when commercial catfish culture was rapidly expanding in the southeastern United States. The etiology of the epizootics remained unknown until Fijan (1968) isolated the virus responsible for the disease. The disease has been observed only in cultured channel catfish and other ictalurids appear resistant (Wolf 1988). The introduction of the disease into California likely occurred as a result of the importation of channel catfish for aquaculture because ictalurids are not indigenous to this state. Likewise, the introduction of the disease to Honduras was apparently associated with importation of channel catfish fry (Wolf 1988).

Regulatory Attempts to Minimize Disease Dissemination

The deleterious effects of an introduced pathogen are often economic as well as biological in nature. Costs involved with introduction of a nontreatable pathogen are often long-term and include the expense of eradication or containment, the loss of markets, and the negative impact on tourism if a sport fishery is involved (Rohovec 1983). With a view to minimizing these risks, numerous countries have instituted regulations that control the importation of fish.

Four Regional Fishery Bodies of the Food and Agriculture Organization, the European Inland Fisheries Advisory Commission (EIFAC), the Indo-Pacific Fisheries Council, the Commission for Inland Fisheries of Latin America, and the Committee for the Inland Fisheries of Africa, have addressed the issue of new introductions of fish and fish eggs (Welcomme 1986). The EIFAC has been the most active with the adoption of the International Code of Practice in conjunction with the International Council for the Exploration of the Sea. The Code addresses the need for disease-free brood stock, quarantine of imports, and periodic fish inspections (Sindermann 1984). This Code has been used with varying success, partly because certain countries have not reacted uniformly to the issue of fish introductions (Sindermann 1986). Recently, this Code has been combined with a review-and-decision model for evaluating proposed introductions of exotic fish species (Kohler and Stanley 1984; EIFAC 1984). This model helps to define the benefits and risks associated with proposed fish introductions and addresses the need to ensure that adequate precautions are taken to prevent the introduction of fish-disease agents. Individual countries have also adopted specific regulations governing the importation of fish. There is little uniformity regarding both the content of these various policies and the extent to which they are enforced. Generally, though, they often deal with specific families of fish that are important to aquaculture, e.g., salmonids. Often specific pathogens are categorized based on the severity of the disease and whether the pathogen already exists in the country. This classification serves to distinguish those pathogens allowed to enter a country and which are not. In addition regulations govern the action taken when a certain pathogen is found in the country.

Regulations governing fish importation into the United States are primarily at the regional and state government level and follow this general pattern of categorizing various fish pathogens. Federal law governing the importation of fish is addressed in the Lacey Act (Title 50 amendment) which mandates imported salmonids to be free of viral hemorrhagic septicemia virus and *M. cerebralis.* An important feature of successful import regulations is the standardization of procedures used to detect specific fish pathogens. This has been accomplished for Canada and the United States with specific documents that outline recommended procedures (United States Department of Interior 1968; Amos 1985; Department of Fisheries and Oceans 1984). It is also important that fish-disease inspections be done by qualified specialists. Fish-health workers are approved by the U.S. Fish and Wildlife Service in order to conduct Title 50 examinations and for inspections of fish being shipped into and, within Canada, inspectors are required to be approved by the Department of Fisheries and

Oceans. In addition, the American Fisheries Society has a certification program that recognizes Fish Health Inspectors and Fish Pathologists. Several states require that fish-health inspections be done by certified fish pathologists.

Conclusion

While there have been definite socioeconomic benefits derived from certain introductions and transfers of fish, there have also been many examples of deleterious results such as the dissemination of certain serious fish parasites, bacteria, and viruses. Most of these deleterious introductions and transfers have occurred in the absence of any fish-disease policies and often before there was an adequate understanding of the seriousness of the diseases. Recent advances in the field of fish health have allowed the fish diagnostician and inspector to detect fish-disease agents at a much greater level of sensitivity. In contrast, the development of useful fish-disease policies has occurred much more slowly. Policies should be developed and implemented that promote cooperation between regulatory agencies, aquacultural interests, and fishery resource managers. These policies should be sufficiently adaptable to incorporate new information regarding specific diseases and newly developed and proven diagnostic procedures. Finally, they should be realistic in addressing the legitimate reasons for fish introductions and transfers. The implementation of good fish-disease policies requires the commitment of significant resources. From the public agencies, adequately trained human resources will be needed to enforce policies and inspect fish stocks that are candidates for introduction or transfer. From aquaculture interests, public and private, there should be a significant investment in development of specific disease-free stocks.

Acknowledgments

This work was sponsored by Oregon Sea Grant through NOAA Office of Sea Grant, Department of Commerce, under

Grant No. 87-89 (FY 88) NA 85AA-D-SG095 (Project No. R/FSD-10). The authors thank C. Pelroy for typing the manuscript. Oregon Agricultural Experiment Station Technical paper No. 8788.

Literature Cited

Amos, K.H . 1985. Procedures for the detection and identification of certain fish pathogens, 3rd edition. American Fisheries Society Fish Health Section, Corvallis, Oregon.

Arkush, K.D., G. Bovo, P. de Kinkelin, J.R. Winton, W.H. Wingfield and R.P. Hedrick. 1989. Biochemical and antigenic properties of the first isolates of infectious hematopoietic necrosis virus from salmonid fish in Europe. J. Aquat. Anim. Health 1:148-153.

Besse, P. and P. de Kinkelin. 1965a. Sur l'existence en France de la necrose pancreatique de la truite arc-en-ciel *(Salmo gairdneri)*. Bull. Acad. Vet. Fr. 38:185-190.

Besse, P. and P. de Kinkelin. 1965b. La necrose pancreatique des alevins arc-en-ciel. Piscicult. Fr. 2:16-19.

Bovo, G., G. Giorgetti, P.E.V. Jorgensen and N.J. Olesen. 1987. Infectious haematopoietic necrosis: first detection in Italy. Bull. Eur. Assoc. of Fish Pathol. 7:124.

Bullock, G.L., H.M. Stuckey and E.B. Shotts, Jr. 1978. Enteric redmouth bacterium: comparison of isolates from different geographic areas. J. Fish Dis. 1:351-356.

Bullock, G.L., R.R. Rucker, D. Amend, K. Wolf and H.M. Stuckey. 1976. Infectious pancreatic necrosis: transmission with iodine-treated and non-treated eggs of brook trout *(Salvelinus fontinalis)*. J. Fish. Res. Board Can. 33:1197-1198.

Burrows, R.E., L.A. Robinson and D.D. Palmer. 1951. Tests of hatchery foods for blueback salmon *(Oncorhynchus nerka)* 1944-1948. United States Fish and Wildlife Service Special Scientific Report Fisheries 59. Washington, D.C.

Busch, R.A. and A.J. Lingg. 1975. Establishment of an asymptomatic carrier state infection of enteric redmouth disease in rainbow trout *(Salmo gairdneri)*. J. Fish. Res. Board Can. 32:2429-2432.

Chen, S.N., C.H. Kou, R.P. Hedrick and J.L. Fryer. 1985. The occurrence of viral infections of fish in Taiwan, p. 313-319. *In* A.E. Ellis (ed.), Fish and shellfish pathology. Academic Press, London.

Dalsgaard, I., J. From and V. Horiyck. 1984. First observation of *Yersinia ruckeri* in Denmark. Bull. Eur. Assoc. of Fish Pathol. 4:10.

De Kinkelin, P. and P. Besse. 1966. Une epizootie de necrose pancreatique dans les salmonicultures françaises. Bull Off. Int. Epizoot. 65:999-1010.

De Kinkelin, P., A.M. Hattenberger, C. Torchy and F. Liefrig. 1987. Infectious haematopoietic necrosis (IHN): first report in Europe, p. 57. *In* Third International European Association of Fish Pathologists Conference, Bergen, Programme and Abstracts.

De la Cruz, J.A., A. Rodriguez, C. Tejedor, E. de Lucas and L. R. Orozco. 1986. Isolation and identification of *Yersinia ruckeri*, causal agent of ERM, for the first time in Spain. Bull. Eur. Assoc. Fish Pathol. 6:43-44.

Department of Fisheries and Oceans. 1984. Fish health protection regulations: manual of compliance. Fisheries Marine Service Miscellaneous Special Publication 31 (Revised). Ottawa, Canada.

EIFAC (European Inland Fisheries Advisory Commission). 1984. Report of the European Inland Fisheries Advisory Commission Working Party on stock enhancement. European Inland Fisheries Advisory Commission Technical Paper 44.

Evelyn, T.P.T., J.E. Ketcheson and L. Prosperi-Porta. 1984. Further evidence for the presence of *Renibacterium salmoninarum* in salmonid eggs and for the failure of povidone-iodine to reduce the intra-ovum infection rate in water-hardened eggs. J. Fish Dis. 7:173-182.

Fernando, C.H. and J.Y. Furtado. 1963. Some studies on helminth parasites of freshwater fishes, p. 5-9. *In* Proceedings of the regional symposium on scientific knowledge of tropical parasites. University of Singapore, Singapore.

Fijan, N. 1968. Progress report on acute mortality of channel catfish fingerlings caused by a virus. Bull Off. Int. Epizoot. 69:1167-1168.

Guenther, R.W., S.W. Watson and R.R. Rucker. 1959. Etiology of sockeye salmon "virus" disease. U. S. Fish and Wildlife Service Special Scientific Report Fisheries 296. Washington, D.C.

Hah, Y.-C., S.W. Hong, M.H. Kim, J.L. Fryer and J.R. Winton. 1984. Isolation of infectious hematopoietic necrosis virus from goldfish (*Carassius auratus*) and chum salmon *(Oncorhynchus keta)* in Korea. Korean J. Microbiol. 22:85-90.

Halliday, M. M. 1976. The biology of *Myxosoma cerebralis:* the causative organism of whirling disease of salmonids. J. Fish Biol. 9:339-357.

Heckmann, R.A., P.D. Greger and J.E. Deacon. 1986. The Asian tapeworm, *Bothriocephalus acheilognathi,* infecting endangered fish species from the Virgin River, Utah, Nevada, and Arizona. Fish Health Sect. Am. Fish. Soc. Newsl. 14(1):5.

Hedrick, R.P., W.D. Eaton, J.L. Fryer, Y.C. Hah, J.W. Park and S.W. Hong. 1985. Biochemical and serological properties of Birnaviruses isolated from fish in Korea. Fish Pathol. 20:463-468.

Hnath, J.G. 1983. Whirling disease, p. 223-229. *In* F.P. Meyer, J.W. Warren and T.G. Carey (eds.), A guide to integrated fish health management in the Greak Lakes basin. Great Lakes Fishery Commission, Ann Arbor, Michigan.

Hofer, B. 1903. Uber die Drehkrankheit der Regenborenforelle. Allg. Fischztg 28:7-8.

Hoffman, G.L. 1970a. Intercontinental and transcontinental dissemination and transfaunation of fish parasites with emphasis on whirling disease *(Myxosoma cerebralis)*, p. 69-81. *In* S.F. Snieszko (ed.), A symposium on diseases of fishes and shellfishes. American Fisheries Society, Washington, D. C.

Hoffman, G.L. 1970b. Whirling disease of trout and salmon caused by *Myxosoma cerebralis* in the United States of America. Riv. Ital. Piscicol. Ittiopatol. 5:29-31.

Hoffman, G.L. 1976. The Asian tapework, *Bothriocephalus gowkongensis,* in the United States, and research needs in fish parasitology, p. 84-90. *In* Proceedings of 1976 fish farming conference and annual convention Catfish Farmers of Texas. Texas A & M University, College Station, Texas.

Hoffman, G.L. 1980. The Asian tapeworm continues to travel. Fish Health Sec. Am. Fish. Soc. Newsl. 8(3):3.

Hoffman,G.L. 1981. Recently imported parasites of baitfishes and relatives, p. 45-46. In Third annual proceedings Catfish Farmers of America Research Workshop. Las Vegas, Nevada.

Hoffman, G.L. and G. Schubert. 1984. Some parasites of exotic fishes, p. 233-261. *In* W.R. Courtenay, Jr. and J.R. Stauffer, Jr. (eds), Distribution, biology and management of exotic fishes. Johns Hopkins University Press, Baltimore, Maryland.

Horsch, C.M. 1987. A case history of whirling disease in a drainage system: Battle Creek drainage of the upper Sacramento River basin, California. J. Fish Dis. 10:453-460.

Hunter, V.A., M.D. Knittel and J.L. Fryer. 1980. Stress-induced transmission of *Yersinia ruckeri* infection from carriers to recipient steelhead trout, *Salmo gairdneri* Richardson. J. Fish Dis. 3:467-472.

Johnson, K.A., J.E. Sanders and J.L. Fryer. 1979. *Ceratomyxa shasta* in salmonids. United States Fish and Wildlife Service Fish Disease Leaflet 58. Washington, D.C.

Kohler, C.C. and W.R. Courtenay, Jr. 1986. American Fisheries Society position on introductions of aquatic species. Fisheries 11:39-42.

Kohler, C.C. and J.G. Stanley. 1984. Implementation of a review and decision model for evaluating proposed introductions of aquatic organisms in Europe and North America. Eur. Inland Fish. Advis. Comm. Tech. Pap. 42:541-549.

Korting, W. 1974. Bothriocephalosis in the carp. Vet. Med. Rev. 2:165-171.

Lesel, R., M. Lesel, F. Gavini and A. Vanillaume. 1983. Outbreak of enteric redmouth in rainbow trout in France. J. Fish Dis. 6:385-387.

Lorz, H.V., A. Amandi, C.R. Banner and J.S. Rohovec. 1989. Detection of *Myxobolus (Myxosoma) cerabralis* in salmonid fishes in Oregon. J. Aquat. Anim. Health. 1:217-221.

MacCrimmon, H.R. 1971. World distribution of rainbow trout *(Salmo gairdneri).* J. Fish. Res. Board Can. 28:663-704.

MacCrimmon, H.R. and J.S. Campbell. 1969. World distribution of brook trout *(Salvelinus fontinalis).* J. Fish. Res. Board Can. 26:1699-1725.

MacCrimmon, H.R. and T.L. Marshall. 1968. World distribution of brown trout, *Salmo trutta.* J. Fish. Res. Board Can. 25:2527-2548.

McAllister, P.E. and X. Reyes. 1984. Infectious pancreatic necrosis virus: isolation from rainbow trout, *Salmo gairdneri* Richardson, imported into Chile. J. Fish Dis. 7:319-322.

McArdle, J.F. and C. Dooley-Martyn. 1985. Isolation of *Yersinia ruckeri* Type I (Hagerman Strain) from goldfish *Carassius auratus* (L.). Bull. Eur. Assoc. Fish Pathol. 5:10-11.

Michel, C., B. Faivre and P. de Kinkelin. 1986. A clinical case of enteric redmouth in minnows *(Pimephales promelas)* imported in Europe as bait-fish. Bull. Eur. Assoc. Fish Pathol. 6:97-99.

M'Gonigle, R.H. 1941. Acute catarrhal enteritis of salmonid fingerlings. Trans. Am. Fish. Soc. 70:297-303.

Molner, K. 1970. An attempt to treat fish bothriocephalosis with devermin. Acta Veterinaria Academiae Scientiarum Hungaricae 30:325-331.

Molner, K. 1982. Parasite range extension by introduction of fish to Hungary. Twelfth meeting of European Inland Fisheries Advisory Commission (EIFAC), Symposium on stock enhancement in the management of freshwater fisheries. Budapest, Hungary.

Parisot, T.J., W.T. Yasutake and G.W. Klontz. 1965. Virus diseases of the salmonidae in western United States. I. Etiology and epizootiology. Ann. N. Y. Acad. Sci. 126:502-519.

Roberts, M.S. 1983. A report of an epizootic in hatchery reared rainbow trout, *Salmo gairdneri,* at an English trout farm caused by *Yersinia ruckeri.* J. Fish Dis. 6:551-552.

Rohovec, J.S. 1983. Development of policies to avoid the introduction of infectious diseases among populations of fish and shellfish, p. 371-373. *In* P.T. Arana (ed.), Proceedings of the international conference on marine resources of the Pacific. Viña del Mar, Chile.

Rucker, R.R. 1966. Redmouth disease of rainbow trout *(Salmo gairdneri).* Bull. Off. Int. Epizoot. 65:815-830.

Rucker, R.R., W.J. Whipple, J.R. Parvin and C.A. Evans. 1953. A contagious disease of salmon, possibly of virus origin. Fish Bull. 54:35-46.

Sanders, J.E. and J.R. Barros. 1986. Evidence by the fluorescent antibody test for the occurrence of *Renibacterium salmoninarum* among salmonid fish in Chile. J. Wildl. Dis. 22:255-257.

Sanders, J.E. and J.L. Fryer. 1981. Bacterial kidney disease of salmonid fish. Ann. Rev. Microbiol. 35:273-298.

Sanders, J.E., J.L. Fryer and R.W. Gould. 1970. Occurrence of the myxosporidan parasite, *Ceratomyxa shasta,* in salmonid fish from the Columbia River basin and Oregon coastal streams, p. 133-141. *In* S.F. Snieszko (ed.), A symposium on diseases of fishes and shellfishes. American Fisheries Society, Washington, D.C.

Sano, T. 1966. Epizootics in fingerling rainbow trout. Fish Pathol. 1:37-46.

Sano, T. 1971. Studies on viral diseases of Japanese fishes-I. Infectious pancreatic necrosis of rainbow trout: First isolation from epizootics in Japan. Bull. Jpn. Soc. of Sci. Fish. 37:495-498.

Sano, T. 1976. Viral diseases of cultured fishes in Japan. Fish Pathol. 10:221-226.

Sindermann, C.J. 1984. Diseases in marine aquaculture. Helgol. Meeresunters. 37:505-532.

Sindermann, C.J. 1986. Strategies for reducing risks from introductions of aquatic organisms: a marine perspective. Fisheries 11:10-15.

Smith, I.W. 1964. The occurrence and pathology of Dee disease. Department of Agriculture and Fisheries of Scotland. Freshwater and Salmon Fisheries Research 34: 12 p. Edinburgh, Scotland.

Sparboe, O., C. Koren, T. Hastein, T. Poppe and H. Stenwig. 1986. The first isolation of *Yersinia ruckeri* from farmed Norwegian salmon. Bull. Eur. Assoc. Fish Pathol. 6:41-42.

United States Department of Interior. 1968. Approved procedure for determining absence of viral hemorrhagic septicemia and whirling disease in certain fish and fish products. United States Fish and Wildlife Service Fish Disease Leaflet 9. Washington, D.C.

Watson, S.H. 1954. Virus diseases of fish. Trans. Am. Fish. Soc. 83:331-341.

Watson, S.W., R.W. Guenther and R.R. Rucker. 1954. A virus disease of sockeye salmon: interim report. U. S. Fish and Wildlife Service Special Scientific Report Fisheries 138. Washington, D.C.

Welcomme, R.L. 1986. International measures for the control of introductions of aquatic organisms. Fisheries 11:4-9.

Wobeser, G. 1973. An outbreak of redmouth disease in rainbow trout *(Salmo gairdneri)* in Saskatchewan. J. Fish. Res. Board Can. 30:571-575.

Wolf, K. 1966. The fish viruses. Adv. Virus Res.12:35-101.

Wolf, K. 1988. Fish viruses and fish viral diseases. Cornell University Press, Ithaca, New York.

Wolf, K. and M.E. Markiw. 1984. Biology contravenes taxonomy in the Myxozoa: new discoveries show alternation of invertebrate and vertebrate hosts. Science 225:1449-1452.

Wolf, K.S., S.F. Snieszko and C.E. Dunbar. 1959. Infectious pancreatic necrosis, a virus-caused disease of fish. Excerpta Medica 13:228. Abstract.

Wood, E.M., S.F. Snieszko and W.T. Yasutake. 1955. Infectious pancreatic necrosis in brook trout. Arch. Pathol. 60:26-28.

CHAPTER 3

Dispersal of Genetically Altered and Unaltered Microbial Agents

Challenges and Opportunities for Marine Biotechnology in Environmental Bioremediation

RITA R. COLWELL

Abstract: Recent developments in genetic engineering and biotechnology have opened up new opportunities for waste control and toxic waste degradation *in situ*. Specific considerations are necessary for use of engineered organisms in marine systems, since freshwater microorganisms do not function optimally in marine systems. Marine bacteria have been employed for treatment of oil spills occurring in coastal and ocean waters with reasonable success. Isolation of oil-degrading microorganisms has been effective and can be accomplished with relative ease. New approaches have been taken to amplify and enhance strain capability for biodegradation of hydrocarbons, employing the techniques of genetic engineering. Successes to date have been good. Furthermore, marine bacteria have been isolated which are capable of degrading a variety of toxic chemicals. The genetics of these organisms is being studied, and engineering organisms for rapid clean-up of toxic chemical spills in the marine environment now appears possible. An important factor to consider in developing marine bacteria for environmental application is that those bacteria which occur naturally at the site to be remediated should be selected as candidate strains for genetic engineering. By employing autochthonous organisms and amplifying their genes for degradative pathways or by insertion of genes coding for degradation of specific chemicals and using the techniques of genetic engineering, it should be possible to enhance *in situ* degradation and accomplish bioremediation. A useful hypothesis to test is that, upon depletion of the toxic chemical "nutrient," the engineered organism will resume occupancy of its appropriate ecological niche. Microcosms have been employed showing that, indeed, such is the case. Mesocosm studies would be useful to gather additional data prior to application in the natural environment. The potential of biotechnology, applied to marine and freshwater systems for bioremediation, waste control, and targeted biodegradation is significant. However, careful microcosm and mesocosm testing, as well as ample containment trials should be done, before large-scale field application is undertaken.

Introduction

The principles of biotechnology have been applied by humans for centuries — for example, in the rotation of leguminous crops for soil fertilization, selective breeding of fish to create progeny of higher reproductive capacity and faster growth, and in classical waste treatment. The new biotechnology involves genetic engineering, genetic manipulation, bioengineering, and/or applied genetics. Much controversy has centered on the application of organisms altered by the introduction of recombinant DNA molecules, as well as organisms modified by cell fusion, transformation, and transduction. Use of mutagenic agents has also been included in the public debate concerning release of genetically engineered organisms to the environment. When looking back over the past fifteen years, one cannot help but be astonished at the rapidity of the progress that has been made. Fifteen years ago, the first reports concerned expression of ribosomal genes of *Xenopus* cloned into *Escherichia coli* that could apparently be detected. Today, we are accustomed to seeing not only *E. coli*, but a host of other procaryotic and eucaryotic cells successfully harnessed for the production of important products on an industrial scale. All has been achieved under guidelines and regulations that, over the last fifteen years, have been reduced in rigor on the basis of experience. There is no instance in which the technology has actually harmed the health of any of the workers involved in recombinant DNA (rDNA) research. Thus, it is truly a unique application of science for useful purpose. Because of very precise genetic alterations made possible by rDNA techniques, the opportunities, especially in bioremediation, are almost limitless.

So what is the debate all about? It is really on three major fronts. The first is a scientific debate. Will the introduction of rDNA organisms (genetically engineered microorganisms or "GEMs") into the environment cause harm? Available data document the usefulness of the introduction of microorganisms capable of biodegradation of hydrocarbons and xenobiotics that are produced by traditional genetic methods. There is no consensus right now among the views of molecular biologists who are well accustomed to inducing genetic change, the views of ecologists who have, in fact, recorded problems associated with introduction of exotic species, e.g., the kudzu vine, gypsy moth, and gray squirrels, and the views of pathologists and medical microbiologists as to what defines a pathogen. What is emerging, however, as a very clear consensus is that the end product designed for release is significant, but not the method for obtaining the product. The engineered microorganism or enzyme that is produced and not the method (protoplast fusion, genetic recombination, or other method of molecular genetic engineering) should be reviewed for regulation. We should not lose sight of this very sensible and fundamental point in the debate concerning recombinant microorganisms for release in bioremediation.

The second aspect of public concern with release of genetically engineered organisms is a regulatory debate. The goal should be balanced regulation that doesn't stifle progress, but is not so lax that we create the microbial equivalent of kudzu. Finally, and perhaps the most important, is the public-perception aspect of the GEM release debate. According to a recent report released by the National Academy of Sciences (Everybody Counts 1989), the public does not understand the highly technological milieu of recombinant DNA. How, then, do we explain rDNA in a way that the public can understand its beneficial aspects, since the negative aspects are conjured quite successfully? We must all participate in this debate and work to dispel the myth that recombinant microorganisms, per se, are a danger. Scientists and engineers have a responsibility to explain their research and its benefits. We must discuss, not dismiss, these fears.

What about bioremediation? The potential of biotechnology has just begun to be recognized. Cloning of genes involved in the degradation of toxic organic components and metabolism of heavy metals is of primary interest.

The use of microorganisms, or their metabolites to treat toxic chemicals to render them harmless, goes back to 1914, when researchers were isolating strains efficient in degrading noxious sewage components and reducing the volume of organic matter. There is a long history of introduction of microorganisms into the environment, and generally a very safe history.

Today, however, the interest is on microorganisms that degrade some of the most toxic of chemicals that human beings produce. One of the best known released toxic chemicals is 2,4,5 T (2,4,5-trichorophenoxyacetic acid) which has been used extensively as a component of agent orange in Vietnam, and as a herbicide in various countries. The compound has created toxicological problems (Grant 1979), but a possible solution for the very slow rate of biodegradation is the isolation of a strain of *Pseudomonas cepacia* shown to degrade as much as 95% of the 2,4,5 T in soil within one week (Chatterjee et al. 1982).

Let me offer some of the challenges from the environmental, or release aspect: the issues — once the organism(s) are engineered (either classically or by means of rDNA) and are functioning in the environment — include survival, persistence, dispersion, genetic transfer to autochthonous organisms from engineered organisms, and their effect on natural cycles and ecosystems. The presence of plasmids in estuarine and marine bacteria has been recognized since 1975, when plasmids in Chesapeake Bay bacteria were characterized. Of the bacterial isolates tested, 40% contained plasmids. Furthermore, Guerry (1977) was able to transfer plasmids from *E. coli* to *Vibrio parahaemolyticus*, an important finding because it showed very clearly that *E. coli* enterotoxigenic strains entering Chesapeake Bay via sewage could transfer plasmid-mediated traits, e.g., metal resistance (R Plasmids), antiobiotic resistance, or toxin production, to indigenous organisms, such as *V. parahaemolyticus*, an autochthonous Chesapeake Bay vibrio. The transfer rate was found to be rather low, about 1 in 10^5; however, demonstration of intergeneric transfer was a significant finding.

The number of heterotrophic bacteria in Chesapeake Bay water is ca. 10^5 ml^{-1}. Thus, it is highly probable that genetic transfer occurs regularly among Chesapeake Bay bacteria and other estuarine bacteria.

In a series of studies over a 10-year period at coastal and

deep-ocean sites, plasmid incidence was found to be exceptionally high for sewage effluent bacteria and low, though detectable, for bacteria in clean water samples. Interestingly, the molecular weights of plasmids carried in sewage effluent were bimodal in distribution, with some very large plasmids detected in marine isolates and many strains carrying small plasmids (Baya et al. 1986). At the clean water site, the smaller, lower molecular weight plasmids were observed. Summarizing, a control site near Ocean City showed bacteria containing few plasmids. Introduction of terrestrial organisms, via sewage effluent, clearly raises the potential for introduction of genetic material.

Sizemore (1977) studied the incidence of resistance of R plasmids in marine bacteria collected from surface waters of the Puerto Rico trench. The largest number of resistance plasmid-bearing strains were isolated, however, from harbors and coastal waters. At the deep-water site, there were few bacteria carrying plasmids, but they were detectable; however, deep-sea sediment samples showed increased numbers of plasmids. Furthermore, we were able to observe, by enrichment procedures employing toxic chemicals, even at the deepest sites, e.g., from sediment samples collected from the Abyssal Plain off the coast of South Africa, the presence of plasmids in bacteria growing in the presence of toxic chemicals.

The assumption is that bacteria in the aquatic environment can exchange genetic information, *in situ*. Data for freshwater systems are convincing. Bacteria in sewage effluent show transfer of genetic material (Grabow et al. 1974), including antibiotic resistance, between strains of *E. coli* suspended in cellulose dialysis bags in a South African river when the temperature was about 20 degrees. The frequency of transfer in the river system was about 3 per 10 cells. The actual transfer of plasmid genetic material in that very early study was not proven, but rather phenotypic changes were monitored. Subsequently, membrane chambers were used and the actual transfer of genetic information demonstrated.

Circumstantial evidence for intergeneric plasmid transfer by bacteria also was shown by Wortman and Colwell (1988) for bacteria isolated from the gut of deep-sea amphipods. Identical plasmids were detected in a *Vibrio* and a *Pseudomonas* isolated from the gut contents of an amphipod, collected from the Bay of Biscay at a deep- sea site of ca. 4,000 meters. Thus, even under high hydrostatic conditions, low temperature, and high salinity of the deep sea, genetic transfer may take place.

Thus, the considerations regarding release of genetically engineering organisms — fate, movement, dispersal, survival in the environment — are all key issues. And in the ocean, if one releases a genetically engineered microorganism, there is no recall. A notice cannot be issued requesting the bacteria be brought back for "overhauling and tinkering," i.e., trait improvement, if they don't function as expected or if the bacteria spread too far in the ocean. So, the kinds of releases for which we prepare ourselves must be those well studied in the laboratory in microcosms and small-scale mesocosms.

Let me illustrate some potential ecological effects, extrapolating what can occur when one releases microorganisms and toxic chemicals into the environment. In the case cited, industrial effluents from pharmaceutical industries in Puerto Rico were collected in large storage tanks. Studies of a dump-site area, accomplished over a 10-year period in the Puerto Rico trench, were done. Using a variety of methods, the microbial populations in the water were monitored. On the east side of the dump site, 40 miles north of Arecibo, Puerto Rico, the specific activity of the microorganisms was surprisingly high. The assumption had been that the dumped material would sweep into the water column, be diluted, and wash away. It turned out that from the results of the microbiological studies, the wastes moved northward. Subsequently, the path of the waste flow via the ocean currents was confirmed using drogues.

On plate count (freshwater) media, freshwater bacteria were detected, apparently in large numbers from the wastes. Further from the dump site, the number of marine bacteria that were enumerated increased and the number of freshwater bacteria decreased.

The distribution of Gram-positive and Gram-negative organisms also proved very important. Offshore, at the east end of the transect of the dump site, few Gram-positive bacteria were detectable. Within the dump-site area, the percent of Gram-positive bacteria was significantly increased and the number of Gram-negative bacteria decreased. Thus, a change in the microbial flora occurred as a result of the dumping and microbial community structure change was discernible (Singleton et al. 1985). At the east end of the transect, in the dump-site area, the numbers of *Pseudomonas* and *Vibrio* spp., typical genera of marine bacteria, decreased and the number of Gram-positive cocci increased.

A transect was run for bacteriological samples collected from the Caribbean Sea to the Sargasso Sea. Within the dump site, the marine bacteria numbers were lower. The point to be made is that changes in the microbial community structure, i.e., species composition, were induced by the introduction of chemicals at the dump site area.

Another interesting and similar set of observations was made at the Campeche Bank in the Gulf of Mexico at the site of the Campeche blowout. Hydrocarbon-degrading bacteria were present in very low numbers in surrounding water, but enriched at the site of the oil release. Plasmids in marine bacteria from the area of the oil spill were detectable. Thus, there was a genetic response of the microorganisms, with enhancement of oil degraders (Leahy et al. 1990b). The capacity of bacteria to degrade oil and toxic chemicals offers great promise in bioremediation *in situ*. Several caveats must be mentioned, however. First, the oil-degrading microorganisms isolated from the site of the oil spill should be candidates for engineering for enhanced degradation. Such bacteria can be added back, in large numbers, to seed the oil-spill site. When the oil is completely de-graded, these microorganisms, being naturally occurring organisms, will resume their compositional structure in the bacterial communities.

Thus, microorganisms added to treat toxic spills or to enhance degradation of wastes which accumulate in such areas should be those organisms isolated from the site that are part of a natural living flora. Engineering them to degrade the oil or chemical spilled is a much more environmentally sensible procedure to follow (Leahy and Colwell 1990a).

The major effect of ocean-dumping activities observed in our studies was an alteration or restructuring of the composition of the culturable bacterial community. The nonculturable components were also alive and functioning, though not able to be recovered by techniques available in microbiological laboratories at the present time.

What is the fate of genetically-altered bacteria introduced into the ocean? Will they persist for a long period of time? Studies done by Timmis et al. (1973) showed that microorganisms can be engineered to contain genes coding for pathways for breakdown of specific compounds. Thus, a single strain may carry a full complement of genes for complete biodegradation. By molecular genetic methods, one can construct in the laboratory, pollutant-degrading, genetically engineered strains.

The next question is, then, will these organisms persist in the environment? Timmis et al. (1973) demonstrated from microcosm studies that genetically engineered organisms do persist and function in sewage sludge microcosms. The *Pseudomonas* strain capable of degrading substituted benzoates within the very complex ecosystem of an activated sludge system proved successful in application. Extrapolating to the environment, the data strongly suggest engineered organisms can be successfully used for treatment of spills and for degradation of toxic chemicals.

One must consider, however, seasonality of bacterial species, a feature not well appreciated by a nonmicrobiologist. Also important is the issue of addition of engineered organisms to the environment. Seasonal effects on, and competition of, the introduced organism with naturally occurring species are only two of many questions to be answered.

Many bacteria such as *E. coli* and *Salmonella* spp., when discharged into the marine environment, will go into a viable but nonculturable stage such that they cannot be detected by standard methods, e.g., heterotrophic plate counts or, most probable liquid broth counts. By direct detection methods, however, such bacteria can be detected and ennumerated (Colwell et al. 1983). Results of these studies have shown that these bacteria retain the ability to infect animals and, potentially, human beings (Grimes et al. 1986). Therefore, yet another aspect of the release of the genetically engineered microorganisms to the marine environment must be considered the survival of these forms in the environment in the non-culturable form. Genetic methods, i.e., gene probes, and monoclonal antibody methods, must be used to detect organisms released to the environment to ensure their detection and effective monitoring.

What is the message to convey? It is that we have the capacity to engineer organisms to degrade toxic chemicals. We know that genetic transfer, accomplished in the laboratory, does take place in the natural environment. Persistence, lateral genetic transfer, and potential interference or inhibition of the natural cycles of the ocean are examples of issues which must be considered before there is any massive release of genetically engineered microorganisms into the ocean.

Acknowledgment

This study was supported in part by Cooperative Agreement No. CR817791.01 between the U.S. Environmental Protection Agency and the University of Maryland and the National Science Foundation, Grant No. BSR-8806509.

Literature Cited

Baya, A.M., P.R. Brayton, V.L. Brown, D.J. Grimes, E. Russek-Cohen, and R.R. Colwell. 1986. Coincident plasmids and antimicrobial resistance in marine bacteria isolated from polluted and unpolluted Atlantic ocean samples. Appl. Environ. Microbiol. 51:1285-1292.

Chattergee, D.K., J.J. Kilbane, and A.M. Chakrabarty. 1982. Biodegradation of 2,4,5 trichloroacetic acid in soil by pure culture of *Pseudomonas capasia*. Appl. Environ. Microbiol. 44: 514-518.

Colwell, R.R. 1983. Biotechnology in the Marine Sciences. Science 222:19-24.

Everybody Counts Report: A Report to the Nation on the Future of Mathematics Education. 1989. Mathematics and Sciences Education Board of the National Research Council.

Grant, W.F. 1979. The genotoxic effect of 2, 4, 5 – T. Mutat. Res. 65: 83-119.

Grabow, W.O., O.W. Prozesky and L.S. Smith. 1974. Drug resistant coliforms call for review of water quality standards. Water Res. 8: 1-9.

Grimes, D.J., and R.R. Colwell. 1986. Viability and virulence of *E. coli* suspended in membrane chamber in semi-tropical ocean water. FEMS Microbiol. Letter 34:161-165.

Guerry, P., and R.R. Colwell. 1977. Isolation of cryptic plasmid deoxyribonucleic acid from Kanagawa-positive strains of *Vibrio parahaemolyticus.* Infect. and Immun. 16:328-334.

Leahy, J., and R.R. Colwell. 1990a. Microbial degradation of hydrocarbons in the environment. Microbiol. Rev. 54:305-315.

Leahy, J.G., C.C. Somerville, K.A. Cunningham, A.G. Adamantiades, J.J. Byrd, and R.R. Colwell. 1990b. Hydrocarbon mineralization in sediments and plasmid incidence in sediment bacteria from the Campeche bank. Appl. and Environ. Microbiol. 56:1565-1570.

Singleton, F.L., D.J. Grimes, H.S. Xu, and R.R. Colwell. 1985. Microbiological studies of ocean dumping, p. 187-208. *In* Kester et al. (ed.), Wastes in the Ocean, Vol. 5, Deep-Sea Waste Disposal. John Wiley & Sons, Inc.

Sizemore, R., and R.R. Colwell. 1977. Plasmids carried by antibiotic-resistant marine bacteria. Antimicrob. Agents Chemother. 12:373-382.

Timmis, K. and U. Winkler. 1973. Isolation of covalently closed circular deoxyribonucleic acid from bacteria which produce exocellular nuclease. J. Bacteriol. 113: 508-509.

Wortman, A.T., and R.R. Colwell. 1988. Frequency and characteristics of plasmids in bacteria isolated from deep-sea amphipods. Appl. Environ. Micorbiol. 54:1284-1288.

Impacts and Fates of Microbial Pest-Control Agents In the Aquatic Environment

WILLIAM E. WALTON AND MIR S. MULLA

Abstract: Four groups of microbial pest-control agents show potential as entomopathogens for pestiferous and disease-transmitting dipterans which utilize aquatic environments: viruses, protists, fungi, and bacteria. Currently, viral and protistan pathogens are not attractive commercially and, therefore, either are not used in pest-control programs or are applied in very restricted habitats. The disadvantages of viruses and protists for large-scale pest-control include (1) obligate parasitism and often complex life cycles, (2) low or variable infectivities for their original and alternate dipteran hosts, (3) environmental factors which limit their usage in large-scale pest-control programs, and (4) expensive and labor-intensive in vivo production. Entomopathogenic viral isolates and protists do not pose a significant concern to aquaculture because of their host specificity and limited usage as microbial pest-control agents.

The fungi and bacteria are the most promising microbial pest-control agents. Large-scale pest-control programs that utilize fungal pathogens also suffer from many of the aforementioned problems and, additionally, fungal pathogens often are slow acting and are facultative parasites that exhibit low infectivities for their original hosts. The attractiveness of fungi as pest-control agents is increasing because of new developments in culture methodologies and a better understanding of fungal life cycles and field efficacy. Currently, bacterial agents such as *Bacillus thuringiensis* (H-14) are used most widely. Bacterial pest-control agents are host specific, do not persist in aquatic ecosystems in environmentally significant quantities, and are commercially attractive. At current recommended application rates and by normal modes of contact, both bacterial and fungal pest-control agents are safe to nontarget organisms, particularly those organisms in aquaculture.

Introduction

A variety of microbial pathogens are used widely, or have the potential, to control pestiferous organisms that utilize aquatic environments during their life cycles. Although the adult stages of mosquitoes, black flies, and closely related taxa are often important vectors of diseases that afflict humans and livestock, the aquatic life-cycle stages of these dipteran pests are controlled frequently via microbial agents. As promising microbial insecticides and biological control agents are replacing conventional chemical control with pesticides, the impacts and fates of these microbial agents in the aquatic environment are of increasing interest to the aquaculturist.

Four groups of microbial pathogens are used currently, or show potential, as pest-control agents for dipteran pests: (1) viruses, (2) protistans (or protozoans), (3) fungi, and (4) bacteria. An extensive literature exists for each group; hence, we discuss only briefly the status of these groups as microbial pest-control agents. Reviews of microbial insecticides and biological control agents have been published elsewhere (Jamnback 1973; Chapman 1974, 1985; Davidson 1982; Mulla 1985, 1989; Lacey and Undeen 1986; Lacey and Mulla 1989), and the reader is referred there for more extensive treatments of the subject. In this chapter, we also describe the effects of the microbial pathogens on the target organisms and on nontarget organisms, and discuss the fates of microbial pest-control agents in the aquatic environment. Although microbial pest-control agents also are used and have been studied extensively in terrestrial ecosystems, our focus is only on those microbial pest-control agents which are applied purposely in aquatic environments.

Viruses

Several virus groups have been isolated from mosquitoes and black flies (Federici 1973, 1985; Lacey 1982). Viral isolates from mosquitoes are more common than from black flies (Lacey and

Undeen 1986) and can be grouped into two general categories: occluded and nonoccluded viruses.

In the occluded viruses, large proteinaceous occlusion bodies are produced late in the infection and encase the virus particles (progeny). Nuclear polyhedrosis viruses (baculoviruses) and cytoplasmic polyhedrosis viruses (reoviruses) are more common in mosquitoes than are entomopoxviruses (Federici 1985). The relatively large-sized occluded viruses are associated with the larval digestive tract. For black flies, only cytoplasmic polyhedrosis virus (Reoviridae) has been isolated (Lacey 1982).

The second virus group, the nonoccluded viruses, does not produce occlusion bodies; however, they often form paracrystalline arrays of virons (Federici 1985). Nonoccluded viruses have been isolated from several mosquito genera and are associated with the epidermis, imaginal tissue, and the fat bodies of the larva. The relatively large-sized iridoviruses and the smaller desonucleosis viruses (parvoviruses) have been isolated from mosquitoes. The former group is more common (Federici 1985). Iridoviruses also have been isolated from black flies (Lacey 1982).

The minute viral pathogens are obligate, intracellular parasites and have limited host ranges (Federici 1985; Payne 1988). The most promising viral pest-control agents, the baculoviruses, differ biochemically from viruses found in vertebrates, plants, and microorganisms (Payne 1988). Also, viral isolates are often genus- or species-specific and their efficacy is influenced by environmental factors such as the deleterious effects of ultraviolet radiation from sunlight (Payne 1988). Despite their prevalence in dipterous pests, no virus is used widely as a microbial pest-control agent (Federici 1985; Lacey and Undeen 1986). Large-scale control programs using entomopathogenic viruses have been hindered because viruses (1) often exhibit low infectivity for their original and alternate mosquito hosts (Federici 1985) and (2) are obligate intracellular parasites that must be produced in mosquito larvae or insect cell cultures (Payne 1988). Therefore, no virus has been produced economically in sufficient quantities to control pests or vectors of human diseases. Currently, entomopathogenic viral isolates do not pose a significant concern to aquaculture because of their limited usage as microbial pest-control agents.

Protista

Two groups of protistan pathogens have been identified as potential microbial control agents for mosquitoes: microsporidians and ciliates (Lacey and Undeen 1986). Microsporidian infections are very common in mosquito larvae; every well-studied mosquito species has at least one microsporidian parasite (Hazard 1985). Microsporidians are categorized into four types based on their developmental patterns and life cycles (Hazard 1985).

Hazard (1985) concluded that current varieties of types I and II microsporidians, such as *Nosema algerae* (type I) and *Vavraia culicis* (type II), show little potential as mosquito-control agents. Whereas varieties in the relatively more virulent types III and IV show more promise as mosquito-control agents, their development as biological control agents is slowed by inadequate understanding of the modes of horizontal transmission and their complex life cycles. Some microsporidian life cycles may include an alternate or an intermediate copepod host (Andreadis 1985; Sweeney et al. 1985; Chen and Barr 1988). Recent work suggests that type IV microsporidians, such as species of *Amblyospora, Culicospora, Culicosporella,* and several other genera, are the most promising microsporidian biological control agents (Hazard 1985; Chen and Barr 1988). However, these protists are not used commonly as microbial control agents. To date, most studies of these microsporidian control agents have been restricted to the laboratory (Sweeney et al. 1985, 1988, 1989).

In studies involving several types of microsporidia, it was found these protists often exhibit relatively low to moderate infectivity in nature (Maddox et al. 1977; Anthony et al. 1978; Andreadis 1989), do not persist in the environment (Undeen and Alger 1975; Anthony et al. 1978), and spores from the inoculum and from infected individuals settle rapidly (Lacey and Undeen 1986). Despite transovarial transmission of some microsporidians in mosquitoes, this phenomenon often is restricted to one generation (Lord et al. 1981) and, therefore, does not perpetuate the pathogen (Andreadis and Hall 1979). However, Andreadis (1989) found that *Amblyospora connecticus*, a naturally occurring microsporidian parasite of the brown saltmarsh mosquito (*Aedes cantator*), was transmitted transovarially for multiple generations, overwintered, and successfully reinfected the mosquito population the following spring.

Large-scale control programs using microsporidians also are hindered by expensive and labor-intensive in vivo production. Because microsporidians are expensive to produce, and their life cycles are inadequately understood, it is unlikely that microsporidians will be developed in the near future as microbial pest-control agents for hematophagous Nematocerans (Lacey and Undeen 1986).

Among the Ciliophora, two genera show the most potential as mosquito-control agents. *Tetrahymena pyriformis* is worldwide in distribution and can tolerate a broad range of environmental conditions; however, *Tetrahymena* is not widely used as a protistan pathogen (Clark 1985). *Tetrahymena* lacks a resistant resting cyst stage and infectivity is limited to certain strains (Corliss and Coats 1976; Clark 1985). Although, under optimum conditions, large quantities of this protist can be produced, concentrated, and stored, further testing in natural situations is required (Clark 1985).

The genus, *Lambornella*, is receiving increasing attention as a mosquito-control agent. This ciliate has an active host-seeking stage, a desiccation-resistant cyst, relatively high infection levels in nature, and self-dispersal via infected adult mosquitoes (Anderson et al. 1986; Egerter et al. 1986). *Lambornella clarki* can live as a parasite or as a free-living form that feeds on bacteria and other microorganisms (Wasburn et al. 1988). Depending on environmental conditions, it is relatively persistent in its natural treehole environments (Anderson et al. 1986). However, secondary infections by a normally saprophytic fungus (*Pythium flevonse*) sometimes kill the ciliate and its mosquito host (*Aedes sierrensis*) (Washburn et al. 1988). By reducing the ciliate and its host populations, the fungal infections decrease the recycling potential of this agent in nature.

Lambornella undergoes a morphological transformation within a few days after treeholes flood with water. Anderson et al. (1986) found that attachment cysts were formed when the newly formed spherical cells attached to the mosquito cuticle. Within 24 to 36 hours of attachment, each ciliate made a hole in the larval cuticle and entered the hemocoel. Parasitic amplification occurred within the hemocoel and mosquitoes succumbed to ciliatosis within three weeks (Washburn et al. 1988).

In California, *L. clarki* currently is being studied as a control agent for treehole mosquitoes (Eldridge 1988). However, given the current, and projected, limited use, and its distinct habitat preferences, *Lambornella* poses no threat to aquaculture. While *Lambornella* is common and persists in treeholes, it certainly is not ubiquitous (Washburn and Anderson 1986) and may have somewhat limited dispersal capabilities.

Fungi

Approximately 13 genera of fungi have been identified as showing potential to control mosquitoes (Roberts and Panter 1985). The most promising fungi are *Coelomomyces, Culicinomyces,* and *Lagenidium* (Lacey and Undeen 1986). However, fungal control agents suffer from many of the same problems for large-scale control efforts as do the viral and protistan agents. The fungi often are sensitive to environmental conditions such as water temperature and salinity. Many fungal pathogens do not grow and germinate at high and/or low temperatures and in saline water. Second, epizootics via inocula are often variable and unpredictable. In addition, as compared to the bacterial control agents, fungi are relatively slow acting and many groups are facultative parasites which do not readily infect dipterous pests.

Large-scale production also is a problem. The culture, fermentation, and harvesting methods are not commercially attractive. Infective stages are often fragile and, in some groups, they sink readily from the surface water and away from potential hosts. Additionally, life cycles are often complex and poorly understood.

However, new fungal strains and culture methodologies may overcome the aforementioned problems (Kerwin and Washino 1988).

Coelomomyces

Coelomomyces (Chytridomycetes, Blastocladiales) is an obligate parasite (Lacey and Undeen 1986) that infects a large number of species in several mosquito genera, particularly *Aedes*, *Anopheles*, *Armigeres*, and *Culex* (Federici et al. 1985). However, in most mosquito genera, only 1 to 3 species are known hosts for this fungus. An attractive feature of this fungus is that *Coelomomyces* presumably is dispersed by infected female mosquitoes which are infected during the late larval stages and do not succumb to mycosis prior to pupation and adult emergence (Lacey and Undeen 1986). The fungus is introduced into larval developmental sites when infected females attempt to oviposit. Despite the prevalence of this fungus in mosquitoes, *Coelomomyces* is not used in large-scale pest-control efforts.

The life cycle of *Coelomomyces* is complex and is discussed in detail by Couch and Bland (1984) and Federici et al. (1985). Briefly, in a generalized life cycle, mosquito larvae are infected by the biflagellate zygote that is formed by the fusion of uniflagellate gametes which emerge from either copepod or ostracod hosts. The zygote typically encysts on the intersegmental membranes and penetrates the larval cuticle via a germ tube. Within the larval hemocoel, mycelia are formed and eventually produce resting sporangia. Larval death usually occurs within 5 to 10 days, presumably from a depletion of body nutrients (Federici et al. 1985). As the larval cadaver degenerates, large numbers (10,000 to 50,000) of resting sporangia are released into the environment. The resting sporangia germinate and undergo meiosis, resulting in haploid meiospores. The meiospores infect the appropriate crustacean host in a manner analogous to that observed in mosquito larvae. Following infection of the crustacean's hemocoel, gametogenesis occurs and the fungal gametes are released into the environment after the death of the crustacean host.

Four major problems impede large-scale usage of *Coelomomyces*. First, *Coelomomyces* often requires an obligate alternate or intermediate crustacean (copepod or ostracod) host (Whisler et al. 1974, 1975; Federici et al. 1985). Second, little is known about the life cycles and infectivities in field situations. Third, culture methods are involved and not attractive commercially. Infective stages are difficult to maintain and exhibit reduced activity after 24 hours (Federici et al. 1985). Last, the 70 or so *Coelomomyces* species associated with mosquitoes do not provide significant or predictable levels of mosquito-control (Lucarotti et al. 1985; Federici et al. 1985). Federici et al. (1985) suggested that, at present, this fungus is probably most useful when combined with other pest-control methodologies. Additionally, further work is needed to determine the effects of *Coelomomyces* on nontarget organisms (Lacey and Undeen 1986).

Culicinomyces

Unlike *Coelomomyces* infections which begin by encystment on the larval cuticle, *Culicinomyces* (Hyphomycetes) infects mosquitoes via the digestive tract. Following ingestion by mosquito larvae, conidia of *Culicinomyces* typically adhere to and germinate in the foregut or the hindgut (Lacey and Undeen 1986). Sweeney (1979a) reported that infections also occur through the anal papillae when mosquito larvae are subjected to high spore concentrations. Larvae die between 2 days to > 1 week after the hemocoel is penetrated and invaded by hyphae (Sweeney 1983). Larval death also may occur within 2 days after ingesting high concentrations of conidia without subsequent proliferation of hyphae (Panter and Russell 1984; Lacey and Undeen 1986). Following larval death, the hyphae swell and rupture the cadaver's cuticle. Conidiophores grow exteriorly and produce the infective stage, the conidium (Roberts and Panter 1985).

Culicinomyces is amenable to mass culture because large numbers of conidia can be produced by surface culture on artificial media (Lacey and Undeen 1986). However, several problems currently inhibit its use in large-scale pest-control programs. The drawbacks of *Culicinomyces* include the reduced infectivities of the stored product, high effective dosages, short persistence in nature, and tendencies of current formulations to sink, thereby reducing contact with target species (Lacey and Undeen 1986).

Culicinomyces has a broad mosquito host range (Sweeney et al. 1973; Couch et al. 1974; Sweeney 1975; Russell et al. 1983), and in the laboratory, has persisted for more than 100 days at 14°C (Frances et al. 1984). However, in nature and at higher temperatures (25°C) in the laboratory, *Culicinomyces* showed little or no persistence (Sweeney and Panter 1977; Frances et al. 1984). Sweeney (1981) and Sweeney et al. (1983) found no evidence of residual activity of *Culicinomyces* in artificial and natural habitats. Besides a tendency for natural infections to occur during cooler seasons, *Culicinomyces* exhibited low infectivities in polluted environments (Lacey and Undeen 1986). Despite its broad environmental tolerances, another hyphomycetes species, *Tolypocladium cylindrosporum*, also did not exhibit significant residual activity in crab holes (Gardner et al. 1982, 1986).

Culicinomyces efficacy is affected by the species and age of the mosquito (Lacey and Undeen 1986). Susceptibility to *C. clavisporus* may be related to feeding mode in the mosquito larvae. Since the infective stages of this fungus settle out of the water surface, surface feeding species may be less susceptible than are species that feed deeper in the water column (anophelines < *Aedes aegypti* or *Culex quinquefasciatus*) (Cooper and Sweeney 1982; Lacey and Undeen 1986). Earlier instars are more susceptible to *Culicinomyces* than are older larvae (Sweeney 1983; Panter and Russell 1984).

Culicinomyces clavisporus is pathogenic to larvae of the Culicidae, and some species in the Chironomidae, Ceratopogonidae, Ephydridae, Syrphidae, Simuliidae, and Chaoboridae (Sweeney 1979b, Knight 1980). However, other dipteran groups, such as the Tipulidae and Psychodidae, are not susceptible (Sweeney 1979b). This fungus is not pathogenic to other insect groups, aquatic vertebrates (amphibians, reptiles, and *Gambusia*), freshwater shrimps and terrestrial vertebrates (Sweeney 1975; Egerton et al. 1981; Mulley et al. 1981). Although a localized reaction occurred in homeotherms that were injected with *C. clavisporus*, viable conidia were not reisolated (Mulley et al. 1981) and homeothermic vertebrates were not susceptible to large oral doses of the fungus (Egerton et al. 1981).

Lagenidium

Lagenidium giganteum (Oomycetes: Lagenidiales) is a facultative parasite that exhibits specificity for larvae of many mosquito species (Lacey and Undeen 1986). It has been isolated from mosquitoes throughout the world. The biflagellate zoospore of *Lagenidium* infects mosquito larvae through the mouth and cuticle (McCray 1985). Following penetration of the digestive tract by the spores, hyphae proliferate in the larval hemocoel. Sporangial formation and larval death are often simultaneous (McCray 1985).

Unlike *Coelomomyces* in which sexual reproduction occurs outside the mosquito larva, in *Lagenidium*, asexual and sexual reproduction may occur within an host individual and also within a single hypha (McCray 1985). Zoosporogenesis occurs within vesicles that are formed terminally on the exit tubes arising from the sporangia. After asexual reproduction, the vesicle wall degenerates and infective zoospores are released into the environment.

Sexual reproduction follows the differentiation of hyphal segments into either antheridia or oogonia. A thick-walled oospore is formed after the antheridial protoplast enters an adjacent oogonium (McCray 1985). Upon germination, the previously dormant oospore releases infective zoospores.

In some habitats, *Lagenidium* appears to function as a classical biological control agent: it provides long-term control following a single application (Kerwin and Washino 1987, 1988). Kerwin and Washino (1986) found that one application of fungal asexual stages suppressed *Culex tarsalis* and *Anopheles freeborni* for an entire season. *Lagenidium* recycled during the next year, and multiseason control is therefore possible (Kerwin and Washino 1988). In California, this fungus has been tested in rice fields, roadside ditches, and irrigation pastures (Eldridge 1988).

In natural and artificial habitats, *Lagenidium* has been inoculated successfully as sexual (oospores) and asexual (myecilium) stages. However, this fungus does have several problems typical of fungal pathogens. First, environmental factors influenced the efficacy of the fungus. High (> 35°C) and low (< 15°C) temperatures inhibited fungus activity (Fetter-Lasko and Washino 19838; Jaronski and Axtell 1983a, 1983b). Cool temperatures (< 14°C) prolonged fungal development and decreased zoospore infectivity (Guzman and Axtell 1987). Jaronski and Axtell (1982) found that zoosporogenesis was inhibited by environmental factors such as high chemical oxygen demand (COD) and ammonia levels. In California, infectivity also was reduced in organically rich habitats (Kerwin and Washino 1988). *Lagenidium giganteum* also was affected adversely by salinity (Merriam and Axtell 1982).

In addition to environmental constraints, levels of control in nature were variable. The presence and relative abundance of zoospores were highly variable in field tests (Guzman and Axtell 1987). Since recycling of the fungus is related to host densities and species (Fetter-Lasko and Washino 1983; Kerwin and Washino 1984), persistence of *Lagenidium* epizootics throughout the mosquito-breeding season in continually flooded habitats often requires the continual addition of mosquito larvae. However, overwintering and persistence in intermittently flooded habitats has been observed (Glenn and Chapman 1978; Kerwin and Washino 1988).

Last, as compared to bacterial agents, *Lagenidium* culture and harvesting is relatively expensive and labor intensive. Infectivity of stored products was variable in both the dried oospore (Jaronski et al. 1983; Kerwin et al. 1986) and myecilial stages (McCray et al. 1973; Kerwin and Washino 1986; Su et al. 1986). Recent advances in stored products, such as the encapsulation of infective stages (Axtell and Guzman 1987), and in culture methods (Jaronski and Axtell 1984; Kerwin and Washino 1987) may reduce these problems.

Currently, fungal microbial control agents are not used widely in aquatic pest-control programs. However, they may be used more in the future. For example, an experimental use permit for large-scale testing of *Lagenidium* in California has been requested from the U.S. Environmental Protection Agency (Eldridge 1988). For nontarget species under aquaculture, and the aquaculturalist, *Lagenidium* is noninfective and safe. This fungus has been tested against rotifers; crustaceans such as copepods, cladocerans, ostracods, fairy shrimp, and crayfish; gastropod molluscs; worms; assorted aquatic insects; amphibians and fish (McCray et al. 1973; McCray 1985). Only one nontarget organism was infected: the predatory midge, *Chaoborus* (Brown and Washino 1977, 1979). *Chaoborus* is related closely to mosquitoes. In addition, extensive testing has been carried out on terrestrial biota, including monitoring humans working with the fungus (McCray 1985; Siegel and Shadduck 1987; Kerwin et al. 1988). No internal (lungs) or external infections were observed (McCray 1985).

Bacteria

Bacterial pathogens in the genus *Bacillus* are the most widely used microbial pathogen. Particular *Bacillus* species and strains are very effective control agents for dipterous pests and are amenable to commercial production (Lacey and Undeen 1986). *Bacillus thuringiensis* variety *israelensis* or B.t.i. has proven successful for controlling mosquito and black fly populations in aquatic environments worldwide (Lacey 1985; Mulla 1985; Lacey and Undeen 1986; Lacey and Mulla 1989). In California, this pathogen has been used successfully against mosquitoes in numerous biotypes which range in size from small standing waters to large wildlife refuge ponds, in particulate concentration from relatively unpolluted, clearwater ponds to relatively polluted, dairy wastewater lagoons, and in setting from rural, agriculture to urban and periurban sites (Mulla 1985; Eldridge and Federici 1988).

Several *B. thuringiensis* serotypes are toxic to mosquito larvae. *Bacillus thuringiensis* var. *israelensis* also is known as serotype 14 or H-14. Serotypes or serovars are based on comparison of antibodies to flagellar (or "H") antigens of the bacteria. Mosquitoes also are susceptible to strains of other serotypes such as *B. thuringiensis* var. *darmstadiensis* (serotype 10), *B. thuringiensis* var. *morrisoni* (serotype 8a 8b) (Lacey and Undeen 1986), and *B. thuringiensis* var. *kyushuensis* (serotype 11a 11b) (Ohba and Aizawa 1979). However, *B. thuringiensis* (H-14) is the most commonly used variety (Gaugler and Finney 1982; Undeen and Lacey 1982; Lacey 1985).

Whereas *B. thuringiensis* (H-14) was isolated from infected mosquito larvae in Israel (Goldberg and Margalit 1977), a second *Bacillus* species, *B. sphaericus*, is common worldwide (Payne 1988). These bacteria are aerobic, spore-forming saprophytes that live in many soil and aquatic habitats (Davidson 1982). *Bacillus sphaericus* is not registered currently for general use in the United States; however, it too has been tested experimentally against mosquitoes in a variety of habitats (Mulla 1985, 1989; Eldridge and Federici 1988).

In both *Bacillus* species, spores are found in spherical bodies that are produced during sporulation. Spores are resistant to a variety of environmental conditions including relatively high temperatures, dessication, and some chemical agents (Eldridge and Federici 1988), but not to UV radiation (Burke et al. 1983; Kirschbaum 1985). During sporulation, *Bacillus* spp. also produce a structure known by several names: the delta-endotoxin, proteinaceous crystal, or parasporal body. Within this crystal reside the protein precursors that, after degradation, are responsible for the toxic activity against dipterous pests (Baumann et al. 1985; Broadwell and Baumann 1986, 1987; Eldridge and Federici 1988; Federici et al. 1989).

Some strains of *B. thuringiensis* produce other minor toxins. One of these other toxins, the beta-exotoxin, has a wider range of toxic activity than does the delta endotoxin (Eldridge and Federici 1988). The results from numerous studies summarized in Lacey (1985) and Lacey and Undeen (1986) have shown that the beta-exotoxin is toxic to mammals by injection, but not by normal means of external contact or by ingestion. However, strains of *B.*

thuringiensis that produce this toxin cannot be used as insecticides in the United States (Eldridge and Federici 1988).

Within the parasporal body are several regions which differ in their electron densities and are known as inclusions. Bacilli in the *thuringiensis* group exhibit 2 to 4 inclusions (Eldridge and Federici 1988; Federici et al. 1989). *B. thuringiensis* (H-14) has three inclusions, whereas, *B. sphaericus* has only one inclusion. Recent research suggests that there are probably four separate proteins associated with the parasporal body and these proteins differ in mass.

In *B. thuringiensis* (H-14), Federici et al. (1989) have tentatively assigned a 27 kDa protein to the least electron dense "large" inclusion. A 65 kDa protein is thought to reside within the moderately electron-dense "bar" inclusion, and the 128 and 135 kDa proteins have been assigned tentatively to the highly electron dense "high-density" inclusion (Federici et al. 1989). In addition, when the paraspore is lysed, new proteins are formed as breakdown products of those listed above. It has been suggested that the 65, 128 and 135 kDa proteins of *B. thuringiensis* (H-14) are the precursors of the polypeptides that are toxic to mosquitoes and, currently, it appears that none of these proteins alone is as toxic per unit weight as is the mixture of them found in the parasporal body (Eldridge and Federici 1988).

Baumann et al. (1985) and Broadwell and Baumann (1986, 1987) found that the *B. sphaericus* inclusion contained 43, 63, 110 and 125 kDa proteins. They suggested that the 43 and the 110 kDa proteins were toxic to mosquito larvae. A 40 kDa polypeptide was appreciably more toxic to cells of *Culex quinquefasciatus* than was its 43 kDa precursor.

Laboratory studies have shown that susceptibilities to *Bacillus* toxins differ among mosquito species and that the toxicities differ among formulations of a particular *Bacillus* strain. Lacey (1985), Singer (1985), and Davidson (1985) summarized the results from studies on *B. thuringiensis* (H-14) and *B. sphaericus* against mosquitoes. Similar results were evident in field studies using the two *Bacillus* species (Lacey 1985; Mulla 1985, 1986, 1989).

In general, *Culex* species were more susceptible to low concentrations of bacterial spores than were *Anopheles* and *Aedes* larvae (Lacey and Singer 1982; Singer 1985; Mulla 1986; Mulla et al. 1986), and some *Culex* and *Anopheles* species were more susceptible to *B. sphaericus* than they were to *B. thuringiensis* (H-14) (Mulla et al. 1986; Payne 1988). Additionally, *B. sphaericus* has a narrower host range than does *B. thuringiensis* (H-14) and is less toxic (Eldridge and Federici 1988), or is not toxic, at operational application rates to filter-feeding Diptera which are related closely to mosquitoes. For example, black flies (Lacey and Undeen 1986) and some chironomid midges (Mathavan and Valpandi 1984; Ali and Nayar 1986) are not susceptible to *B. sphaericus*. However, like mosquitoes, susceptibilities of black flies and chironomids to *B. thuringinensis* (H-14) differ among species and among B.t.i. formulations (Gaugler and Finney 1982; Undeen and Lacey 1982; Ali and Nayar 1986).

Currently, it is thought that three digestive tract characteristics cause the differential susceptibilities observed among mosquito species and result in the safety of these bacterial pest-control agents to nontarget organisms. First, an organism must have an alkaline digestive tract. The toxic proteins, which were discussed above, are derived from the solubilization of the parasporal body proteins by enzymes that are active only under alkaline conditions (Eldridge and Federici 1988). The anterior midgut of mosquito larvae is highly alkaline ($pH > 10$) and the posterior midgut is less so (pH 7-8; Dadd 1975; Davidson 1988).

Second, the susceptible organism also must exhibit an enzyme complement suitable for the degradation of the endotoxin crystal. Although, *Aedes aegypti*, *A. triseriatus*, and closely related species have alkaline digestive tracts and the appropriate enzymes which degrade the *Bacillus* parasporal body, the bacterial toxin does not bind to cultured cells or the midguts of these resistant species (Davidson et al. 1987; Davidson 1988). Third, at least for *B. sphaericus*, the susceptible organism also must exhibit the appropriate glycoprotein receptors in the digestive tract that permit the toxin to enter cells via receptor-mediated endocytosis

(Davidson 1988). Davidson (1988) found that the bacterial toxin bound to sharply delineated regions of the posterior midgut and the proximal lobes of the gastric caecum in susceptible *Culex quinquefasciatus* larvae.

The ultrastructural and histopathological events that follow the ingestion of *B. thuringiensis* (H-14) or *B. sphaericus* toxins differ slightly (J.-F. Charles, pers. comm.). Typically, following ingestion of the toxin, the cells of the larval gut epithelium swell, distend, and contain many lytic vacuoles (Lacey and Undeen 1986). The cells either lyse and are sloughed into the gut lumen (*B. thuringiensis* H-14) or simply separate from one another (*B. sphaericus*). In both cases, larval death presumably is caused by the complete loss of osmotic regulation (Lacey and Undeen 1986). The timing of larval death obviously depends on the toxin concentration.

Several factors affect the efficacy of mosquito-control with bacterial insecticides. As was discussed previously, the species of mosquito is important. The age of the mosquito larvae also is important. Younger instars are more susceptible to the bacterial toxins than are the older larval instars (Lacey and Undeen 1986). Additionally, larval feeding behavior and the availability of the toxin in the feeding zone are important factors determining activity (Lacey 1985; Mulla 1985). For example, besides differences in gut receptor sites, the surface-feeding behavior of anophelines and young *Aedes* larvae may reduce further their susceptibility to the bacterial toxins. By feeding in the surface zone, these mosquitoes are less likely to ingest bacterial toxins which settle from the water's surface than are culicine larvae which browse on vegetation surfaces and the substrata (Mulla 1985; Lacey and Undeen 1986). Rishikesh et al. (1983) suggested that toxin efficacy also is reduced via ingestion and biodegradation by nontarget organisms.

Availability of food, larval density, pollutants, vegetative cover, and environmental factors such as solar radiation, water temperature, and ionic content influence the activity of bacterial control agents in the field (Mulla et al. 1984a, Lacey 1985; Mulla 1985; Lacey and Undeen 1986). For example, in eutrophic environments, larval diving rate and depth are less than in relatively oligotrophic habitats (Lacey and Undeen 1986). Mosquito larvae concentrate their feeding in the upper, food-rich strata under eutrophic conditions; hence, larvae are less likely to ingest the settled, bacterial toxins than they would under food-poor conditions (Lacey and Undeen 1986). Also, toxin settling rate is enhanced in turbid, eutrophic, or in polluted, waters (Van Essen and Hembree 1982; Mulla 1985). The rate of *Bacillus* application needed for satisfactory control of mosquito larvae is two to five-fold greater in turbid, eutrophic water than in clear, oligotrophic water (Mulla et al. 1982, 1984b; Mulla 1985).

Some *Bacillus* formulations do not penetrate thick vegetative cover (Mulla 1985). Granular and pelletized formulations often are required where thick, emergent vegetation is present. Besides the inherent toxicological and insecticidal potency differences among these formulations and strains, efficacy is influenced additionally by a variety of physicochemical factors. For example, toxin activity is reduced at low temperature (Mulla et al. 1986), perhaps as a consequence of lower larval feeding rates. Sunlight also influences toxicities and persistence since UV light rapidly degrades the *Bacillus* spores (Mulla et al. 1988; Payne 1988). In addition to many of the factors discussed above, conditions affecting downstream carry and larval contact rates are important in lotic habitats (Lacey and Undeen 1986).

One problem with bacterial insecticides is that the spores settle out of the water surface and repeated applications are necessary (Davidson et al. 1984; Mulla 1985; Mulla et al. 1982, 1984a, 1984b, 1988). Mulla (1985) concluded that standard flowable concentrate and water-dispersable formulations of *B. thuringiensis* (H-14) provided effective initial control, but after 7 to 14 days, control declined to undetectable levels. Davidson et al. (1984) found that *B. sphaericus* spores in the sediments remain viable for about 4 weeks as long as water was present. However, no *Bacillus* activity was found in subsequent floodings. The fact that spores sink and are degraded by ultraviolet radiation reduces the likelihood that they will be dispersed from typical mosquito developmental sites which have been inoculated with B.t.i. or *B. sphaericus*. Payne (1988) concluded that *B. thuringiensis* was very effective in enclosed environments such as granaries, but in nature, inoculations generally failed to spread. Although studies have attempted to detect toxins in sediments of habitats treated with *B. thuringiensis* (H-14) and *B. sphaericus*, to date, no definitive results have been obtained to show that toxins remain in the sediments in environmentally significant quantities (Eldridge and Federici 1988).

Some recycling of *Bacillus* spores has been reported, particularly in small containers. Laboratory and field studies have documented toxin recycling in habitats utilized by container-breeding mosquitoes, such as used tires and small vessels (Ignoffo et al. 1981; Aly 1983; Larget-Theiry 1984; Aly et al. 1985; Zaritsky and Kawaled 1986; Kramer in press). Also, recycling has been documented in some natural habitats such as treeholes, roadside ditches, dairy wastewater lagoons, and other habitats (Mulligan et al. 1978; Hertelin et al. 1979; Singer 1980, 1985; Dossou-Yovo and Hougard 1987; Mulla et al. 1988). Hertelin et al. (1979) found that *B. sphaericus* spores were present 9 months after application in a roadside ditch in Florida. In Illinois, Singer (1980, 1985) found that detritus remained insecticidal for 7 to 13 months after application. Kramer (in press) found that the abundance of tire-breeding *Culiseta incidens* in California was reduced significantly for 4 to 6 weeks posttreatment.

Mulla (1985) argued, in terms of larval control, that studies which have documented the presence of spores in the sediments do not provide a true indication of recycling in nature. Although sediments may remain insecticidal, their toxicity is evident only after resuspension that does not occur typically in nature. In the case of dairy wastewater lagoons, water-management practices continually resuspend the spore-laden sediments (Mulla et al. 1988). Additionally, unlike *B. thuringiensis*, some *B. sphaericus* strains are able to grow saprophytically in heavy-polluted water (WHO 1985; cited in Payne 1988). In small containers in the laboratory and small-sized natural habitats, it appears that cannibalism of *Bacillus*-infected cadavers is the mechanism by which the recycling takes place (Aly 1983; Larget-Thiery 1984; Zaritsky and Kawaled 1986; Kramer in press). If uninfected larvae do not, or cannot, ingest infected conspecifics, larval control declines rapidly.

It is unlikely that *Bacillus* pest-control agents currently pose any threat to aquaculture. Both *B. thuringiensis* (H-14) and *B. sphaericus* have been tested extensively against many invertebrate and vertebrate groups. The results from these studies have been summarized by Lacey (1985), Singer (1985), Mulla (1989) and Lacey and Mulla (1989). For example, during field testing in standing waters, nontarget taxa were not reduced by *Bacillus* spp. (Garcia et al. 1980, 1981; Miura et al. 1980, 1982; Ali et al. 1981; Purcell 1981; Sebastian and Brust 1981; Mulla et al. 1982, 1984a; Mulla 1986, 1989). Laboratory studies also have shown that (1) mosquito predators were not affected by ingesting *Bacillus*-intoxicated mosquito larvae (Mulligan and Schaefer 1982; Mathavan and Velpandi 1984; Olejnicek and Maryskova 1986; Aly and Mulla 1987; Mathavan et al. 1987; Mulla 1989), (2) nonpredatory nontarget taxa were not affected by *Bacillus* toxins (Schnetter et al. 1981; Mathavan and Velpandi 1984; Mulligan and Schaefer 1982), and (3) susceptible dipterans, which are closely related to mosquitoes, were often 13- to 75-fold more tolerant than were mosquito larvae (Ali et al. 1981; Mulla 1989). Therefore, at current application rates, and via natural forms of contact such as external contact or ingestion, *B. thuringiensis* (H-14) and *B. sphaericus* are safe and nontoxic to nondipteran taxa (Lacey and Undeen 1986).

Some *B. thuringiensis* (H-14) toxicity was reported for some stream insects and for brook trout, but the toxicities have been attributed to the xylene component of the particular formulation and not to the endotoxin (Fortin et al. 1986). Subsequent studies have confirmed this result (see Lacey and Mulla 1989). We did not detect any reduction in fish populations when we combined mosquitofish, *Gambusia* affinis, with *B. sphaericus* treatments (Walton et al., unpublished results).

While the nontarget invertebrate taxa in the studies listed above were primarily co-occurring insects, several studies failed to detect any effect of *Bacillus* toxins on crustaceans and molluscs (Schnetter et al. 1981; Mulla et al. 1982, 1984b; Mulligan and Schaefer 1982; Mathavan and Velpandi 1984; Reish et al. 1985). Holck and Meek (1987) found that the crayfish *Procambarus clarkii* was less susceptible than were the larvae of three mosquito species. The 96-hr LC 50's of 25- to 40-mm crayfish were 2,000 to 7,000 times those of mosquito larvae and 500 times the maximum labeled rates for field applications of these bacterial agents to control mosquitoes.

In summary, *Bacillus* pest-control agents exhibit several advantageous characteristics. First, *Bacillus* is not an obligate parasite of mosquitoes. Therefore, the culture and mass production of the bacterium is much easier and less expensive than for other microbial pathogens that are obligatory mosquito parasites. Second, insecticidal activity is relatively stable when products are stored at room temperatures (Lacey and Undeen 1986). Third, *Bacillus* toxins exhibit specific pathenogenicities and are safe to nontarget organisms at current application rates and by common modes of contact. Only some dipteran groups that are related to mosquitoes and black flies are susceptible to bacterial pest-control agents; albeit, at much higher concentrations than those required to control pestiferous and disease-transmitting species. Last, in nature, the toxin is typically short lived and is degraded rapidly.

While bacterial pest-control agents are safe to most nontarget organisms and do not persist in the environment, efforts continue to increase the host-specificity, insecticidal activity, and efficacy of *Bacillus* formulations. In order to increase *Bacillus* toxicity and host-specificity, future research is likely to focus on biotechnological aspects such as the characterization of the paraspore proteins, the toxin's mode of action, and genetic alteration of the toxin's protein precursors (Eldridge and Federici 1988). There also is an interest to insert the toxin genes into other organisms, particularly prokaryotes (Payne 1988). The prokaroytic organism presumably will recycle the toxin and maintain it in the feeding zone of the target organisms.

Future research also will focus on the development and refinement of asporular, floating, and slow-release formulations (Lacey et al. 1984; Apperson et al. 1986; Mulla 1985; Novak et al. 1985). Such efforts will enhance the efficacy of *Bacillus* in the field, reduce the number of larvicide applications, and, for asporular formulations, address a concern often raised against bacterial pest-control agents: the fate of *Bacillus* spores in the environment. Research is also warranted on developing formulations that maintain the toxins in the feeding zones of the target species.

Conclusion

Four groups of microbial pest-control agents show potential as entomopathogens for pestiferous and disease-transmitting insects which utilize aquatic environments. Viruses and protists either are not used in large-scale aquatic pest-control programs or are applied in very restricted habitats such as the ciliate *Lambornella clarki* in treeholes. In general, these pathogens exhibit low or variable infectivities for their dipteran hosts and, presently, commercial production is not economically attractive. The bacteria and the fungi are the most promising microbial pest-control agents.

Currently, *Bacillus thuringiensis* (H-14) is used most often to control dipteran pests, does not persist in aquatic ecosystems in environmentally significant quantities, and is commercially attractive. However, the attractiveness of fungi for large-scale pest-control programs is increasing because of new developments in culture methodologies and a better understanding of fungal life cycles and field efficacy. At current recommended application rates and by normal modes of contact, both bacteria and fungal agents (e.g., *Lagenidium giganteum*) are safe to nontarget organisms, particularly those in aquaculture. A more serious threat to aquaculture is that posed by geographic transfers of populations or by the introduction of exotic and nonendemic species which harbor infectious diseases or outcompete endemic species.

Literature Cited

Ali, A., R.A. Baggs and J.P. Stewart. 1981. Susceptibility of some Florida chironomids midges and mosquitoes to various formulations of *Bacillus thuringiensis* serovar. *israelensis*. J. Econ. Entomol. 74:672-677.

Ali, A. and J.K. Nayar. 1986. Efficacy of *Bacillus sphaericus* Neide against larval mosquitoes (Diptera: Culicidae) and midges (Diptera: Chirnomidae) in the laboratory. Florida Entomol. 69:685-690.

Aly, C. 1983. Feeding behavior of *Aedes vexans* larvae influencing efficacy of *Bacillus thuringiensis* var. *israelensis* toxin. Bull. Soc. Vector Ecol. 8:94-100.

Aly, C. and M.S. Mulla. 1987. Effect of two microbial insecticides on aquatic predators of mosquitoes. Zeits. Angew. Entomol. 103:113-118.

Aly, C., M.S. Mulla and B.A. Federici. 1985. Sporulation and toxin production by *Bacillus thuringiensis* var. *israelensis* in cadavers of mosquito larvae (Diptera: Culicidae). J. Invertebr. Pathol. 46:251-258.

Anderson, J.R., D.E. Egerter and J.O. Washburn. 1986. The biology and biological control potential of *Lambornella clarki* (Ciliophora: Tetrahymenidae), an endoparasite of the western treehole mosquito, *Aedes sierrensis*. Proc. Pap. 54th Annu. Conf. Calif. Mosq. Vector Control Assoc. 54:149-150.

Andreadis, T.G. 1985. Experimental transmission of a microsporidian pathogen from mosquitoes to an alternate copepod host. Proc. Nat. Acad. Sci. (USA) 82:5574-5577.

Andreadis, T.G. 1989. Infection of a field population of *Aedes cantator* with a polymorphic microsporidian, *Amblyospora connecticus*, via release of the intermediate host, *Acanthocyclops vernalis*. J. Am. Mosq. Control Assoc. 5:81-89.

Andreadis, T.G. and D.W. Hall. 1979. Significance of transovarial infections of *Amblyospora* sp. (Microspora: Thelohaniidae) in relation to parasite maintenance in the mosquito *Culex salinarius*. J. Invertebr. Pathol. 34:152-157.

Anthony, D.W., K.E. Savage, E.I. Hazard, S.W. Avery, M.D. Boston and S.W. Oldacre. 1978. Field tests with *Nosema algerae* Vavra and Undeen (Microsporida, Nosematidae) against *Anopheles albomanus* Weidemann in Panama. Misc. Publ. Entomol. Soc. Am. 11:17-28.

Apperson, C.S., E.E. Powell and F. Van Essen. 1986. Evaluation of a sustained release formulation of *Bacillus thuringiensis* (H-14) for control of woodland *Culex* mosquitoes. J. Am. Mosq. Control Assoc. 2:376-378.

Axtell, R.C. and D.R. Guzman. 1987. Encapsulation of the mosquito pathogen *Lagenidium giganteum* (Oomycetes: Lagenidiales) in calcium alginate. J. Am. Mosq. Control Assoc. 3:450-459.

Baumann, P., B.M. Unterman, L. Baumann, A.H. Broadwell, S.J. Abbene and R.D. Bowditch. 1985. Purification of the larvicidal toxin of *Bacillus sphaericus* and evidence for high-molecular weight precursors. J. Bacteriol. 163:738-764.

Broadwell, A.H. and P. Baumann. 1986. Sporulation-associated activation of *Bacillus sphaericus* larvicide. Appl. Environ. Microbiol. 52:758-764.

Broadwell, A.H. and P. Baumann. 1987. Proteolysis in the gut of mosquito larvae results in further activation of the *Bacillus sphaericus* toxin. Appl. Environ. Microbiol. 53:1333-1337.

Brown, J.K. and R.K. Washino. 1977. Developments in research with the Clear Lake gnat *Chaoborus astictopus* in relation with fungus *Lagenidium giganteum.* Proc. Pap. 45th Annu. Conf. Calif. Mosq. Vector Control Assoc. 45:105.

Brown, J.K. and R.K. Washino. 1979. Evaluating the fungus *Lagenidium giganteum* for the biological control of the Clear Lake gnat, *Chaoborus astictopus,* in an agricultural pond in Lake County, Calif. Proc. Pap. 47th Annu. Conf. Calif. Mosq. Vector Control Assoc. 47:37.

Burke, W.F., K.O. McDonald and E.W. Davidson. 1983. Effect of UV light on spore viability and mosquito larvicidal activity of *Bacillus sphaericus* 1593. Appl. Environ. Microbiol. 46:954-956.

Chapman, H.C. 1974. Biological control of mosquito larvae. Annu. Rev. Entomol. 19:33-59.

Chapman, H.C., editor. 1985. Biological control of mosquitoes. Am. Mosq. Control Assoc. Bull. No. 6, Fresno, Calif.

Chen, W.-J. and A.R. Barr. 1988. Development of the mosquito microsporidian parasite *Amblyospora californica* in an alternate copepod host, *Acanthocyclops vernalis*. Proc. Pap. 56th Annu. Conf. Calif. Mosq. Vector Control Assoc. 56:146-152.

Clark, T.B. 1985. *Tetrahymena* and *Lambornella* (Protozoa), pages 56-58. *In* H.C. Chapman (ed.), Biological control of mosquitoes. Am. Mosq. Control Assoc. Bull. No. 6, Fresno, Calif.

Cooper, R. and A.W. Sweeney. 1982. The comparative activity of the Australian and United States strains of *Culicinomyces clavisporus* bioassayed in mosquito larvae of three different genera. J. Invertebr. Pathol. 40:383-387.

Corliss, J.O. and D.W. Coats. 1976. A new cuticular cyst-producing tetrahymenid ciliate, *Lambornella clarki* n. sp., and the current status of ciliatosis in culicine mosquitoes. Trans. Am. Microsc. Soc. 95:725-739.

Couch, J.N. and C.E. Bland. 1984. The genus *Coelomomyces*. Academic Press, New York.

Couch, J.N., S.V. Romney and B. Rao. 1974. A new fungus which attacks mosquitoes and related Diptera. Mycologia 66:374-379.

Dadd, R.H. 1975. Alkalinity within the midgut of mosquito larvae with alkaline active digestive enzymes. J. Insect Physiol. 21:1847-1853.

Davidson, E.W. 1982. Bacteria and the control of arthropod vectors of human and animal disease, p. 285-315. *In* E. Kurstak (ed.), Microbial and viral pesticides. Marcel Dekker, New York.

Davidson, E.W. 1985. *Bacillus sphaericus* as a microbial control agent for mosquito larvae, p. 213-226. *In* M. Laird and J.W. Miles (eds.), Integrated mosquito-control methodologies. Volume 2. Academic Press, London.

Davidson, E.W. 1988. Binding of the *Bacillus sphaericus* (Eubacteriales: Bacillaceae) toxin to midgut cells of mosquito (Diptera: Culicidae) larvae: relationship to host range. J. Med. Entomol. 25:151-157.

Davidson, E.W., C. Shellabarger, M. Meyer and A.L. Beiber. 1987. Binding of the *Bacillus sphaericus* mosquito larvicidal toxin to cultured insect cells. Can. J. Microbiol. 33:982-989.

Davidson, E.W., M. Urbina, J. Payne, M.S. Mulla, H. Darwazeh, H.T. Dulmage, and J.A. Correa. 1984. Fate of *Bacillus sphaericus* 1593 and 2362 spores used as larvicides in the aquatic environment. Appl. Environ. Microbiol. 47:125-129.

Dossou-Yovo and J.-M. Hougard. 1987. Persistence and recycling of *Bacillus sphaericus* 2362 spores in *Culex quinquefasciatus* breeding sites in West Africa. Appl. Microbiol. Biotech. 25:341-345.

Egerter, D.E., J.R. Anderson and J.O. Washburn. 1986. Dispersal of the parasitic ciliate *Lambornella clarki*: implications for ciliates in the biological control of mosquitoes. Proc. Nat. Acad. Sci. (USA) 83:7335-7339.

Egerton, J.R., J.W. Hartley, R.C. Mulley and A.W. Sweeney. 1981. Susceptibility of laboratory and farm animals and two species of duck to the mosquito fungus *Culicinomyces* sp. Mosq. News 38:260-263.

Eldridge, B.F. 1988. Conventional chemical pesticides for mosquito-control: past and future. Proc. Pap. 56th Annu. Conf. Calif. Mosq. Vector Control Assoc. 56:91-98.

Eldridge, B.F. and B.A. Federici. 1988. Bacterial mosquito larvicides: Present status of knowledge and future directions for research. Proc. Pap. 56th Annu. Conf. Calif. Mosq. Vector Control Assoc. 56:117-127.

Federici, B.A. 1973. Virus pathogens of mosquitoes and their potential use in mosquito-control. Proceedings of the International Seminar on Mosquito-control, University of Quebec, Trois Rivieres, Quebec, Canada. pp. 93-105.

Federici, B.A. 1985. Viral pathogens, p. 62-74. *In* H.C. Chapman (ed.), Biological control of mosquitoes. Am. Mosq. Control Assoc. Bull. No. 6, Fresno, California.

Federici, B.A., P. Luthy and J.E. Ibarra. In Press. The parasporal body of *Bacillus thuringiensis* subsp. *israelensis*: Structure, protein composition, and toxicity. *In* H. de Barjac and D. Sutherland (eds.), Bacterial control of mosquitoes and black flies: biochemistry, genetics, and applications of *Bacillus thuringiensis* and *Bacillus sphaericus*. Rutgers University Press, New Brunswick, New Jersey.

Federici, B.A., P.W. Tsao and C.J. Lucarotti. 1985. *Coelomomyces* (Fungi), p. 75-86. *In* H.C. Chapman (ed.), Biological control of mosquitoes. Am. Mosq. Control Assoc. Bull. No. 6, Fresno, California.

Fetter-Lasko, L.L. and R.K. Washino. 1983. In situ studies on seasonality and recycling pattern in California of *Lagenidium giganteum* Couch, an aquatic fungal pathogen of mosquitoes. Environ. Entomol. 12:635-640.

Fortin, C., D. Lapointe and G. Charpentier. 1986. Susceptibility of brook trout (*Salvelinus fontinalis*) fry to a liquid formulation of *Bacillus thuringiensis* serovar. *israelensis* (Tecknar®) used for black fly control. Can. J. Fish. Aquat. Sci. 43:1667-1670.

Frances, S.P., R.C. Russell and C. Panter. 1984. Persistence of the mosquito pathogen fungus *Culicinomyces* in artificial aquatic environments. Mosq. News 44:321-324.

Garcia, R., B. Des Rochers and W. Tozer. 1980. Studies on the toxicity of *Bacillus thuringiensis* var. *israelensis* against organisms found in association with mosquito larvae. Proc. Pap. 48th Annu. Conf. Calif. Mosq. Vector Control Assoc. 48:33-36.

Garcia, R., B. Des Rochers and W. Tozer. 1981. Studies on *Bacillus thuringiensis* var. *israelensis* against mosquito larvae and other organisms found in association with mosquito larvae. Proc. Pap. 48th Annu. Conf. Calif. Mosq. Vector Control Assoc. 49:25-29.

Gardner, J.M., C.M. Chang and J.S. Pillai. 1982. The effects of salinity and temperature on the growth of three strains of the mosquito pathogenic fungus *Tolypocladium cylindrosporum* Gams. WHO/VBC/82.849, World Health Organization, Geneva, Switzerland. 5 p.

Gardner, J.M., R.C. Ram, S. Kumar and J.S. Pillai. 1986. Field trials of *Tolypocladium cylindrosporum* against larvae of *Aedes polynesiensis* breeding in crab holes in Fiji. J. Am. Mosq. Control Assoc. 2:292-295.

Gaugler, R. and J. Finney. 1982. A review of *Bacillus thuringiensis* var. *israelensis* (serotype 14) as a biological control agent for black flies (Simuliidae). Misc. Publ. Entomol. Soc. Am. 12:1-17.

Glenn, F.E. and H.C. Chapman. 1978. A natural epizootic of the aquatic fungus *Lagenidium giganteum* in the mosquito *Culex territans*. Mosq. News 38:522-524.

Goldberg, L.G. and J. Margalit. 1977. A bacterial spore demonstrating rapid larvicidal activity against *Anopheles sergentii, Uranotaenia unguiculata, Culex univitattus, Aedes aegypti* and *Culex pipiens*. Mosq. News 37:355-358.

Guzman, D.R. and R.C. Axtell. 1987. Population dynamics of *Culex quinquefasciatus* and the fungal pathogen *Lagenidium giganteum* (Oomycetes: Lagenidiales) in stagnant water pools. J. Am. Mosq. Control Assoc. 3:442-449.

Hazard, E.I. 1985. Microsporidia (Microspora) (Protozoa), p. 51-55. *In* H.C. Chapman (ed.), Biological control of mosquitoes. Amer. Mosq. Assoc., Bull. No. 6, Fresno, California.

Hertelein, B.C., R. Levy and T.W. Miller, Jr. 1979. Recycling potential and selective retrieval of *Bacillus sphaericus* from soil in a mosquito habitat. J. Invertebr. Pathol. 33:217-221.

Holck, A.R. and C.L. Meek. 1987. Dose-mortality responses of crawfish and mosquitoes to selected pesticides. J. Am. Mosq. Control Assoc. 3:407-411.

Ignoffo, C.M., C. Garcia, M.J. Kroha, T. Fukuda and T.L. Couch. 1981. Laboratory tests to evaluate the potential efficacy of *Bacillus thuringiensis* var. *israelensis* for use against mosquitoes. Mosq. News 41:85-93.

Jamnback, H. 1973. Recent developments in control of black flies. Annu. Rev. Entomol. 18:281-304.

Jaronski, S.T. and R.C. Axtell. 1982. Effects of organic water pollution on the infectivity of the fungus *Lagenidium giganteum* (Oomycetes: Lagenidiales) for larvae of *Culex quinquefasciatus* (Diptera: Culicidae): Field and laboratory evaluation. J. Med. Entomol. 19:255-262.

Jaronski, S.T. and R.C. Axtell. 1983a. Persistence of the mosquito fungal pathogen *Lagenidium giganteum* (Oomycetes: Lagenidiales) after introduction into natural habitats. Mosq. News 43:332-337.

Jaronski, S.T. and R.C. Axtell. 1983b. Effects of temperature on infection, growth and zoosporogenesis of *Lagenidium giganteum,* a fungal pathogen of mosquito larvae. Mosq. News 43:42-45.

Jaronski, S.T., R.C. Axtell, S.M. Fagan and A.J. Domnas. 1983. In vitro production of zoospores by the mosquito pathogen *Lagenidium giganteum* (Oomycetes: Lagenidiales) on solid media. J. Invertebr. Pathol. 41:305-309.

Jaronski, S.T. and R.C. Axtell. 1984. Simplified production system for the fungus *Lagenidium giganteum* for operational mosquito-control. Mosq. News 44:377-381.

Kerwin, J. and R.K. Washino. 1984. Cyclic nucleotide regulation of oosporogenesis by *Lagenidium giganteum* and related fungi. Exp. Mycol. 7:109-115.

Kerwin, J.L. and R.K. Washino. 1986. Ground and aerial application of the sexual and asexual stages of *Lagenidium giganteum* (Oomycetes: Lagenidiales) for mosquito-control. J. Am. Mosq. Control Assoc. 2:182-189.

Kerwin, J.K. and R.K. Washino. 1987. Ground and aerial application of asexual stage of *Lagenidium giganteum* for control of mosquitoes associated with rice culture in the Central Valley of California. J. Am. Mosq. Control Assoc. 3:59-64.

Kerwin, J.L. and R.K. Washino. 1988. Isolation of a new strain of *Lagenidium giganteum* and implications for control of floodwater and other rapidly developing mosquito species. Proc. Pap. 56th Annu. Conf. Calif. Mosq. Vector Control Assoc. 56:182-184.

Kerwin, J.L., D.A. Dritz and R.K. Washino. 1988. Nonmammalian safety tests for *Lagenidium giganteum* (Oomycetes: Lagenidiales). J. Econ. Entomol. 81:158-171.

Kerwin, J.L., C.A. Simmons and R.K. Washino. 1986. Oosporogenesis by *Lagenidium giganteum* in liquid culture. J. Invertebr. Pathol. 47:258-270.

Kirschbaum, J.B. 1985. Potential implication of genetic engineering and other biotechnologies to insect control. Annu. Rev. Entomol. 30:51-70.

Knight, A.L. 1980. Host range and temperature requirements of *Culicinomyces clavisporus.* J. Invertebr. Pathol. 36:423-425.

Kramer, V.L. In press. Efficacy and persistence of *Bacillus sphaericus, Bacillus thuringiensis* var. *israelensis* and methoprene against *Culiseta incidens* (Diptera: Culicidae) in tires. J. Econ. Entomol.

Lacey, L.A. 1982. Viral pathogens of vector *Nematocera* and their potential for microbial control, p. 428-436. *In* Proc. 3rd Internat. Colloq. Invertebr. Pathol., University of Sussex, Brighton, England.

Lacey, L.A. 1985. *Bacillus thuringiensis* Serotype H-14 (Batcteria), p. 132-158. *In* H.C. Chapman (ed.), Biological control of mosquitoes. Am. Mosq. Control Assoc. Bull. No. 6, Fresno, California.

Lacey, L.A. and M.S. Mulla. In press. Safety of *Bacillus thuringiensis* var. *israelensis* and *Bacillus sphaericus* to non-target organisms in the aquatic environment. *In* M. Laird, E.W. Davidson and L.A. Lacey (eds.), Safety aspects of microbial insecticides. CRC Press.

Lacey, L.A. and S. Singer. 1982. Larvicidal activity of new isolates of *Bacillus sphaericus* and *Bacillus thuringiensis* (H-14) against anopheline and culicine mosquitoes. Mosq. News 42:537-543.

Lacey, L.A. and A.H. Undeen. 1986. Microbial control of black flies and mosquitoes. Annu. Rev. Entomol. 31:265-296.

Lacey, L.A., M.J. Urbina and C.M. Heitzman. 1984. Sustained release formulations of *Bacillus sphaericus* and *Bacillus thuringiensis* (H-14) for control of container breeding *Culex quinquefasciatus*. Mosq. News 44:26-32.

Larget-Thiery, I. 1984. Simulation studies on the persistence of *Bacillus thuringiensis* H-14. WHO/VBC/84.906, World Health Organization, Geneva, Switzerland. 8 p.

Lord, J.C., D.W. Hall and E.A. Ellis. 1981. Life cycle of a new species of *Amblyospora* (Microspora: Amblyosporidae) in the mosquito *Aedes taeniorhynchus*. J. Invertebr. Pathol. 37:66-72.

Lucarotti, C.J., B.A. Federici and H.C. Chapman. 1985. Progress in the development of *Coelomomyces* fungi for use in integrated mosquito-control programmes, p. 251-268. *In* M. Laird and J.M. Miles (eds.), Integrated mosquito-control methodologies. Volume 2. Academic Press, London.

Maddox, J.V., N.E. Alger, A. Ahmad and M. Aslamkhan. 1977. The susceptibility of some Pakistan mosquitoes to *Nosema algerae* (Microsporidia). Pakistan J. Zool. 9:19-22.

Mathavan, S. and A. Velpandi. 1984. Toxicity of *Bacillus sphaericus* strains to selected target and non-target aquatic organisms. Indian J. Med. Res. 80:653-657.

Mathavan, S., A. Velpandi and J.C. Johnson. 1987. Sub-toxic effects of *Bacillus sphaericus* 1593 M on feeding, growth and reproduction of *Laccotrephes griseus* (Hemiptera: Nepidae). Exp. Biology 46:149-153.

McCray, E.M. 1985. *Lagenidium giganteum* (Fungi), p. 87-98. *In* H.C. Chapman (ed.), Biological control of mosquitoes. Am. Mosq. Control Assoc. Bull. No. 6, Fresno, California.

McCray, E.M., Jr., D.J. Womeldorf, R.C. Husbands and D.A. Eliason. 1973. Laboratory and field tests with *Lagenidium* against California mosquitoes. Proc. Pap. 41st Annu. Conf. Calif. Mosq. Vector Control Assoc. 41:123-128.

Merriam, T.L. and R.C. Axtell. 1982. Salinity tolerance of two isolates of *Lagenidium giganteum* (Oomycetes: Lagenidiales), a fungal pathogen of mosquito larvae. J. Med. Entomol. 19:388-393.

Miura, T., R. Takahashi and F. Mulligan, III. 1980. Effects of the bacterial mosquito larvicide *Bacillus thuringiensis* serotype H-14 on selected aquatic organisms. Mosq. News 40:619-622.

Miura, T., R. Takahashi and F. Mulligan, III. 1982. Impact of the use of candidate bacterial mosquito larvicides on some selected aquatic organisms. Proc. Pap. 49th Annu. Conf. Calif. Mosq. Vector Control Assoc. 49:45-48.

Mulla, M.S. 1985. Field evaluation and efficacy of bacterial agents and their formulations against mosquito larvae, p. 227-250. *In* M. Laird and J.W. Miles (eds.), Integrated mosquito-control methodologies, Volume 2. Academic Press, London.

Mulla, M.S. 1986. Efficacy of the microbial agent *Bacillus sphaericus* Neide against mosquitoes (Diptera: Culicidae) in southern California. Bull. Soc. Vector Ecol. 11:247-254.

Mulla, M.S. In Press. Activity, field efficacy and use of *Bacillus thuringiensis* H-14 against mosquitoes. *In* H. de Barjac and D.J. Southerland (eds.), Bacterial control of mosquitoes and black flies: biochemistry, genetics, and applications of *Bacillus thuringiensis* and *Bacillus sphaericus*. Rutgers University Press, New Brunswick, New Jersey.

Mulla, M.S., H. Axelrod, H.A. Darwazeh and B.A. Matanmi. 1988. Efficacy and longevity of *Bacillus sphaericus* 2362 formulations for control of mosquito larvae in dairy wastewater lagoons. J. Am. Mosq. Control Assoc. 4:448-452.

Mulla, M.S., H.A. Darwazeh and C. Aly. 1986. Laboratory and field studies on new formulations of two microbial control agents against mosquitoes. Bull. Soc. Vector Ecol. 11:255-263.

Mulla, M.S., H.A. Darwazeh, E.W. Davidson and H.T. Dulmage. 1984a. Efficacy and persistence of the microbial agent *Bacillus sphaericus* against mosquito larvae in organically enriched habitats. Mosq. News 44:166-173.

Mulla, M.S., H.A. Darwazeh, E.W. Davidson, H. Dulmage and S. Singer. 1984b. Larvicidal activity and field efficacy of *Bacillus sphaericus* strains against mosquito larvae and their safety to non-target organisms. Mosq. News 44:336-347.

Mulla, M.S., B.A. Federici and H.A. Darwazeh. 1982. Larvicidal efficacy of *Bacillus thuringiensis* serotype H-14 against stagnant-water mosquitoes and its effects on non-target organisms. Environ. Entomol. 11:788-795.

Mulley, R.C., J.R. Egerton, A.W. Sweeney and W.J. Hartley. 1981. Further tests in mammals, reptiles and an amphibian to delineate the host range of the mosquito fungus *Culicinomyces* sp. Mosq. News 41:528-531.

Mulligan, F.S., III and C.H. Schaefer. 1982. Integration of a selective mosquito-control agent *Bacillus thuringiensis* serotype H-14, with natural predator populations in pesticide-sensitive habitats. Proc. Pap. 49th Annu. Conf. Calif. Mosq. Vector Control Assoc. 49:19-22.

Mulligan, F.S., III., C.H. Schaefer and T. Miura. 1978. Laboratory and field evaluation of *Bacillus sphaericus* as a mosquito-control agent. J. Econ. Entomol. 71:774-777.

Novak, R.J., D.J. Gubler and D. Underwood. 1985. Evaluation of slow-release formulations of Abate and *Bacillus thuringiensis* for the control of *Aedes aegypti* in Puerto Rico. J. Am. Mosq. Control Assoc. 1:449-453.

Ohba, M. and K. Aizawa. 1979. A new subspecies of *Bacillus thuringiensis* H-14 possessing 11a: 11b flagellar autigenic structure: *Bacillus thuringiensis* subsp. *kyushuensis*. J. Invertebr. Pathol. 33:387-388.

Olejnicek, J. and B. Maryskova. 1986. The influence of *Bacillus thuringiensis* var. *israelensis* on the mosquito predator *Notonecta glauca*. Folia Parasitol. 33:279-280.

Panter, C. and R.C. Russell. 1984. Rapid kill of mosquito larvae by high concentrations of *Culicinomyces clavisporus* conidia. Mosq. News 44:242-244.

Payne, C.C. 1988. Pathogens for the control of insects: where next? Philos. Trans. R. Soc. Lon. B 318:225-248.

Purcell, B.H. 1981. Effects of *Bacillus thuringiensis* var. *israelensis* on *Aedes taeniorhynchus* and some non-target organisms in the salt marsh. Mosq. News 41:476-484.

Reish, D.J., J.A. LeMay and S.L. Asato. 1985. The effect of BTI (H-14) and methoprene on two species of marine invertebrates from southern California estuaries. Bull. Soc. Vector Ecol. 10:20-22.

Rishikesh, N., H.D. Burges and M. Vandekar. 1983. Operational use of *Bacillus thuringiensis* serotype H-14 and environmental safety. WHO/VBC/83.871, World Health Organization, Geneva, Switzerland.

Roberts, D.W. and C. Panter. 1985. Fungi other than *Coelomomyces* and *Lagenidium*, p. 99-109. *In* H.C. Chapman (ed.), Biological control of mosquitoes. Am. Mosq. Control Assoc. Bull. No. 6, Fresno, California.

Russell, R.C., C. Panter and P.I. Whelan. 1983. Laboratory studies on the pathogenicity of the mosquito fungus *Culicinomyces* to various species in their natural waters. Gen. Appl. Entomol. 15:53-63.

Schnetter, W., S. Engler, J. Morawcsik and N. Becker. 1981. Wirksamkeit von *Bacillus thuringiensis* var. *israelensis* gegen Stechmuckenlarven und Nontarget-Organismen. Mitt. Dtsch Ges. allg. angew. Entomol. 2:195-202.

Sebastien, R.J. and R.A. Brust. 1981. An evaluation of two formulations of *Bacillus thuringiensis* var. *israelensis* for larval mosquito-control in sod-lined simulated pools. Mosq. News 41:508-511.

Siegel, J.P. and J.A. Shadduck. 1987. Some observations of the safety of the entomopathogen *Lagenidium giganteum* to mammals. J. Econ. Entomol. 80:994-997.

Singer, S. 1980. *Bacillus sphaericus* for the control of mosquitoes. Biotech. Bioengin. 22:1335-1355.

Singer, S. 1985. *Bacillus sphaericus* (Bacteria), p. 123-131. *In* H.C. Chapman (ed.), Biological control of mosquitoes. Am. Mosq. Control Assoc. Bull. No. 6, Fresno, California.

Su, X., D.R. Guzman and R.C. Axtell. 1986. Factors affecting storage of mycelial cultures of the mosquito fungal pathogen *Lagenidium giganteum* Couch (Oomycetes: Lagenidiales). J. Am. Mosq. Control Assoc. 2:350-354.

Sweeney, A.W. 1975. The insect pathogenic fungus *Culicinomyces* in mosquitoes and other hosts. Aust. J. Zool. 23:59-64.

Sweeney, A.W. 1979a. Infection of mosquito larvae by *Culicinomyces* sp. through anal papillae. J. Invertebr. Pathol. 33:249-251.

Sweeney, A.W. 1979b. Further observations on the host range of the mosquito fungus *Culicinomyces*. Mosq. News 39:140-142.

Sweeney, A.W. 1981. Preliminary field tests of the fungus *Culicinomyces* against mosquito larvae in Australia. Mosq. News 41:470-476.

Sweeney, A.W. 1983. The time-mortality response of mosquito larvae infected with fungus *Culicinomyces*. J. Invertebr. Pathol. 42:162-166.

Sweeney, A.W., R.D. Cooper, B.E. Medcraft, R.C. Russell, M. O'Donnell and C. Panter. 1983. Field tests of the mosquito fungus *Culicinomyces* against the Australian encephalitis vector *Culex annulirostris*. Mosq. News 43:290-297.

Sweeney, A.W., S.L. Doggett and G. Gullick. 1989. Laboratory experiments on infection rates of *Amblyospora dyxenoides* (Microsporidia: Amblyosporidae) in the mosquito *Culex annulirostris.* J. Invertebr. Pathol. 53:85-92.

Sweeney, A.W., M.F. Graham and E.I. Hazard. 1988. Life cycle of *Amblyospora dyxenoides* sp. nov. in the mosquito *Culex annulirostris* and the copepod *Mesocyclops albicans.* J. Invertebr. Pathol. 51:46-57.

Sweeney, A.W., E.I. Hazard and M.F. Graham. 1985. Intermediate host for an *Amblyospora* sp. (Microspora) infecting the mosquito *Culex annulirostris.* J. Invertebr. Pathol. 46:98-102.

Sweeney, A.W., D.J. Lee, C. Panter and L.W. Burgess. 1973. A fungal pathogen for mosquito larvae with potential as a mosquito insecticide. Search 4:344-345.

Sweeney, A.W. and C. Panter. 1977. The pathogenicity of the fungus *Culicinomyces* to mosquito larvae in a natural field habitat. J. Med. Entomol. 14:495-496.

Undeen, A.H. and N.E. Alger. 1975. The effects of the microsporidian, *Nosema algerae,* on *Anopheles stephensi.* J. Invertebr. Pathol. 22:258-265.

Undeen, A.H. and L.A. Lacey. 1982. Field procedures for the evaluation of *Bacillus thuringiensis* var. *israelensis* (serotype 14) against black flies (Simuliidae) and nontarget organisms in streams. Misc. Publ. Entomol. Soc. Am. 12:25-30.

Van Essen, F.W. and S.C. Hembree. 1982. Simulated field studies with four formulations of *Bacillus thuringiensis* var. *israelensis* against mosquitoes: residual activity and effect of soil constituents. Mosq. News 42:66-72.

Washburn, J.O. and J.R. Anderson. 1986. Distribution of *Lambornella clarki* (Ciliophora: Tetrahymenidae) and other mosquito parasites in California treeholes. J. Invertebr. Pathol. 43:296-309.

Washburn, J.O., D.E. Egerter, J.R. Anderson and G.A. Simmons. 1988. Density reduction in larval mosquito (Diptera: Culicidae) populations by interactions between a parasitic ciliate (Ciliophora: Tetrahymenidae) and an opportunistic fungal (Oomycetes: Pythiaceae) parasite. J. Med. Entomol. 25:307-314.

Whisler, H.C., S.L. Zebold and J.A. Schemanchuk. 1974. Alternate host for the mosquito parasite *Coelomomyces.* Nature (London) 251:715-716.

Whisler, H.C., S.L. Zebold and J.A. Shemanchuk. 1975. Life history of *Coelomomyces psorophorae.* Proc. Nat. Acad. Sci. (USA) 72:693-696.

WHO. 1985. Informal consultation on the development of *Bacillus sphaericus* as a microbial control larvicide. TDR/BCV/ *SPHAERICUS*/85.3, World Health Organization, Geneva, Switzerland.

Zaritsky, A. and K. Khawaled. 1986. Toxicity in carcasses of *Bacillus thuringiensis* var. *israelensis*-killed *Aedes aegypti* larvae against scavenging larvae: implications to bioassay. J. Am. Mosq. Control Assoc. 2:555-559.

Distribution of Microbial Agents in Marine Ecosystems as a Consequence of Sewage-Disposal Practices

THOMAS K. SAWYER

Abstract: Microbial agents of human or animal origin, and present in water and sediment, are reliable indicators of sewage pollution in coastal marine waters. Certain viruses, bacteria, fungi, and protozoans commonly present in sewage wastes may survive for periods of days to years, and serve as useful indicators for short- or long-term monitoring purposes. Fecal coliform bacteria, for example, are indicative of recent sewage pollution while spore-forming *Clostridia* serve as indicators of both recent or long-term accumulation of sewage wastes. Viruses belonging to the polio, echo, and cocksackie groups also serve as indicators of sewage pollution in coastal and offshore environments. Spores of terrestrial fungi, including slime molds, may fail to grow in marine sediments but, nevertheless, remain viable for long periods and serve as indicators of sewage and dredge-spoil contamination. Cyst-forming freshwater and soil amoebae belonging to *Acanthamoeba, Hartmannella,* and *Vahlkampfia* have been cultured from sewage contaminated sediments. The amoebae are especially useful since they are unaffected by sediment type, temperature, salinity, or water depth. Nutrient enrichment brought about by wastewater and solid wastes may stimulate the growth of native coastal plant and animal species to population densities that upset the balance of natural community structures, *i.e.*, algal blooms, dinoflagellate blooms, etc. Further studies are needed to better evaluate and manage events that lead to outbreaks of disease, stress, and anoxic bottom water conditions.

Introduction

Sewage pollution of coastal and offshore waters often leads to the closure of commercially valuable shellfish beds and recreational waters. Furthermore, there is increasing evidence that diseases of fish and shellfish are significantly higher in polluted environments than in those otherwise judged to be clean. Efforts to monitor the introduction of microbial agents into marine ecosystems have included studies on enteric viruses (Goyal 1989), bacteria (Babinchak et al. 1977; Cabelli et al. 1982; Grimes & Colwell 1989), and pathogenic protozoa (Sawyer et al. 1982, 1987, 1989; O'Malley et al. 1982). Sindermann (1972) and others have published extensively on microbial diseases of fish and shellfish. Attempts to measure the effects of environmental pollution on animal health include efforts to determine the cause or causes of periodic mass mortalities, changes in species composition and abundance, and the prevalence of disease or mortality (Sindermann 1972). Other concerns include the sanitary quality of seafood and seafood products, alterations of benthic habitats, closure of shellfish-producing areas, and contamination of coastal mariculture facilities with increased land use.

The present report includes a brief overview of some of the known effects of sewage pollution on fish and shellfish, and briefly summarizes unpublished studies on the distribution of potentially pathogenic amoebae in three coastal ecosystems, Cape Cod, Massachusetts, Hempstead Bay, Long Island, New York, and the Yaquina River, Oregon. The amoebae belong to the genus *Acanthamoeba,* form highly resistant cysts, and have been cultured from municipal sewage sludge. Several species survive and grow at temperatures of 37°C or higher and have been identified from patients suffering from brain disease or eye infections. Results of sediment studies are discussed with respect to the frequent association of *Acanthamoeba* with enteric bacteria recovered from bottom sediments.

Methods

Nineteen shoreline sediment samples were collected in Massachusetts (Cape Cod, Nantucket Island, and Plymouth) in September 1985 (Tables 1 and 2); 17 samples were taken in Hempstead Bay, Long Island, New York, in May 1984 (Tables 3 and 4), and 11 were taken from the Yaquina River and Yaquina Bay, Oregon, in September 1984 (Tables 5 and 6). Long Island

Table l. Stations sampled for *Acanthamoeba* in Massachusetts, September 1985.

No.–Location	Type	Comment
1 - Herring River, Dennisport	Coastal river	Closed to shellfishing
2 - Hyannis	Marina ramp	Sand and duckweed
3 - Nantucket Island	Sandy beach	Closed to swimming
4 - Nantucket Island	Sandy beach	Sewage outfall pipe
5 - Nantucket Island	Beach bulkhead	Sand and seaweed
6 - Mashpee River	Coastal river	Polluted river
7 - Chatham	Sandy beach	Residential inlet
8 - Chatham	Boat ramp	Town harbor
9 - Orleans	River shoreline	Closed to shellfishing
10 - Wellfleet	Boat ramp	Town harbor
11 - Provincetown	Sandy beach	Race beach
12 - Provincetown	Harbor/beach	Town beach
13 - Nickerson Pond	Freshwater	Campground
14 - Barnstable	Harbor	Bulkhead sand/mud
15 - Sandwich	Dry salt marsh	Sand dunes
16 - Plymouth Rock	Plymouth memorial	Washed sand
17 - Plymouth Rock	Boat ramp	At restaurant
18 - Falmouth	Boat ramp	Town harbor
19 - Woods Hole	Boat ramp	Town harbor

Table 2. Distribution of *Acanthamoeba* in sediments of Cape Cod, Massachusetts, September 1985 (*=potential pathogens).

No. -Location	Positive Cultures	Species Identified
	Sandy Beaches with Wave Action	
3-Nantucket	1/6	*A. polyphaga, A. hatchetti**, *A. terricola*
4-Nantucket	1/6	*A. polyphaga*
7-Chatham	1/6	*A. hatchetti**
11-Provincetown — Ocean	0/6	-
Positive Cultures	3/24 (13%)	
	Harbors (Marinas and Ramps)	
2-Hyannis	4/6	*A. polyphaga, A. hatchetti**, *A. rhysodes**, *A. tubiashi, A.* sp.*
5-Nantucket	2/6	*A. polyphaga, A. hatchetti**, *A.* sp.*
10-Wellfleet	6/6	*A. polyphaga, A. hatchetti**, *A.* sp.* *A. castellanii, A. terricola*
12-Provincetown	1/6	*A. polyphaga, A. hatchetti**, *A.* sp.*
16-Plymouth Rock	2/6	*A. polyphaga, A. hatchetti**, *A. lenticulata*
17-Plymouth — Ramp	2/6	*A. polyphaga, A. hatchetti** *A. rhysodes**
18-Falmouth	4/6	*A. polyphaga, A. hatchetti**, *A. astronyxis*
19-Woods Hole	4/6	*A. polyphaga, A. hatchetti**, *A.* sp.* *A. rhysodes**, *A. lenticulata**
Positive Cultures	25/48 (52%)	
	Rivers and Creeks — Running Inland from Coast	
1-Dennisport — Herring R.	6/6	*A. polyphaga, A. hatchetti** *A. rhysodes**, *A. castellanii*
6-Mashpee River	5/6	*A. polyphaga, A. hatchetti**, *A.* sp.*
8-Chatham	1/6	*A. polyphaga, A. hatchetti**
9-Orleans	5/6	*A. polyphaga, A. hatchetti**, *A. rhysodes**, *A.* sp.*
14-Barnstable — Marsh	3/6	*A. polyphaga, A. hatchetti**, *A. rhysodes**
15-Sandwich — Marsh	6/6	*A. polyphaga, A. hatchetti**
Positive Cultures	26/36 (72%)	
	Freshwater Pond	
13-Nickerson Pond	4/6	*A. polyphaga, A. hatchetti**, *A. rhysodes**, *A. castellanii, A.* sp.*
Total — all positive cultures = 58/114 (51%)		

Table 3. Stations sampled for *Acanthamoeba* in Hempstead, Long Island, New York, May 1984.

No.-Location	Type	Comment
188-Point Lookout	Bay	On shellfish closure line
123-Meadow Island to Jones Island	Bay	On shellfish closure line
127-Deep Creek Meadow	Bay	Near closure line
130-Jones Beach Causeway	Bay	At sewage treatment plant
20W-South Line Island	Bay	Near closure line
176-Great Island — East Bay	Bay	Near closure line
31.2-North Line Island	Bay	Near closure line
34.2-Squaw Island — South Oyster Bay	Bay	Near closure line
173-Broad Creek Island	Bay	Closed area
171-West Crow Island	Bay	Closed area
JC-Jones Creek	Creek*	Shallow water
CC- Carman Creek	Creek*	Shallow water
NC-Narraskatuck Creek	Creek*	Shallow water
MC-Massapequa Creek	Creek*	Shallow water
SC-2-Seaford Creek	Creek*	Houses on pilings
IC-Island Creek	Creek*	Shallow water

*Shallow water creek sediments were brown to black mud and silt.

Table 4. Distribution of *Acanthamoeba* in sediments of Hempstead Bay, Long Island, New York, May 1984 (*=potential pathogens).

No. -	Location	Positive Cultures	Species Identified
	Open Bay Stations		
20W -	South Line Island	2/6	*A. polyphaga, A. hatchetti**
31.2 -	North Line Island	0/6	-
34.2 -	Squaw Island	2/6	*A. polyphaga, A. hatchetti*, A. rhysodes*, A. castellanii*
118 -	Point Lookout**	0/6	-
123 -	Meadow Island to Jones Island**	0/6	-
127 -	Deep Creek Meadow	0/6	-
130 -	Jones Beach Causeway Sewage Rx Plant	1/6	*A. hatchetti*, A. rhysodes**
171 -	West Crow Island	0/6	-
173 -	Broad Creek Island**	2/6	*A. polyphaga, A. hatchetti* A. rhysodes*, A. lenticulata**
176 -	Great Island	1/6	*A. polyphaga, A. rhysodes**
Positive Cultures		8/60 (13%)	
	Inshore Creek Stations		
CC -	Carman Creek	2/6	*A. hatchetti**
IC -	Island Creek	6/6	*A. polyphaga, A. hatchetti*, A. rhysodes*, A. lenticulata**
IC-2 -	Belmore Creek	4/6	*A. polyphaga, A. hatchetti*, A. rhysodes*, A. lenticulata**
JC -	Jones Creek	5/6	*A. polyphaga, A. hatchetti*, A. rhysodes*, A. castellanii, A. lenticulata*, A. astronyxis*
MC -	Massapequa Creek	1/6	*A. polyphaga, A. hatchetti*, A. lenticulata**
NC -	Narrashatucka Creek	5/6	*A. polyphaga, A. hatchetti*, A. rhysodes**
SC-2 -	Seaford Creek	4/6	*A. polyphaga, A. hatchetti*, A. rhysodes*, A. castellanii, A. astronyxis*
Positive Cultures		27/42(64%)	

**Bay stations located in closed areas or on shellfish closure lines.

Table 5. Stations sampled for *Acanthamoeba* in Yaquina River/Bay, Oregon, September 1984.

No.-Location	Type	Comment
1 - Above Toledo	Boat ramp	Campground near Toledo
2 - Mill Creek at Toledo	Narrow creek	Junction with Yaquina River
3 - Toledo	Dock	Across from sewage lagoons
4 - Toledo	Dock	Center of town below lagoons
5 - Toledo	Shoreline	Below city limits
6 - Yaquina River	Shoreline	Halfway to Newport
7 - Yaquina River	Shoreline	At Queen Oyster Co.
8 - Yaquina Bay	Shoreline	Boat marina, camping trailers
9 - Yaquina Bay	Shoreline	Sawyer's Landing-marina
10 - Port of Newport	Bulkhead	Yacht basin
11 - Yaquina Bay	Shoreline	At EPA/NOAA building

Note: The Toledo sewage treatment plant was designed for 1 MGD (million gallons per day) to serve approximately 3,000 persons. Stations were sampled along a transect from above Toledo (freshwater) down river to Newport, Oregon (seawater from Pacific Ocean).

Table 6. Distribution of *Acanthamoeba* in sediments from Yaquina River and Yaquina Bay, September, 1984 (*=potential pathogens).

No. - Location	Positive Cultures	Species Identified
	Low salinity river stations (<10 o/oo)	
1 - Above Toledo	5/6	*A. polyphaga, A. hatchetti**, *A. tubiashi, A.* sp.*
2 - Mill Creek — at Toledo	3/6	*A. polyphaga, A. hatchetti**, *A.* sp.*
3 - Toledo — boat dock	6/6	*A. polyphaga, A. hatchetti**, *A. tubiashi*
4 - Toledo — town dock	5/6	*A. polyphaga, A. hatchetti**
5 - Toledo — below town	6/6	*A. polyphaga, A. hatchetti**, *A. castellanii*
Positive Cultures	25/30 (83%)	
	Mid- to High-Salinity River/Bay Stations (11-30 ppt)	
6 - Yaquina R. — buoy #37	5/6	*A. polyphaga, A. hatchetti**, *A. tubiashi, A. castellanii*
7 - Queen Oyster Co.	5/6	*A. polyphaga, A. hatchetti**, *A. rhysodes**
8 - Yaquina R. — marina	6/6	*A. polyphaga, A. hatchetti**, *A. rhysodes**
9 - Sawyer's Landing	1/6	*A. polyphaga*
10 - Port of Newport	4/6	*A. polyphaga, A. hatchetti**
11 - Yaquina Bay — mouth	5/6	*A. polyphagha, A. hatchetti**
Positive Cultures	26/36 (72%)	

stations included samples from Hempstead Bay, Great South Bay, and South Oyster Bay and are referred to collectively as Hempstead Bay stations. Shoreline sediments were collected with sterile wooden tongue depressors, placed in sterile plastic dishes, and stored under refrigeration until streaked on 6 replicate agar plates as described by Sawyer and Bodammer (1983); deepwater Long Island stations were sampled with a Ponar bottom grab. Cultures positive for *Acanthamoeba* were subcultured and incubated at 37-39°C to test for ability to grow at mammalian body temperature. Bacteriological most probable numbers (MPNs) were determined for Hempstead Bay seawater at the same time that bottom sediments were collected, and 4 months before the Oregon sediments were sampled. Similar determinations were not made on the Massachusetts samples. Bacteriological data were provided by personnel at the Food and Drug Administration, Northeast Technical Service Unit, North Kingston, Rhode Island. Water column salinity was not determined during the sampling periods. However, FDA personnel measured salinity in May 1984, several months prior to the sediment studies in Oregon in September 1984.

Results

Massachusetts Bottom Sediments. Among the 19 stations sampled in Massachusetts, 14 were from Cape Cod and included sandy beaches, harbors, and boat ramps, saline rivers and creeks, and a freshwater pond (Nickerson Pond); all 3 Nantucket stations were sandy beaches, and 2 Plymouth sediments were from a boat ramp and from behind Plymouth Rock (Table 1). Sandy beaches subject to wave and wind action yielded the smallest number of positive cultures (13%), followed by harbors, marinas, and ramps (52%), and by rivers and creeks (72%) (Table 2). The Plymouth Rock station yielded 2/6 positive cultures, and Nickerson Pond yielded 4/6 (Table 2). One sandy beach station on Nantucket Island was closed to public bathing, and a station at Orleans was closed for shellfishing except for scallops. The open sandy beach at Race Beach, Provincetown, was the only station negative for amoebae. Nine species of *Acanthamoeba* were identified from the Massachusetts sediments.

Hempstead Bay, New York, Bottom Sediments. Among 17 stations sampled in the Hempstead Bay area, 10 were from open waters of the Bay and 7 were from shallow water creeks flowing into Long Island residential areas (Table 3). Bay stations had the smallest number of positive cultures (13%), while creek stations the highest (64%). Only 5 of 10 Bay stations were positive for amoebae while all 7 creek stations were positive (Table 4). Bacterial MPNs taken from the Bay stations over a 10-day period (May 1-10, 1984) showed that all stations were contaminated by coliform and fecal coliform bacteria at one time or another. All Bay stations were located within a half mile of shellfish closure lines in Great South Bay, South Oyster Bay, and Hempstead Bay. Seven of the 10 stations exceeded acceptable total coliform levels for harvesting shellfish, and 3 exceeded acceptable levels for fecal coliforms. Fecal coliform MPNs of over 100 were rare in Bay waters. Creek stations were not sampled over the full 10-day period but showed a maximum total coliform MPN of 35,000 and a maximum fecal coliform MPN of 3,300. The number of cultures yielding amoebae from creek stations (27/42), compared to Bay stations (8/60), showed that there was an association between bacterial MPNs and the frequency with which stations were positive for amoebae. Seven species of *Acanthamoeba* were identified from the Long Island, New York, stations.

Yaquina River and Bay, Oregon, Bottom Sediments. Eleven sediment samples were taken from the Yaquina River system in September 1984. Stations were located several miles above the town of Toledo and extended to the mouth of the river at Newport, Oregon, on the Pacific coast (Table 5). Salinity determinations provided by FDA showed that in May 1984, salinities at high tide ranged from < 1.0 ppt at Toledo to 30.5 ppt at the mouth of the Bay. Sources of pollution were considered to be livestock, sporadic housing along the shoreline, minor tributaries, and the

Toledo sewage treatment plant. All stations along the river and bay yielded one or more species of *Acanthamoeba* and 77% of the cultures were positive (Table 6). Seven of 11 stations had temperature-tolerant potentially pathogenic species. Bacteriological data provided by FDA showed that among 12 stations sampled at intervals from Toledo to the mouth of the Bay, maximum fecal coliform MPNs at high tide ranged from 7.8–240. The association between fecal bacteria and the presence of potentially pathogenic amoebae was particularly evident in the Oregon sediments; 10 of the 11 stations yielded amoebae from 3-5 cultures in each set of 6 replicates. Seven species of *Acanthamoeba* were identified from the Oregon stations (Table 6).

Discussion

Studies reported here further document the value of cyst-forming species of *Acanthamoeba* as indicators for inland or coastal sources of sewage wastes that often lead to the closure of commercially valuable shellfish beds and/or recreational waters. The amoebae are of further interest since they may be found in highly polluted environments distant from sources of sewage wastes and negative for culturable fecal coliforms (Sawyer et al. 1989). Sites for the present studies were selected to compare (1) a northeastern shoreline that supports commercially valuable shellfish, (2) mid-Atlantic deeper water bays supporting harvestable numbers of shellfish, and (3) a Pacific coast river system ranging from fresh to salt water and supporting a commercially valuable and recreational shellfish population. The three geographically separated sites have areas that are open to shellfishing, temporarily closed during rainy seasons, or permanentaly closed due to sewage contamination. Recovery rates for *Acanthamoeba* (% positive cultures), and species diversity were found to vary with station depth and the severity of contamination by sewage bacteria. Sediment characteristics also were found to influence the recovery of amoebae. Shellfish closure areas in Hempstead Bay with high water column bacterial MPNs sometimes were negative for both amoebae and bacteria when bottom sediments were silt-free hard sands subject to the scouring action of tides and currents.

Dispersal of Potentially Pathogenic Amoebae. Species of *Acanthamoeba* that do not grow at mammalian body temperature include *A. polyphaga, A. castellanii, A. astronyxis,* and *A. terricola,* and they are almost universally distributed in soil, water, dust, leaf mold, river mud, etc. By contrast, temperature-tolerant species capable of causing disease in humans or animals, including *A. culbertsoni, A. hatchetti, A. rhysodes, A. lenticulata,* and *A.* sp., are most often isolated from municipal sewage sludge, sewage contaminated water or sediment, or thermally-polluted water. Extensive sediment studies in the Gulf of Mexico and New York Bight at stations away from sources of sewage or dredge-spoil pollution were consistently negative for *Acanthamoeba* (Sawyer 1980). By contrast, sediments taken at the New York 12-mile sewage disposal site, the 4-mile dredge spoil site, and the 40-mile Philadelphia-Camden site (PDS) were routinely positive for the amoebae (Sawyer 1980; O'Malley et al. 1982; Sawyer et al. 1982). Four strains of *Acanthamoeba* recovered from the New York or PDS disposal site were found to kill experimentally infected laboratory mice (Daggett et al. 1982). Bacteriological studies in the New York Bight apex (Babinchak et al. 1977) and at the PDS (O'Malley et al. 1982) showed that both sites were contaminated with sewage-associated coliform bacteria; stations with the highest coliform MPNs were either closed to shellfishing or failed to support shellfish growth. Unpublished follow-up studies have shown that sediments from the PDS were negative for fecal coliform bacteria within one year after the cessation of ocean dumping, and negative for *Acanthamoeba* three years after the cessation of dumping. Thus, the association between sewage contamination of the sea bottom, and the presence of pathogenic amoebae was demonstrated at both the nearshore New York disposal site and the offshore Philadelphia-Camden site.

Results of the present study showed that resistant, cyst-forming species of *Acanthamoeba* are excellent indicators for the severity of estuarine and coastal sewage contamination on both east and west coasts of the United States. *Acanthamoeba polyphaga* is recognized as a widely distributed species in both clean and contaminated sediments while temperature-tolerant strains of *A. hatchetti, A. lenticulata,* and *Acanthamoeba* sp., and others are routinely found in environments subject to recent thermal and/or sewage contamination.

Enteric Viruses Associated with Sewage Pollution. Goyal et al. (1984) found human enteric viruses (cocksackie, echo, polio), and unidentified viruses, in sediments from New York Bight and PDS stations that also were positive for *Acanthamoeba* and/or enteric bacteria. Goyal (1989) also found cocksackie B3 virus at the PDS 17 months after the cessation of dumping, indicating that this virus may survive in the environment for longer periods of time than other enteric viruses. Earlier studies by Goyal et al. (1979) and LaBelle et al. (1980) showed that enteric viruses could be isolated from waters that met microbiological standards for harvesting shellfish, and from shellfish when water samples were negative for virus. They emphasized the fact that bacteriological standards for determining the safety of shellfish for human consumption do not fully consider the occurrence of enteroviruses. Lewis et al. (1985) studied sediments and beach sands near a sewer outfall in Australia and also recovered polio 2, coxsackie B4 and B5, and other unidentified viruses. They did not always find a significant relationship between enteric bacteria and enteric viruses, and reported that fecal coliforms are not always reliable indicators of viral pollution. Lewis et al. (1986) also found that viruses could be recovered from sediments 19.5 km downstream from a sewage outfall during periods of rainfall, but for only 4 km during periods of normal flow. Human enteric viruses, fecal bacteria and potentially pathogenic amoebae are of concern to human health and are associated with considerable economic loss when harvestable waters do not meet sanitary standards. Dispersal of the pathogens is widespread in shallow coastal bays and rivers but may be limited in offshore waters as influenced by ocean disposal practices.

Enteric Bacteria Associated with Sewage Pollution. Babinchak et al.(1977) made extensive studies on the distribution of fecal bacteria at or near the New York 12-mile sewage disposal site and reported fecal coliform MPNs of over 100 at 45/87 test stations; similar MPNs were recorded 37 km from the center of the dumpsite. O'Malley et al. (1982) recovered total coliforms and fecal coliforms 37 km to the northeast and southwest of the center of the Philadelphia-Camden ocean disposal site, and concluded that an area of 1,190 km^2 was affected by ocean disposal practices. The dispersal of enteric bacteria in waters of the open ocean is influenced by tides, currents, storms, and winds and the size of particulate matter of bottom sediments. Dispersal of fecal coliforms away from shoreline outfalls has been reported to vary from 5-8 km.

Yde and deMaeyer-Cleempoel (1980) found indicator fecal bacteria at a distance of 5 km from the Belgian coastline with no significant differences occurring at water depths of 7-33 m. Loutit and Lewis (1985) collected water and sediment samples from a 50 km^2 area around a coastal sewage outfall in New Zealand and recovered fecal bacteria up to 8 km from the outfall. They found the distribution of bacteria to be affected by currents and wind, and recovered bacteria from sediments even when water samples were negative. Grimes and Colwell (1989) studied wastewater from a treatment plant near Barceloneta, Puerto Rico, that discharged pharmaceutical and chemical wastes, and wastes from a food-processing plant. They recovered *Salmonella, Vibrio, Shigella,* and *Campylobacter,* and reported a fecal coliform index of 1.2×10^8 100 ml^{-1}, with the highest numbers at 10-40-m depth near the mouth of the outfall. They found that the numbers of fecal bacteria were high enough to close the coastal waters to swimming, surfing, shellfish, and other recreational activities. Most studies on the dispersal of sewage-associated bacteria have employed total coliforms, fecal coliforms, and fecal streptococci as indicator species. Ongoing studies with more resistant spore-forming bacteria such as *Clostridium perfringens,* and viable but non-culturable *Eschericia coli,* however, clearly show that the dispersal of pathogens away from their source is more extensive than presently appreciated.

Dispersal of Other Sewage-Associated Microorganisms. Crewe and Owen (1981) provided a list of pathogens, including protozoans, trematodes, cestodes, and nematodes of human and veterinary interest found in sewage from England, Wales, and Ireland. They concluded that most of the parasites are of little concern to human health except in areas of activity near sources of sewage discharges or outfalls. Burge and Marsh (1978) listed some of the major disease-causing pathogens found in sewage, including bacteria, viruses, protozoa, and intestinal helminths. They stated that most disease outbreaks involve night soil, raw sewage, or wastewaters used in growing vegetables to be eaten raw (typhoid fever, cholera, ascariasis, amoebiasis, dysentery, enteric fevers, etc.). Intestinal protozoans and helminths do not appear to be of major concern in coastal waters of the United States, but very little is known of their role in human disease in underdeveloped countries. Kott and Kott (1970) studied the viability of *Entamoeba histolytica* cysts exposed to seawater and found that they only survived for several days. The mean number of cysts was 42 in 10 liters for influent samples, and 7 in 10 liters in effluents. The authors suggested that dilution factors are sufficient to minimize the hazard of amoebic infection in bathers or swimmers; however, there is increasing concern for potential pathogens that may cause serious health problems in persons with deficient immune systems or other predisposing conditions.

Stewart and Brown (1969) isolated a myxobacterium (*Cytophaga* N-5) from sewage that killed green and blue green algae. Goldstein (1973) discussed the physiology, ecology, and taxonomy of thraustochytrid marine Phycomycetes are reportedly common in polluted littoral waters, including certain species that could be isolated from hydrogen sulfide-rich sediments. He also reviewed some of the literature concerning the recovery of a herpes-like virus from an estuarine *Thraustochytrium* species, and the recovery of virus-like particles from the oyster pathogen, *Labyrinthomyxa marina*. Sawyer and Meyer (1977) recovered a marine yeast-like fungus, *Sterigmatomyces halophilus*, from the mantle fluid of the Japanese oyster, *Crassostrea gigas*. The fungus previously was recovered from air samples at Key Biscayne, Florida, from the Indian Ocean at depths ranging up to 1,997 m, and from a case of human keloid blastomycosis in Brazil. Atwell and Colwell (1981) recovered a variety of terrestrial Acinomycetes from sediments in New York Harbor and the New York Bight. The authors attributed unusually high bacterial counts in ocean sediments to dredge-spoil disposal practices, and found that certain species were useful markers for the dispersal of such spoils. Burge and Marsh (1978) have discussed the public-health significance of some fungi and thermophilic actinomycetes present in sludge and sewage composts. Thus, there are a variety of microorganisms other than viruses, bacteria, and protozoans that are useful for estimating the dispersal of sewage wastes away from outfalls or ocean disposal sites.

Effects of Sewage-Associated Pathogens on Fish and Shellfish

Studies on wild-caught fish and shellfish, as well as those raised in aquaculture or mariculture facilities, have shown that sewage-associated bacteria may be responsible for serious economic losses. *Edwardsiella tarda,* a cause of enteritis in humans, is known to cause disease in both eels and fish (Wakabayashi and Egusa 1973; Kusada et al. 1976). Amandi et al. (1982) isolated *E. tarda* from fall chinook salmon, *Oncorhynchus tshawytscha,* in Oregon. The authors stated that the bacteria have been isolated from a variety of mammals, birds, reptiles, and amphibians, including sea gulls and sea mammals along the Oregon coast. Thus, fecal matter from shore-dwelling birds and mammals, as well as from sources of sewage pollution, must be considered when investigating outbreaks of disease or mortality.

Disease and mortality in coho salmon, *O. kisutch,* and other salmonids due to *Clostridium botulinum,* has been reported on the Pacific coast (Huss and Eskildsen 1974; Eklund et al. 1984), and serological studies have shown that fish may demonstrate antibodies to a variety of pathogenic bacteria. Janssen and Meyers

(1968) found antibodies to *Pasteurella pestis, P. pseudotuberculosis, Salmonella paratyphi A, Shigella flexneri, Proteus vulgaris, Pseudomonas aeruginosa, Escherichia coli B, Aeromonas hydrophila,* and *A. shigelloides,* in white perch, *Morone americanus,* captured in Chesapeake Bay, Maryland. Robohm et al. (1979) tested three species of marine fish captured in the New York Bight, finding that antibody titers against several bacterial species in two of them were higher in polluted waters than in unpolluted waters. Results of their studies suggested that in the polluted New York Bight apex had an increased level and diversity of bacteria during warm summer months. Fin rot disease in the New York Bight apex also was highest in fish captured in geographical areas located near major metropolitan domestic and industrial development (Ziskowoski et al. 1987).

Significant acreage along the shorelines and coasts has been closed to shellfishing (clams, oysters, scallops) because of sewage pollution and associated human pathogens. Leonard and Harkness (this volume) report that 52% of the mid-Atlantic harvest-limited waters are affected by wastewater treatment plants; 44% of the shellfish growing waters near seaports at Charleston, South Carolina, and Savannah, Georgia, and 1.1 million acres (34%) in the Gulf of Mexico are similarly affected. Sawyer et al. (1989) reported that hard clam *Mercenaria mercenaria* beds in Raritan Bay, New York, and New Jersey classified as condemned, closed, or restricted had higher bacterial counts in bottom sediments than in the overlying water. Surface and bottom water had maximum total coliform MPNs of 1,300 compared with sediment MPNs of up to 49,000; surface- and bottom-water fecal coliforms ranged up to 280 compared with sediment maximum of 3,300. Fabrikant (1984) studied the effect of sewage effluent in the clam *Parvilucina tenuisculpta* and found that size and population density increased as sediment organic nitrogen increased. When nitrogen concentrations reached a critical level, however, there was a dramatic decrease in population density due to the toxicity of released compounds. The dispersal of microbial agents, including viruses, bacteria, fungi, and protozoans, away from shellfish closure areas must be considered as a serious threat to shellfish beds and commercial rearing facilities as shorelines are lost with continued land development for industrial and residential purposes. Furthermore, as discussed by Fabrikant (1984), short-term and long-term studies are necessary for measuring the effects of sewage effluents on fish and shellfish health.

Mechanisms for the Distribution of Microbial Agents. Goyal (1984, 1989) reported that enteric viruses may be recovered from sediments when water samples are negative, and some viruses survive for long periods of time when attached to sediment particulates. Sediment transport is an important mechanism for the dispersal of viruses in the marine environment as shown by Goyal (1989) who found coxsackie viruses 13 km southeast of the New York 12-mile disposal site. Goyal (1989) also recovered enteroviruses from the intestine and hepatopancreas of rock crabs *Cancer irroratus* captured in New York waters and the Philadelphia-Camden disposal site. Seidel et al. (1983) experimentally seeded tissue homogenates of the blue crab *Callinectes sapidus* with virus and reported a recovery efficiency of 52% for three enteroviruses and 23% for rotavirus (SA11). Improved isolation and identification methods may show that crustaceans have an important role in the transport of human viruses away from sources of sewage pollution. Babinchak et al. (1982) studied *C. sapidus* from South Carolina waters and obtained fecal coliform MPNs from gills that ranged from 210-4,300. The authors recovered bacteria from gills that were brown to mahogany in color and suggested that crab gills may accumulate high fecal bacteria populations that persist even when the influx of bacteria into some ecosystems is low. Although studies on crabs as transport vectors for bacteria and viruses are limited in number, there are sufficient data to justify further studies, especially in deep-ocean disposal sites. Bullis et al. (1988) have published preliminary data on shell disease in deep sea red crabs *Geryon quinquedens*. The authors cultured *Vibrio alginolyticus, V. campbellii, V. fluvialis, Flavobacter meningosepticum, F. breve,* and *Eschericia coli* from crabs that may have been affected by the dispersion of contaminants from the deepwater 106 Dump Site.

Mahoney et al. (1974) studied Australian antigen in clams *Mercenaria mercenaria* collected in closed areas along the Atlantic coast of Maine. The antigen, a marker for type B hepatitis virus, was identified from 20 of 24 tissue pools, and transmission from clam to clam was demonstrated in laboratory experiments. Continued studies on the antigen in hard clams might show that they are useful markers for the movement of virus away from sources of coastal pollution. Al-Mossawi et al. (1983) used the clam *Circinita callipyga* as a bioindicator for sewater pollution by fecal coliform and *Salmonella* spp. bacteria. The authors studied coastal waters of Kuwait and found salmonellae in clams when water samples were negative; fecal coliform MPNs ranged from approximately 10^3–10^6 g^{-1} and up to 17% of the isolates were resistant to certain antibiotics. Devanas et al. (1980) found that a number of bacteria, including *Vibrio, Neisseria, Pseudomonas,* and *Aeromonas,* were resistant to one or more of 11 different well-known antibiotics, and to cadmium. The resistant bacteria were cultured from sediments collected at sewage, dredge-spoil, and acid-waste disposal sites in the New York Bight apex. Most studies on the distribution of microbial pathogens associated with sewage or other wastes have been made in coastal or nearshore environments. Recent findings suggest that further research is needed to obtain new information on the distribution of pathogens in offshore waters and sediments. Deepwater fish, shellfish, and fine particulate surficial sediments may concentrate microorganisms that otherwise are diluted beyond detectable limits in the water column.

Summary and Conclusions

The distribution of microbial pathogens in the marine environment is influenced by the binding capacity of particulate matter, sediment grain size, climatic conditions, prevailing winds and currents, and transport by invertebrates and vertebrates. More than 110 different viruses may be present in sewage (Goyal 1989), but not all of them are easily cultured and identified in the laboratory. Attempts to plot the distribution of pathogenic bacteria in marine ecosystems also are influenced by the length of time that different species remain both viable and culturable. Resistant or spore-forming bacterial species, although not always indicative of recent contamination, may be cultured from samples that otherwise are negative for less resistant fecal coliforms. Resistant species are valuable indicators of sewage-associated chemical and biological contaminants in sediments distant from sources of pollution. The frequency with which terrestrial cyst-forming potentially pathogenic protozoa and spore-forming fungi may be cultured from bottom sediments is useful for following the progressive accumulation or dispersal of sewage and dredge-spoil contaminants in marine ecosystems. Serological studies and the identification of bacteria and viruses from animal tissues and body fluids provide some insight as to the diversity of pathogenic species entering coastal and estuarine ecoystems from pollution sources. Authorized increases in sewage-treatment-plant discharge permits, and the construction of new plants, must continue to take into account the risk of possible contamination and closure of nearby shellfish beds, mariculture facilities, and recreational areas.

The impact of increased land development and water use emphasizes the need for continued studies on the distribution of sewage-associated pathogens on harvestable resources and human health. Finally, the design of reliable predictive models for assessing the impacts of coastal or ocean waste disposal on natural resources and water quality must rely on statistically valid measurements. Berman (1983) showed that the spatial distribution of heavy metals, fecal steroids (coprastanol), polychlorinated biphenyls, sediment characteristics, current patterns, etc., at ocean waste-disposal sites was related statistically to the recovery of pathogenic amoebae from contaminated sediments. The predictive model proposed by Berman (1983) should be a valuable tool for predicting trends in the effects of ocean dumping on human health, the sanitary quality of fish and shellfish, and the distribution of genetically altered plants and animals in marine ecosystems.

Acknowledgments

The author gratefully acknowledges the staff of F.D.A.'s Northeast Technical Service Unit, Shellfish Sanitation Branch, at North Kingston, Rhode Island, for providing bacteriological data from the Hempstead Bay and Yaquina River systems. Special appreciation is expressed to Dr. Aaron Rosenfield, National Marine Fisheries Service, Oxford, Maryland, for providing the opportunity for the author to participate in the symposium from which this paper derives.

Literature Cited

Al-Mossawi, M.J., M.H. Kadri, A.A. Salem and T.D. Chugh. 1983. The use of clams as bioindicator of fecal pollution in seawater. Water Air Soil Pollut. 20:257-263

Amandi, A., S.F. Hiu, J.S. Rohovec and J.L. Fryer. 1982. Isolation and characterization of *Edwardsiella tarda* from fall chinook salmon *(Oncorhynchus tshawytscha).* App. Env. Microbiol. 43:1380-1384.

Attwell, R.W. and R.R. Colwell. 1981. Actinomycetes in New York harbour sediments and dredging spoil. Mar. Pollut. Bull. 12:351-353.

Babinchak, J.A., J.T. Graikoski, S. Dudley and M.F. Nitkowski. 1977. Distribution of faecal coliforms in bottom sediments from the New York Bight. Mar. Pollut. Bull. 8:150-153.

Babinchak, J.A., D. Goldmintz and G.P. Richards. 1982. A comparative study of autochthonous bacterial flora on the gills of the blue crab, *Callinectes sapidus*, and its environment. Fish. Bull. 80:884-890.

Berman, C.R., Jr. 1983. A statistical model to predict the incidence of pathogenic protozoa (Amoebida:Acanthamoebidae) in ocean sediments using surrogate variables. Ph.D. Dissertation, School of Marine Science, College of William and Mary, Gloucester Point, Virginia.

Bullis, R., L. Leibovitz, L. Swanson and R. Young. 1988. Bacteriologic investigation of shell disease in the deep sea red crab, *Geryon quinquedens.* Biol. Bull. 175:304.

Burge, W.D., and P.B. Marsh. 1978. Infectious disease hazards of landspreading sewage wastes. J. Environ. Qual. 7:1-9.

Cabelli, V.J., A.P. Dufour, L.J. McCabe and M.A. Levin. 1982. Swimming-associated gastroenteritis and water quality. Amer. J. Epidemiol., 115:606-616.

Crewe, W. and R.R. Owen. 1981. The occurrence of sewage-borne pathogens in the UK. J. Water Pollut. Control 5:632-637.

Daggett, P.-M., T.K. Sawyer and T.A. Nerad. 1982. Distribution and possible interrelationships of pathogenic and nonpathogenic *Acanthamoeba* from aquatic environments. Microb. Ecol. 8:371-386.

Devanas, M.A., C.D. Litchfield, C. McClean and J. Gianni. 1980. Coincidence of cadmium and antibiotic resistance in New York bight apex benthic microorganisms. Mar. Pollut. Bull. 11:264-269.

Eklund, M.W., F.T. Poysky, M.E. Peterson, L.W. Peck and W.D. Brunson. 1984. Type E botulism in salmonids and conditions contributing to outbreaks. Aquaculture 41: 293-309.

Fabrikant, R. 1984. The effect of sewage effluent on the population density and size of the clam, *Parvilucina tenuisculpta.* Mar. Pollut. Bull. 15:249-253.

Goyal, S.M. 1989. Virus survival at sewage-sludge disposal sites, p. 57-63. *In* D.W. Hood, A. Schoener and P. Kilho Park (eds.), Oceanic Processes in Marine Pollution, Volume 4: Scientific Monitoring Strategies for Ocean Waste Disposal. Robert E. Krieger Publishing Company, Malabar, Florida.

Goldstein, S. 1973. Zoosporic marine fungi (Thraustochytriaceae and Dermocystidiaceae). Ann. Rev. Microb. 27:13-26.

Goyal, S.M., C.P. Gerba and J.L. Melnick. 1979. Human enteroviruses in oysters and their overlying waters. Appl. Env. Microbiol. 37:572-581.

Goyal, S.M., W.N. Adams, M.L. O'Malley and D.W. Lear. 1984. Human pathogenic viruses at sewage sludge disposal sites in the middle Atlantic region. Appl. Environ. Microbiol. 48:758-763.

Grimes, D.J. and R.R. Colwell. 1989. Ocean discharge of industrial and domestic wastewater at Barceloneta, Puerto Rico: Bacteriological considerations, p.139-148. *In* D.W. Hood, A. Schoener and P. Kilho Park (eds.), Oceanic processes in marine pollution, Volume 4: Scientific Monitoring Strategies for ocean waste disposal. Robert E. Krieger Publishing Company, Malabar, Florida.

Huss, H.H. and U. Eskildsen. 1974. Botulism in farmed trout caused by *Clostridium botulinum* type E. Vet. Med. 26:733-738.

Janssen, W.A. and C.D. Meyers. 1968. Fish: serologic evidence of infection with human pathogens. Science 159:547-548.

Kott, H. and Y. Kott. 1970. Viability of *Entamoeba histolytica* cysts exposed to sea water. Rev. Int. Océanogr. Med. 28-9:85-95.

Kusuda, R., T. Toyoshima, Y. Iwamura and H. Sako. 1976. *Edwardsiella tarda* from an epizootic of mullets *(Mugil cephalus)* in Okitsu Bay. Bull. Jap. Soc. Sci. Fish. 42:271-275.

LaBelle, R.L., C.P. Gerba, S.M. Goyal, J.L. Melnick, I. Cech and G.F. Bogdan. 1980. Relationships between environmental factors, bacterial indicators, and the occurrence of enteric viruses in estuarine sediments. Appl. Env. Microbiol. 39:588-596.

Lewis, G.D., M.W. Loutit and F.J. Austin. 1985. Human enteroviruses in marine sediments near a sewage outfall on the Otago coast. N.Z. J. Mar. Freshwater Res. 19: 187-192.

Lewis, G.D., F.J. Austin and M.W. Loutit. 1986. Enteroviruses of human origin and faecal coliforms in river water and sediments down stream from a sewage outfall in the Taieri River, Otago. N.Z. J. Mar. Freshwater Res. 20:101-105.

Loutit, M.W. and G. Lewis. 1985. Faecal bacteria from sewage effluent in sediments around an ocean outfall. N.Z. J. Mar. Freshwater Res. 19:179-185.

Mahoney, P., G. Fleischner, I. Millman, W.T. London, B.S. Blumberg and I.M. Arias. 1974. Australia antigen: detection and transmission in shellfish. Science 183:80-81.

O'Malley, M.L., D.W. Lear, W.N. Adams, J. Gaines, T.K. Sawyer and E.J. Lewis. 1982. Microbial contamination of continental shelf sediments by wastewater. J. Water Pollut. Cont. Fed. 54:1311-1317.

Robohm, R.A., C. Brown and R.A. Murchelano. 1979. Comparison of antibodies in marine fish from clean and polluted waters of the New York bight: relative levels against 36 bacteria. Appl. Env. Microbiol. 38:248-257.

Sawyer, T.K. 1980. Marine amoebae from clean and stressed bottom sediments of the Atlantic Ocean and Gulf of Mexico. J. Protozool. 27:13-32.

Sawyer, T.K. and S.M. Bodammer. 1983. Marine amoebae (Protozoa: Sarcodina) as indicators of healthy or impacted sediments in the New York bight apex, p. 337-352. *In* I.W. Duedall, B.H. Ketchum, P.K. Park and D.R. Kester (eds.), Wastes in the ocean, Vol. 1: Industrial and sewage wastes in the ocean. Wiley-Interscience, New York.

Sawyer, T.K. and S.A. Meyer. 1977. A nonfilamentous marine fungus, *Sterigmatomyces halophilus,* from mantle fluid of the Japanese oyster, *Crassostrea gigas*. J. Invertebr. Path. 29:395-396.

Sawyer, T.K., E.J. Lewis, M. Galasso, D.W. Lear, M.L. O'Malley, W.N. Adams and J. Gaines. 1982. Pathogenic amoebae in ocean sediments near wastewater sludge disposal sites. J. Water Pollut. Cont. Fed. 54:1318-1323.

Sawyer, T.K., T.A. Nerad, P.-M. Daggett and S.M. Bodammer. 1987. Potentially pathogenic protozoa in sediments from oceanic sewage-disposal sites, p. 183-194. *In* J.M. Capuzzo and D.R. Kester (eds.), Oceanic processes in marine pollution, Volume 1: Biological processes and wastes in the ocean. Robert E. Krieger Publishing Company, Malabar, Florida.

Sawyer, T.K., E.J. Lewis, J. Musselman, W.N. Adams, J.Gaines, L. Chandler and S. Rippey. 1989. Sewage-associated protozoans (Amoebida) and bacteria as indicators of the sanitary quality of commercial shellfish beds, p. 73-81. *In* D.W. Hood, A. Schoener and P. K. Park (eds.), Oceanic processes in marine pollution, Volume 4: Scientific monitoring strategies for ocean waste disposal. Robert E. Krieger Publishing Company, Malabar, Florida.

Seidel, K. M., S.M. Gogal, V.C. Rao and J.L. Melnick. 1983. Concentration of rotavirus and enteroviruses from blue crabs *(Callinectes sapidus)*. Appl. Env. Microbiol. 46:1293-1296.

Sindermann, C.J. 1972. Some biological indicators of marine environmental degradation. Symposium on the fate of the Chesapeake Bay. J. Wash. Acad. Sci. 62:184-189.

Stewart, J.R. and R.M. Brown. 1969. Cytophaga that kills or lyses algae. Science 164:1523-1524.

Wakabayashi, H. and S. Egusa. 1973. *Edwardsiella tarda (Paracolobactrum anguillimortiferum)* associated with pond-cultured eel disease. Bull. Japan Soc. Sci. Fish. 39:931-936.

Yde, M. and S. DeMaeyer-Cleempoel. 1980. Faecal pollution of Belgian coastal water. Mar. Pollut. Bull. 11:108-110.

Ziskowski, J.J., L. Despres-Patanjo, R.A. Murchelano, A.B. Howe, D. Ralph and S. Atran. 1987. Disease in commercially valuable fish stocks in the northwest Atlantic. Mar. Pollut. Bull. 18:496-504.

CHAPTER 4

Dispersal of Genetically Manipulated Macroorganisms

Fish Genetic Engineering: A Novel Approach in Aquaculture

THOMAS T. CHEN
CHAU-MIN LIN
LUCIA IRENE GONZALEZ-VILLASEÑOR
REX DUNHAM
DENNIS POWERS
Z. ZHU

Abstract: As the world population increases rapidly, there is a need to increase protein production. Aquaculture and mariculture possess the greatest potential for the production of animal proteins. In this paper we will discuss the application of recombinant DNA technology and genetic engineering in aquaculture and mariculture.

Introduction

As the world population increases there is an urgent need to develop technologies for increased protein production. Aquaculture and mariculture possess the greatest potential for the production of animal proteins. Traditionally, success in fisheries has been largely dependent upon the natural population of freshwater and marine fishes; however, due to over exploitation by commercial and sportfishing operations and poor restocking programs, the harvest has decreased drastically. In recent years, active promotion of aquaculture has resulted in a significant increase in fish production.

Production of finfish and shellfish has grown significantly in the past decade from advances in disease control, production of vaccines, improvement of genetic stocks, nutrition, and development of effecient culturing and management technologies. Recent advances in recombinant DNA technology and genetic engineering promise to revolutionize aquaculture and mariculture through manipulation of genes that control growth, development, and disease resistance in fish. Using rainbow trout and other economically important fish species as models, we are investigating strategies to enhance the growth rate of fish for aquaculture purposes by manipulating growth hormone genes. While this effort is only beginning, numerous promising advances have already been accomplished. In this paper we shall discuss some of our recent findings to illustrate our general approaches.

Two Rainbow Trout Growth Hormone m-RNA Sequences

Agellon et al. (1986) reported earlier that polyclonal antibodies raised against the growth hormone (GH) of chum salmon cross-reacted specifically with that of rainbow trout. Using the antiserum to chum salmon GH as a probe, several recombinant clones harboring rainbow trout *(Oncorhynchus mykiss)* GH cDNA sequences have been isolated (Agellon and Chen 1986) from a pituitary cDNA bank. One of the clones, pAF51, encodes an immunoreactive polypeptide of 24 KDa, which is slightly smaller than the pre-GH observed in extracts of rainbow trout pituitary glands (Agellon and Chen 1986). Nucleotide sequence analysis revealed that the cDNA insert in pAF51 contained the entire coding sequence of mature GH polypeptide. This cDNA is encoded by GH gene-1 (GH1) mRNA. In comparison with the nucleotide sequence corresponding to the mature chum salmon GH1 polypeptide (Sekine et al. 1985), rainbow trout GH1 cDNA differs from that of chum salmon GH1 by six nucleotides. Since these substitutions are located at the third positions of their respective codons, the two mature GH polypeptides share identical amino acid sequences. On the other hand, when the nucleotide sequence of the mature rainbow trout GH was compared with that of coho salmon, they differ from each other by 9 nucleotides, resulting in a difference of 6 amino acid residues between each other (Gonzalez-Villasenor et al. 1988).

Recently, the nucleotide sequence of rainbow trout GH gene-2 (GH2) mRNA has also been deduced (Agellon et al. 1988a). The sequences of these two mRNA were determined by sequencing the cDNA of GH1 and GH2, the genomic gene of GH2 (Agellon et al. 1988b) and the mRNA directly. GH1 and GH2 mRNA differ from each other by 5 nucleo-tides at 5' untranslated region and 23 nucleotides at the translated region, resulting in 10 amino acid replacements. The majority of the differences between GH1 and GH2 mRNAs resides in the 3' untranslated region. This region is characterized by specific nucleotide changes as well as single and multiple insertions/deletions.

Growth-Promoting Activity of the Biosynthetic Growth Hormone (GH)

The biosynthetic trout GH polypeptide was prepared by expressing GH1 cDNA sequence in *E. coli* cells. It is expressed in the form of a fusion protein, containing 9 amino acid residues from the N-terminus of *E. coli* β-galactosidase and the entire sequence of the mature GH polypeptide. To test the biological activity of this hybrid protein, the hormone was enriched from the pAF51 cell extract and used directly in bioassays.

After acclimation for 2 weeks at 15°C, groups of 16 yearling rainbow trout were injected weekly with the enriched hormone preparation at a dose of 0.2 µg g^{-1}, 1.0 µg g^{-1}, and 2.0 µg g^{-1} body weight for 4 weeks. Two groups of control animals were used: one group was treated with protein fractions prepared from *E. coli* cells containing plasmid pUC8 alone and the other group without injection. Both control and experimental fish were fed to satiation daily and the amount of food consumed was recorded. Biosynthetic GH at a dose each week of 0.2 µg g^{-1} was sufficient to stimulate a significant increase in weight and length (Agellon et al. 1988c). After treatment with biosynthetic GH for 4 weeks at the dose of 1 µg g^{-1} body weight, the weight gain in the experimental group was 2.0 times greater than the control. Increase in body length was also observed in experimental animals. Furthermore, the chemical composition of muscle tissues of hormone-treated fish is indistinguishable from that of the control fish.

The growth-promoting effect of the biosynthetic GH polypeptide was also assessed for rainbow trout fry (Agellon et al. 1988c). In this study, trout fry was first incubated in a hypertonic saline solution (3.5% NaCl) for 2 min. and then transferred to an isotonic saline solution (0.9% NaCl) containing biosynthetic GH at a concentration of 50 mu or 500 µg L^{-1} for 30 min. A significant increase in weight gain was evident in the hormone-treated fry 5 weeks after hormone administration. These results are in agreement with those reported by Sekine et al. (1985), Wagner et al. (1985), and Gill et al. (1985).

Generation of Transgenic Fish With Enhanced Growth Rates

As discussed earlier, fish will grow faster in response to elevated levels of GH in the circulation. This can be achieved by exogenous application of biosynthetic GH. Alternatively, through transfer of additional copies of GH gene into fish, the resulting transgenic fish should produce elevated levels of GH and, consequently, will grow faster. Fast-growing transgenic fish can be developed for aquaculture purposes through manipulation of GH gene-by-gene transfer technology. Basic strategies in constructing such a strain of rapid-growing transgenic fish will involve: (1) isolation of fish GH genes (cDNA or genomic gene); (2) identification of reporter genes or other genetic markers; (3) identification of appropriate promoters to control the expression of the foreign GH gene; (4) construction of chimeric plasmids consisting of a reporter gene and a GH gene fused with an appropriate promoter; (5) introducing the chimeric plasmid into developing embryos by microinjection; (6) identification and characterization of transgenic fish; (7) studying the pattern of inheritance and the stability of the foreign genes in progeny of the transgenic fish; (8)

investigating physiological and environmental factors that will maximize the performance of the transgenic fish; and (9) assessing the impact of transgenic fish in the environment. As a step toward this direction, we have attempted to transfer trout and/or human GH genes into carp and loach by the microinjection technique.

Microinjection of Trout GH cDNA and Human GH Gene Into Carp and Loach Embryos

One of our gene-transfer attempts was introduction of a trout GH cDNA sequence fused to the long terminal repeat (LTR) sequence of avian Rouse sarcoma virus (RSV) into fertilized common carp eggs *(Cyprinus carpio)*. About 10^6 molecules of BamH-I linearized pRSVrtGHcDNA were microinjected into each of the 1,746 common carp embryos at one-cell, two-cell, or four-cell stage. Following incubation for 4 days, approximately 37% of the injected embryos hatched, of which 57% of the fry survived at least 90 days (Table 1).

Table 1. Percent of hatching, survival, and integration of carp embryos microinjected with pRSVrtGHcDNA at different developmental stages.

Embryo stage	Embryos injected	% hatched	% survival at 90 days	Fingerlings analyzed	% integration
One-cell	1034	39.3	52.7	219	9.9
Two-cell	331	33.0	70.6	77	15.6
Four-cell	381	33.0	58.4	73	1.4
Control (non-injected)	569	33.9			

To detect the integration of pRSVrtGHcDNA in the genome of presumptive transgenic carp, genomic DNA was extracted from pectoral fin clips of the 365 survivors and analyzed by dot blot hybridization, using [^{32}P]-labeled RSV-LTR or rtGHcDNA sequence as a probe. Of these 365 fish, 20 were found to be transgenic by dot blot and Southern blot analyses (Table 1, Zhang et al. 1989). This value represents an overall integration of 5.5% which is similar to that observed in zebrafish (5%) (Stuart et al. 1988), but low compared to those of catfish (Dunham et al. 1987), mice (Palmiter et al. 1982), and frogs (Etkin and Pearman 1987), which were 20%, 33%, and 60%, respectively. The number of pRSVrtGHcDNA molecules present in the fin tissue of positive fish, determined by quantitative dot blot hybridization, ranged from 1 to 5 per haploid genome.

The integration of pRSVrtGHcDNA in the genomes of transgenic fish was further confirmed in 10 dot blot positive DNA samples by digestion with restriction enzyme Bam-H1 or Hind-III, and followed by hybridization to both [^{32}P]-labeled RSV-LTR and/or rtGHcDNA. As shown in Fig. 1, discrete bands from genomic DNA samples of fish 24L, 27L, and 131L hybridized to RSV-LTR, rtGHcDNA, or both. In BamH-1-digested DNA samples of fish 24L and 131L, a 5.2 Kb and a 4.3 Kb DNA fragment hybridized to both probes. When the identical DNA samples were digested with the restriction enzyme Hind-III, the pattern of hybridization was somewhat different: a 3.0 Kb DNA fragment hybridized to both RSV-LTR and rtGHcDNA, a 4.3 Kb DNA fragment hybridized to RSV-LTR alone, and a 0.9 Kb DNA fragment hybridized to rtGHcDNA. Since each of these transgenic fish (e.g., 24L and 131L) contains 2 copies of pRSVrtGHcDNA in its genome, the patterns of hybridization presented in Fig. 1 suggest that pRSVrtGHcDNA is integrated in a pattern of single-copy integration at multiple chromosomal sites. In addition, results of Hind-III digestion further suggest that the Hind-III site residing between RSV-LTR and rtGHcDNA has been modified in one of the integrated pRSVrtGHcDNA molecules. In the case of fish 27L, where 5 copies of pRSVrtGHcDNA molecules were integrated, the additional bands of hybridization in the BamH-1 digestion can be accounted for by a single copy of the foreign gene integrated at different chromosomal sites. The differences observed in the BamH-1 digestion suggest that they carry genomic DNA

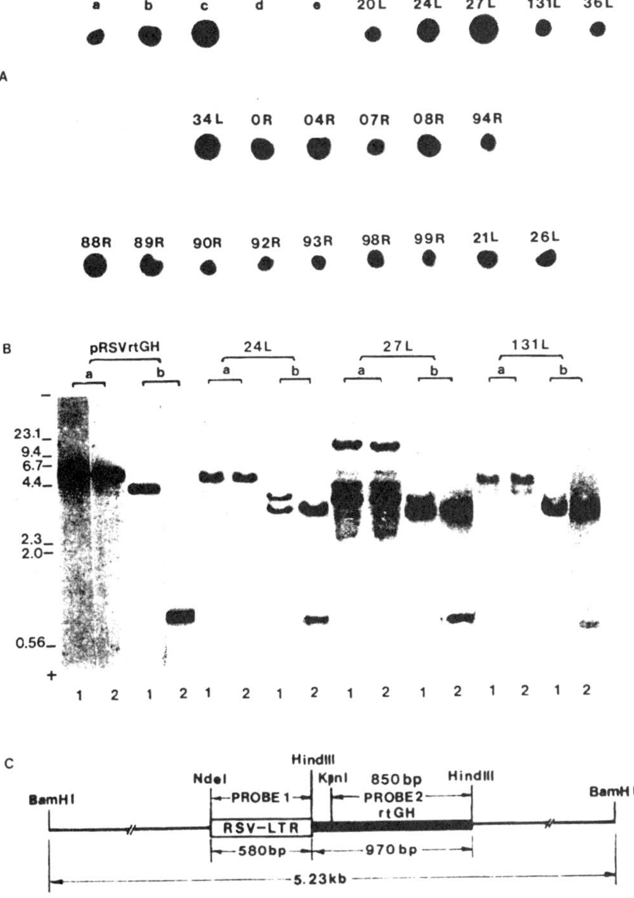

Figure 1. Dot blot and Southern blot analysis of genomic DNA samples isolated from presumptive transgenic fish. A. Genomic DNA (18 mu) isolated from the pectoral fin was denatured and spotted onto nylon membranes along with pRSVrtGHcDNA of 30 pg (a), 60 pg (b), and 150 pg (c) corresponding to 1, 2, and 5 molecules of the integrated gene. Genomic DNA from the nontransgenic carp (d and e) was included as negative controls. The membranes were hybridized to [^{32}P]-labeled pRSVrtGHcDNA. B. Genomic DNA digested with BamH-1 (a) or Hind-III was electrophoresed, transferred to nylon membranes, and then hybridized to either [^{32}P]-labeled probe 1 (1) or probe 2 (2). BamH-1- and Hind-III-digested pRSVrtGHcDNA samples were used as positive controls. C. A 580 bp Nde-I-Hind-III fragment derived from pRSVrtGHcDNA was used as probe 1, and a 850 bp Kpn-I-Hind-III fragment derived from pRSVrtGHcDNA was used as probe 2.

junction fragments of variable sizes. Although foreign DNA sequences introduced into embryos are usually integrated as a head-to-tail concatamer at a single chromosomal site and modified in transgenic individuals (Brinster et al. 1981; Dunham et al. 1987; Wilkie and Palmiter 1987), single- and low-copy integrations exhibiting deletions, modifications, and rearrangements of the inserted foreign sequence have also been observed (Gordon and Ruddle 1985). The results of pRSVrtGHcDNA integration in transgenic carp agree with those reported in other systems.

Gene-transfer studies have also been attempted in other species of carp and loach *(Misgurnus anguillicaudatus)*, using MThGH gene as a model gene (Zhu et al., 1985, 1989). Results of some representative studies are summarized in Table 2. In these studies rates of foreign gene integration are higher than those reported by other laboratories.

Table 2. Integration of MT-hGH gene in carp and loach.

Fish species	Number of fish tested	Number of positive fish	% integration
Loach	10	4	40.0
Crucian carp	64	33	51.6
Red crucian carp	66	28	42.4
Silver crucian carp	23	16	69.6
Mirror carp	89	46	51.7
Red carp	8	7	87.9

Expression of Foreign Gene in Transgenic Fish

Since polyclonal antibodies of chum salmon GH react specifically with trout GH polypeptide, they were used as probes for detecting the expression of rtGHcDNA in transgenic carp by the radioimmunobinding assay (Zhang et al. 1989). Although these antibodies show a partial cross-reactivity with growth hormone of carp, they were rendered specific to trout growth hormone by extensive re-absorption with the pituitary extract of carp.

Since the rtGHcDNA used in the gene-transfer studies does not contain a signal peptide sequence, rainbow trout-growth hormone polypeptide is not expected in the serum of the transgenic fish. We examined the expression of rtGHcDNA in the red blood cells (RBC) of individual transgenic carp instead. As shown in Table 3, all 9 samples tested showed various levels of trout-growth hormone: from 8.0 ng mg^{-1} RBC proteins in fish 20L to 89.1 ng mg^{-1} RBC proteins in fish 131L. There was no correlation between the number of foreign genes integrated and the levels of trout-growth hormone expressed in RBC. These results are in agreement with those observed in transgenic mice (Palmiter et al. 1982).

The body weight of each transgenic carp that received pRSVrtGHcDNA was measured at three months of age. Transgenic individuals derived from embryos microinjected at one-cell stage showed higher weight gain than controls. The mean body weight of these transgenic fish was 143 $\pm$ 31 g, which is 22% larger ($p < 0.05$) than the mean body weight (116 $\pm$ 40 g) of non-transgenic siblings. These results suggest that growth seems to be correlated to the amount of rtGH polypeptide present. Transgenic carp with low rtGH content grew faster than those containing higher rtGH levels. These results are in good agreement with those observed by Agellon et al. (1988c).

Expression of MthGH gene in transgenic silver crucian carp was also detected by radioimmunoprecipitation, and the results are presented in Table 4. At the age of 35 days, the levels of hGH in transgenic individuals ranged from 26.1 to 50 ng for each fish (0.2 g body weight). At the age of 913 days, the levels of hGH in transgenic fish ranged from 1.1 to 4.0 ng ml^{-1} serum.

The body weights of transgenic mirror carp and silver crucian carp were compared to their respective nontransgenic siblings (Table 5). Although there was no difference ($P>0.05$) between the body weights of transgenic mirror carp and their nontrans-

Table 3. Gene copy number and trout growth hormone levels in transgenic carp.

Fish number	pRSVrtGHcDNA molecules*	ngGH mg^{-1} protein in RBC
20L	2	8.0
34L	4	47.7
36L	1	47.5
94R	1	48.8
0R	3	64.2
04R	2	28.6
07R	1	8.9
08R	3	73.8
131L	2	89.1

* copy number per haploid genome.

Table 4. Human growth hormone detected in transgenic silver crucian carp by radioimmunoprecipitation assay.

Age (days)	Number of fish tested	Number fish with hGH	Range of hGH levels
35	20	10	26.1-50 ng/fish (0.2 g body weight)
913	8	3	1.1-4.0 ng ml^{-1} serum

Table 5. Mean body weight of transgenic carp and their control siblings.

	Mirror carp[a]		Silver crucian carp[b]	
	Transgenic	Nontransgenic	Transgenic	Nontransgenic
N	15.0	15.0	28.0	28.0
Mean weight (g)	234.5	210.7	239.3	134.1
SD[c]	51.4	26.8	93.1	25.3
% difference	1.3[d]	78.0[e]		

Age[a] = 153 days[b], Age = 208 days[c], Standard deviation[d]. Transgenic mirror carp were not larger than their nontransgenic siblings (P>0.05)[e], but transgenic silver crucian carp were significantly larger than their nontransgenic siblings (P<0.001).

genic siblings, the transgenic silver crucian carp were 78% larger ($P<0.001$) than their nontransgenic siblings.

Inheritance of Foreign Gene in Transgenic F_1 Generation

P_1 male transgenic individuals (04R, 36L, 131L, and 94R) were crossed to one nontransgenic female in order to study the inheritance of pRSVrtGHcDNA. Of the F_1 progeny analyzed from fish 131L and 94R, 32.3% and 42.3% were found to carry pRSVrtGHcDNA sequence. Although most of the F_1 progeny from fish 36L died, the four survivors inherited the foreign gene sequence. None of the F_1 progeny of fish 04R received the foreign DNA from their father, suggesting that the pRSVrtGHcDNA sequence was not integrated into the germ line of fish 04R. Since transgenic fish are independently derived by injecting the foreign gene at different stages of development, one should expect animals mosaic for rtGHcDNA. The degree of mosaicism in animals determines whether a foreign gene will be present in the germ line, and whether it will be transmitted to the progeny. Since nearly 50% of the progeny derived from fish 94R carry rtGHcDNA, the transformed progenitor cells must be primordial to the entire germ-line. The progeny ratios of 1 transgenic : 3 nontransgenic from fish 131L suggests that his germ-line is mosaic.

The mean body weight of F_1 progeny derived from fish 131L and 94R was measured at three months of age. Positive F_1 progeny of fish 131L were about 20.8% ($p < 0.05$) larger than their nontransgenic siblings, and positive progeny from fish 94R were 40.1% larger than nontransgenic siblings (Table 6). Furthermore, 32% and 46% of these positive progeny were larger than their largest nontransgenic siblings respectively. Results summarized in Table 5 indicate that expression of rtGHcDNA in transgenic fish results in elevated growth rate.

Table 6. Mean weight, range of weight and percent inheritance at 90 days of progeny derived from transgenic common carp 131L and 94R.

	Progeny of fish 131L		Progeny of fish 94R	
	Transgenic	Non-transgenic	Transgenic	Non-transgenic
N	31.0	65.0	11.0	15 0
% inheritance	32.3	42.3		
Mean weight (g)	120.6[a]	99.3[b]	206.0	147.0
SD[c]	17.4	14.7	45.2	48.0
Weight range	95-173[d]	65-129	115-283[e]	67-228
% difference	20.8[f]	40.1[g]		

N[a] = 28[b], N = 38[c], SD = standard deviation[d], 32%, and 46%[e] of transgenic progeny were larger than largest control. Transgenic progeny were larger than nontransgenic progeny at p[f] <0.05 and p[g] < 0.001, respectively.

Future Prospects

Results presented in the previous sections clearly demonstrated that weekly injections of biosynthetic GH dramatically accelerated the normal growth of juvenile rainbow trout. However, administration of GH through repeated injections of individual fish is labor intensive, and subjects the fish to more handling than is ideal. In this regard, the osmotic shock method is a more convenient mode of hormone delivery, and is well suited for large-scale operations. Therefore we believe that the biosynthetic trout GH can be used in aquaculture operations to enhance the growth rates of salmon, trout, and other fish species, provided the following practical conditions have been addressed. A comprehensive study employing intact and hypophysectomized fish, a pure preparation of GH, a convenient and practical means of hormone delivery, and a more detailed dose regimen must all be undertaken to assess the growth-promoting potential of this hormone properly. Furthermore, the effect of both chronic and acute GH treatment, nutrient requirements, and other rearing conditions affecting GH-treated fish should be determined in order to maximize growth.

An alternative application of fish GH in aquaculture is the development of transgenic fish and shellfish strains with enhanced growth rates by the gene-transfer technology. At present we have achieved transfer, expression, and transmission of rtGHcDNA and human GH gene in the carp and loach, and the resulting animals grow considerably faster than their nontransgenic siblings. Nevertheless, generation of a strain of fast-growing transgenic fish for commercial purposes is still in the developmental stages. To realize this goal, more research will be required.

Acknowledgments

This research was supported by grants from NSF (DCB-86-42247) and the Maryland Sea Grant program (R/F 47) to DAP and TTC. This is contribution number 122 of the Center of Marine Biotechnology, University of Maryland, and Journal series number 8-892368P of the Alabama Agricultural Experiment Station.

Abbreviations

GH: growth hormone
cDNA: complementary DNA
LTR: long terminal repeat sequences
RSV: Rouse sarcoma virus
pRSVrtGHcDNA: plasmid containing the long terminal repeat sequence of Rouse sarcoma virus and rainbow trout growth hormone complementary DNA
Bam-H1: restriction endonuclease Bam-H1
Hind-III: restriction endonuclease Hid III
RCB: red blood cells
Kb: kilo-bases
MthGH: promoter of mouse metallothionein gene fused to the coding region of the human growth hormone gene.
Kpn I, Hind III and Nde: restriction enzymes which will digest DNA at specific sites
Puc8: DNA cloning vector in the bacteria

24 Kda: 24 kilo-daltons (molecular weight of polypeptide)
580 bp Nde I-Hind III: DNA fragment with 580 base pairs in length resulting from digestion with restriction enzymes Nde I and Hind III

Literature Cited

Agellon, L.B. and T.T. Chen. 1986. Rainbow trout growth hormone: molecular cloning of cDNA and expression in E. coli. DNA 5:463-471.

Agellon, L.B., T.T. Chen, R.J. van Beneden, R.A. Sonstegard, G.F. Wagner and B.A. McKeown. 1986. Rainbow trout growth hormone: in vitro translation of pituitary RNA and product analysis. Can. J. Fish. Aquat. Sci. 43:1327-1331.

Agellon, L.B., S.L. Davies, D.M. Lin, T.T. Chen and D.A. Powers. 1988a. Rainbow trout has two genes for growth hormone. Molec. Reproduc. Develop. 1:11-17.

Agellon, L.B., S.L. Davies, T.T. Chen and D.A. Powers. 1988b. The nucleotide sequence of a gene encoding rainbow trout growth hormone:implication on evolution of the GH gene structure. Proc. Nat. Acad. Sci. USA 85:5236-5140.

Agellon, L.B., C.J. Emery, J. Jones, S.L. Davies, A.D. Dingle and T.T. Chen. 1988c. Growth hormone enhancement by genetically-engineered rainbow trout growth hormone. Can. J. Fish. Aquat. Sci., 45:146-151.

Brinster, R.L., H.Y. Chen, M.E. Trumbauer, A.W. Senear, R. Warren and R.D. Palmiter. 1981. Somatic expression of herpes thymidine kinase in mice following injection of a fusion gene into eggs. Cell 27:223-231.

Dunham, R.A., J. Eash, J. Askins and T.M. Townes. 1987. Transfer of the metallothionein-human growth hormone fusion gene into channel catfish. Trans. Amer. Fish. Soc. 116:87-91.

Etkin, L.D., and B. Pearman. 1987. Distribution, expression and germ line transmission of exogenous DNA sequences following microinjection into Xenopus laevis. Development 99:15-23.

Gill, J.A., J.P. Stumper, E.M. Donaldson, H.M. Dye, L. Souza, T. Berg, J. Wypych and K. Langley. 1985. Recombinant chicken and bovine growth hormone in cultured juvenile pacific salmon *(Onchorhynchus kisutch)*. Biotechnology 3:643-646.

Gonzalez-Villasenor, L.I., P. Zhang, T.T. Chen and D.A. Powers. 1988. Molecular cloning and sequencing of coho salmon growth hormone cDNA. Gene 65:239-246.

Gordon, J.W., and F.H. Ruddle. 1985. DNA-mediated genetic transformation of mouse embryos and bone marrow — a review. Gene 33:121-136.

Palmiter, R.D., R.L. Brinster, R.E. Manner, M.E. Trunbauer, M.G. Rosenfeld, N.C. Birnberg and R.M. Evans. 1982. Dramatic growth of mice that develop from eggs microinjected with metallothionein-growth hormone fusion gene. Nature, 300:611-616.

Sekine, S., T. Mizukami, T. Nishi, Y. Kuwana, A. Saito, M. Sata, S. Itoh and H. Kawauchi. 1985. Cloning and expression of cDNA for salmon growth hormone in E. Coli. Proc. Nat. Acad. Sci. USA 82:4306-4310.

Stuart, G.W., J.V. McMurray and M. Westerfield. 1988. Replication, integration and stable germ-line transmission of foreign sequences injected into early zebrafish embryos. Development 103:403-412.

Wagner, G.F., R.G. Fargher, J.C. Brown and B.A. McKeown. 1985. Further characterization of growth hormone from chum salmon *(Oncorhynchus keta).* Gen. Comp. Endocrinol. 60:27-34.

Zhang, P., M. Hayat, C. Joyce, L.I. Gonzalez-Villasenor, C.M. Lin, R.A. Dunham, T.T. Chen and D.A. Powers. 1990. Gene transfer, expression and inheritance of pRSV-rainbow trout-GHcDNA in the common carp, Cyprinus carpio. Molec. Reproduc. Develop. 25:3-13.

Zhu, Z., G. Li, L. He and S. Chen. 1985. Novel gene transfer into the fertilized eggs of gold fish (*Carassius auratus* L. 1758). Z. Angew. Ichthyol. 1:31-34.

Zhu, Z., K. Xu, Y. Xie, G. Li and L. He. 1989. A model of transgenic fish. Scientia Sinica. [B2]:147-155.

Environmental Impacts of Inbred, Hybrid and Polyploid Aquatic Species

GARY H. THORGAARD
STANDISH K. ALLEN

Abstract: The recent application of techniques for producing inbred and polyploid fish and shellfish and the use of hybrids in aquaculture and fishery management programs has led to increased concern about the environmental impacts of such forms. These manipulated forms represent a special case of the general problem of environmental impacts of organisms that have been genetically altered by conventional or nonconventional means. Genetically altered organisms may harm natural populations of species by competing with, interbreeding with, or replacing them.

Sterile organisms are least likely to have negative impacts on natural populations. Sterile hybrids or triploids thus seem least likely to have harmful effects when introduced because their genetic impact should be minimal or minimized; however, sterile hybrids or triploids might in some cases interfere with reproduction of natural stocks in nongenetic (e.g., behavioral) ways.

Fertile hybrids have sometimes had very negative impacts on natural populations and should not be used outside of closed systems. Fertile hybrids may, however, provide an opportunity for introducing beneficial genes or chromosome segments into domesticated stocks. Inbred strains have not been widely used in aquaculture or fishery managmeent but would be expected to have negative effects if interbreeding led to decreased genetic diversity.

Introduction

With the development of techniques for genetically manipulating fish and shellfish, there has been increased concern about the potential impacts of such fish on indigenous populations of aquatic species. Inbred and hybrid organisms can be produced using convential breeding methods. Genetic manipulation techniques can be used to rapidly produce highly inbred or polyploid organisms (Thorgaard 1986; Chen et al., this volume). In this paper, we discuss the potential impacts of inbred, hybrid and polyploid aquatic species. While there are valid reasons for concern about the impacts of genetically manipulated forms, the impacts are not necessarily greater than those of organisms generated using conventional methods. In some cases, genetically manipulated organisms may be useful in conservation programs.

Potential Impacts of Inbred Strains

The principal concern about inbred organisms is that they are genetically uniform. If uniform organisms become established in nature, they could replace genetically variable natural populations and establish a "monoculture" which will lack the ability to adapt to changes in the environment and will possibly be highly susceptible to disease organisms. One such example in agriculture was the susceptibility of the U.S. corn crop to southern corn leaf blight in 1970 as a result of the widespread use of corn with a highly uniform cytoplasmic genetic background (National Research Council 1972).

Inbred forms have been produced in aquatic species both through conventional inbreeding and through chromosome set manipulation (gynogenesis and androgenesis). Purposely inbred strains have primarily been used as research tools and have generally not been intended for release into the wild. Such forms should not be released into the wild because of valid concerns about the potential for decreasing genetic variability in wild populations (Streisinger et al. 1981).

Perhaps of greater concern are situations in which unintended inbreeding takes place in cultured species. In species with extremely high fecundities in which it is difficult to obtain large numbers of females at the appropriate stage for spawning (e.g., sturgeon, Conte et al. 1988; striped bass, Smith 1988) there is a strong temptation to use a limited number of parents because the "quota" for offspring can be easily met with just a few parents. This can result in reduced genetic variability in the offspring. When these offspring are re-leased in nature and interbreed with wild populations, the variability and fitness of the wild populations could be reduced.

Reproductive Sterility in Hybrids and Polyploids

Interspecific fish hybrids exhibit a full range of reproductive capability, from fully functional, fertile forms to sterile forms with very little gonad development (Chevassus 1983). It is not safe to assume that a given interspecific hybrid is sterile without clear evidence.

Male triploid fish have been found to have substantial gonad development and to produce aneuploid sperm (Lincoln and Scott 1984; Allen et al. 1986; Benfey et al. 1986). Such sperm is typically reduced in quantity relative to that of diploid males. Male triploid shellfish, however, have sometimes been found to produce haploid sperm (Allen 1987).

Female triploid fish and shellfish generally show very little ovary development (reviewed by Lincoln and Scott 1984). There are apparent exceptions in both fish (e.g., Benfey and Sutterlin 1984) and shellfish (e.g., Komaru and Wada 1989; Allen and Downing, in press) but in general the degree of sterility in triploid females is greater than that in triploid males.

Impacts of Fertile Hybrids

Fertile hybrids can negatively impact natural populations if they interbreed with them. Probably the best example of a fertile fish hybrid interbreeding with and impacting native populations is the fertile hybrid between the rainbow and cutthroat trout. These hybrids have been produced artificially (e.g., Rohrer and Thorgaard 1986) and are also formed in nature by crosses between rainbow and cutthroat trout. Hybridization has been responsible for loss of many valuable native cutthroat trout populations (Behnke 1972; Leary et al. 1984; Trotter 1987; Allendorf and Leary 1988).

Other fertile hybrids could impact natural populations in a similar manner. Hybrids between sauger and walleye (Hearn 1986) and between striped and white bass (Avise and Van Den Avyle 1984; Smith 1988) are both fertile and yet are still being released into the wild by management agencies, in some cases into waters with natural populations of one of the parent species. Fertile hybrids should not be used in such waters because of their potential negative impacts.

Impacts of Polyploids

Triploids have been advocated for use because of their sterility but, as previously discussed, some triploids do produce gametes. Male triploid fish can produce inviable progeny when they interbreed with normal diploid females because they develop gonads and produce limited quantities of aneuploid sperm (Lincoln and Scott 1984; Allen et al. 1986). This could potentially disrupt the spawning of normal fish. Introduction of male triploids would thus not have a permanent genetic impact on the diploid population, but might reduce their spawning success considerably. It might even be possible to use male triploids as a biological control agent as sterile insects have been used (Thorgaard and Allen 1987).

Female triploids, on the other hand, should have a minimal impact on natural populations, particularly if they fail to attempt to spawn (as their hormone levels in some cases suggest, e.g., Lincoln and Scott 1984). They consequently may be preferred over male triploids for use in managmeent programs and aquaculture. Fortunately, techniques for producing all-female populations are available and successful for many species (Hunter and Donaldson 1983; Lincoln and Scott 1983).

Positive Conservation Aspects of Genetic Manipulation

Although "genetic engineering" tends to be perceived as working against conservation, we believe that it can be used in a positive manner in support of conservation programs. Two examples are the use of induced triploidy for sterilization of fish and the use of androgenesis (all-paternal inheritance) for recovering strains from cryopreserved sperm.

Triploidy can prevent introduced fish from successfully interbreeding with indigenous fish or from becoming established. The prime example of this application is with the grass carp, a fish that has great potential for aquatic weed control but which managers in many areas do not want permanently established (Allen and Wattendorf 1987; Thompson et al. 1987). There are numerous other situations in which nonreproducing fish might be desirable; however, it is difficult to produce 100% triploid individuals under normal circumstances. For grass carp, individual fish are tested for ploidy before they can be used in weed-control programs in many areas. Individual testing is not practical for most species; consequently it is difficult to be certain that "triploid" lots in most species will be 100% nonreproducing. An alternative in situations where it is important that no individuals be able to reproduce may be the use of sterile hybrids or interspecific triploid hybrids in which the diploid hybrid is inviable or sterile (Chevassus et al. 1983; Scheerer and Thorgaard 1983; Scheerer et al. 1987).

Androgenesis involves fertilizing irradiated eggs with normal sperm and producing homozygous diploid individuals by suppression of the first cleavage division (Parsons and Thorgaard 1985; Scheerer et al. 1986; May et al. 1988). It can be used for rapidly generating homozygous lines of fish for research programs. Another application of androgenesis is in the recovery of fish strains from cryopreserved sperm (Stoss 1983; Thorgaard et al., 1990). It is not currently possible to cryopreserve eggs or embryos of fish, probably because of their large size (Stoss 1983); however, sperm cryopreservation is relatively successful and, when combined with androgenesis, may allow banking and recovery of valuable or endangered fish strains.

Conclusions

There are reasons to be concerned about impacts of inbred fish on natural populations. A particular hazard may be the unintended inbreeding that is occurring in some cultured species. Fertile interspecific hybrids have been and are being used in inappropriate situations in which they are able to interbreed with natural populations. Although male triploids have the potential to interfere with the spawning and reproduction of normal individuals, female triploids are much less likely to interfere and may be especially valuable where sterile individuals are introduced among reproducing individuals of the same species. Induced triploidy and androgenesis are examples of genetic manipulation techniques that can be used in a positive manner in conservation programs.

Literature Cited

Allen, S.K., Jr. 1987. Reproductive sterility of triploid shellfish and fish. Ph.D. Dissertation, University of Washington, Seattle.

Allen, S.K., Jr. and S.L. Downing. In press. Performance of triploid Pacific oysters, *Crassostrea gigas*. II. Gametogenesis. Can. J. Fish. and Aquat. Sci.

Allen, S.K., Jr., R.G. Thiery and N.T. Hagstrom. 1986. Cytological evaluation of the likelihood that triploid grass carp will reproduce. Trans. Amer. Fish. Soc. 115:841-848.

Allen, S.K., Jr. and R.J. Wattendorf. 1987. Triploid grass carp: status and management implications. Fisheries 12(4):20-24.

Allendorf, F.W. and R.F. Leary. 1988. Conservation and distribution of genetic variation in a polytypic species, the cutthroat trout. Conserv. Biol. 2:170-184.

Avise, J.C. and M.J. Van Den Avyle. 1984. Genetic analysis of reproduction of hybrid white bass x striped bass in the Savannah River. Trans. Amer. Fish. Soc.113:563-570.

Behnke, R.J. 1972. The systematics of salmonid fishes of recently glaciated lakes. J. Fish. Res. Board Can. 29:639-671.

Benfey, T.J. and A.M. Sutterlin. 1984. Growth and gonadal development in triploid landlocked Atlantic salmon *(Salmo salar)*. Can. J. Fisheries Aquat. Sci. 41:1387-1392.

Benfey, T.J., I.I. Solar, G. De Jong and E.M. Donaldson. 1986. Flow-cytometric confirmation of aneuploidy in sperm from triploid rainbow trout. Trans. Amer. Fish. Soc. 115:838-840.

Chevassus, B. 1983. Hybridization in fish. Aquaculture 33:245-262.

Chevassus, B., R. Guyomard, D. Chourrout and E. Quillet. 1983. Production of viable hybrids in salmonids by triploidization. Genet. Select. Evol. 15:519-532.

Conte, F.S., S.I. Doroshov and P.B. Lutes. 1988. Hatchery manual for the white sturgeon. University of California, Division of Agriculture and Natural Resources, Publication 3322, Oakland.

Hearn, M.C. 1986. Reproductive viability of sauger-walleye hybrids. Prog. Fish-Cult. 48:149-150.

Hunter, G.A., and E.M. Donaldson. 1983. Hormonal sex control and its application to fish culture, p. 223-303. *In* W.J. Hoar, D.J. Randall and E.M. Donaldson (eds.), Fish physiology, Vol. 9B, Academic Press, New York.

Komaru, A. and K.T. Wada. 1989. Gametogenesis and growth of triploid scallops, *Chlamys nobilis*. Nipp. Suis. Gakk. 55:447-452.

Leary, R.F., F.W. Allendorf, S.R. Phelps and K.L. Knudsen. 1984. Introgression between westslope cutthroat and rainbow trout in the Clark Fork River drainage, Montana. Proc. Mont. Acad. Sci. 43:1-18.

Lincoln, R.F. and A.P. Scott. 1983. Production of all-female rainbow trout. Aquaculture 30:375-380.

Lincoln, R.F. and A.P. Scott. 1984. Sexual maturation in triploid rainbow trout, *Salmo gairdneri* Richardson. J. Fish Biol. 25:385-392.

May, B., K.J. Henley, C.C. Krueger and S.P. Gloss. 1988. Androgenesis as a mechanism for chromosome set manipulation in brook trout *(Salvelinus fontinalis)*. Aquaculture 27:57-70.

National Research Council. 1972. Genetic vulnerability of major crops. National Academy of Sciences, Washington, D.C.

Parsons, J.E. and G.H. Thorgaard. 1985. Production of androgenetic diploid rainbow trout. J. Heredity 76:177-181.

Rohrer, R.L. and G.H. Thorgaard. 1986. Evaluation of two hybrid trout strains in Henry's Lake, Idaho, and comments on the potential use of sterile triploid hybrids. N. Amer. J. Fish. Manag. 6:367-371.

Scheerer, P.D. and G.H. Thorgaard. 1983. Increased survival in salmonid hybrids by induced triploidy. Can. J. Fish. Aquat. Sci. 40:2040-2044.

Scheerer, P.D., G.H. Thorgaard, F.W. Allendorf and K.L. Knudsen. 1986. Androgenetic rainbow trout produced from inbred and outbred sperm sources show similar survival. Aquaculture 57:289-298.

Scheerer, P.D., G.H. Thorgaard and J.E. Seeb. 1987. Performance and developmental stability of triploid tiger trout *(Salmo trutta* x *Salvelinus fontinalis)*. Trans. Amer. Fish. Soc. 116:92-97.

Smith, T.I.J. 1988. Aquaculture of striped bass and its hybrids in North America. Aquacult. Mag. 14(1):40-49.

Stöss, J. 1983. Fish gamete preservation and spermatozoan physiology, p. 305-350. *In* W.J. Hoar, D.J. Randall and E.M. Donaldson (eds.), Fish physiology, Vol. 9B, Academic Press, New York.

Streisinger, G., C. Walker, N. Dower, D. Knauber and F. Singer. 1981. Production of clones of homozygous diploid zebra fish *(Brachydanio rerio)*. Nature 291:293-296.

Thompson, B.Z., R.J. Wattendorf, R.S. Hestand and J.L. Underwood. 1987. Triploid grass carp production. Prog. Fish-Cult. 49:213-217.

Thorgaard, G.H. 1986. Ploidy manipulation and performance. Aquaculture 57:57-64.

Thorgaard, G.H. and S.K. Allen, Jr. 1987. Chromosome manipulation and markers in fishery management, p. 319-336. *In* N. Ryman and F.M. Utter (eds.), Population genetics and fishery management. University of Washington, Seattle.

Thorgaard, G.H., P.D. Scheerer, W.K. Hershberger and J.M. Myers. 1990. Androgenetic rainbow trout produced using sperm from tetraploid males show improved survival. Aquaculture 85: 245-251.

Trotter, P. 1987. Cutthroat: Native trout of the west. Colorado Associated University Press, Boulder.

Part II
Risk Reduction and Safety

Introduction to Part II

Management of Introductions and Transfers: A Commentary on the Changing Role of the Biologist

Roger Mann

More than ten years ago many of the contributors to this volume sat in an auditorium in Woods Hole, Massachusetts discussing the translocation of the Japanese oyster, *Crassostrea gigas* (Thunberg), from its native oriental zoogeographic range to waters in different regions of the world. Indeed, given its history of movement, the species was considered a suitable test case for discussion of evaluation procedures for introductions and transfers. The proceedings of that meeting were subsequently published (Mann 1979). I wrote the summary overview of the 1979 publication and, in preparing the introduction to this section, it is informative to review the situation as we left it in 1979 and the progress since that time.

Many realities remain unchanged: introductions and transfers will occur; some will be beneficial in both an ecological and economic sense. Future introductions and transfers will be effected, and social and economic pressures to consider movements will continue to increase as a part of both continuing commerce and new ventures. Absolute guarantees of ecological safety cannot be made where movements occur, and all decisions concerning movements are a compromise of biology, economics, politics, philosophy and probably many more disciplines of importance.

For some time the role of the biologist has not been well defined. Sometimes an expert witness, the biologist is trained to design experiments or collect information in a quantitative manner, analyse it statistically and draw appropriate conclusions; suddenly, biologists are thrown into an arena of laypersons with the task of delineating risks and benefits with inadequate data sets. Although several groups had been working on guidelines for assessing requests for and impacts of introductions and transfers (notably the International Council for the Exploration of the Sea, the American Fisheries Society, and the European Inland Fisheries Advisory Commission), their interaction appeared only in a fledgling stage and the biologist was often a minor player in the larger game. Clearly, these groups have, and are moving along, converging courses as their respective efforts in guideline production indicate. Uniformity of opinion among several previously disparate groups speaks to the broad value and importance of their communal recommendations. As can be seen from the papers in this book, the biologist has moved from the periphery to the center in the discussion of introductions, even acting as prime advocate in some instances. This is not to say that the biologist is now adequately equipped to assess all proposed introductions. An expanding data base on all species and environments (both donor and recipient) is essential, together with a clear understanding of the limitations of such information. Consider a zoogeographic range apparently limited by an environmental parameter such as temperature — such data is useful in comparing donor and recipient sites but potentially limited if other range limiting factors, such as salinity or predation are not also considered. Controllable functions, for example, limitation of premeditated movement, must be examined in the light of uncontrollable functions, such as bulk, transoceanic movement of ballast water and entrained species associated with commercial shipping.

Present avenues of commerce have made clandestine movement of species comparatively easy to effect. Control must be made attractive to be effective. An informed public is a powerful ally in this arena. Environmental and ecological activists are present from the high seas to Capitol Hill; they are a powerful political lobby and legal adversary. At the other end of the scale industries utilizing non-native material are image conscious. Negative publicity translates to losses in investor confidence and market share. Self policing makes sense. The research community, also offenders in effecting sometimes careless movements (but curiously often excluded from legislation-controlling movements), would benefit from heightened awareness to and greater self control of movement of species to regions beyond their natural zoogeographic range — perhaps this volume will help. A sharing of responsibility would allow government agencies to assume the role of reviewer and promoter rather than enforcer — a welcome development.

We stand at the edge of new technology which was simply unavailable a decade ago. The assertion that, once released, an introduction was essentially uncontrolled with respect to reproduction and ensuing competition with native species, may be challenged in increasing instances by the application of genetic manipulation to introduced stocks. Cell culture of plant material offers the option of purely vegetative material for introduction. Triploid induction and single sex manipulation of fishes is now available in an increasing number of species. Triploidy induction is practical in a limited number of shellfish species although one hundred percent induction in a population has not yet been achieved on a continuing basis at commercial levels. As the array of tools available increases so do the options for utilization of non-native genetic material in both an ecological and commercial sense. Exciting opportunities await us — we look forward to their development.

Literature Cited

Mann, R. 1979. Exotic species in mariculture. MIT Press, Cambridge, Massachusetts.

Chapter 5

National and Regional Jurisdictions: Activities and Plans

The National Biological Impact Assessment Program

DAVID R. MACKENZIE

Abstract: The need to field-test genetically modified organisms safely is a critical step in the sequence of research leading to the commercialization of biotechnology in agriculture. To address this need the United States Department of Agriculture has established the National Biological Impact Assessment Program (NBIAP) to facilitate the safe field testing of genetically modified organisms. NBIAP fosters safe field testing of genetically modified organisms through a computerized network for information exchange, facilitation of biological monitoring techniques, and by providing support for research in biosafety to develop new field-testing methods and better predictive models.

Introduction

The development of techniques to modify organisms genetically offers immense opportunities to resolve significant problems in agriculture by addressing issues of agricultural sustainability, lowering costs of production, improving the quality of harvested products, providing new or novel products, and protecting crops and livestock from abiotic and biotic stresses. These are just some of the anticipated applications of biotechnology to agriculture.

Along with these research opportunities comes the responsibility for providing adequate protection of public health and the environment. These protection measures are the realm of biosafety and the concerns for biosafety are of importance not only to the scientific community, but to the public at large.

To address biosafety concerns at the federal level, the Domestic Policy Council of the White House established the "Coordinated Framework for the Regulation of Biotechnology," published June 26, 1986, in the Federal Register. The Coordinated Framework uses existing federal laws to regulate the products and articles produced by biotechnology, rather than regulating the research process itself. The Coordinated Framework assumes that there is no fundamental difference between genetically modified organisms and those produced by conventional research methods. Thus, no new federal laws have been enacted. Consequently, several federal agencies are using existing authorities to regulate the products of biotechnology under the Coordinated Framework.

The U. S. Department of Agriculture (USDA) has chosen to meet its biosafety responsibilities for agricultural research through a dual system of review. The first part of this review system uses existing regulations to review and issue permits for field testing or commercial use for certain regulated products and articles. The second system involves scientific reviews, conducted by the USDA, of certain federally funded experiments under guidelines for field tests outside of the laboratory with genetically modified organisms. The first system (product regulation) is administered through the USDA's Marketing and Inspection Service while the scientific reviews are conducted under the authority delegated to Science and Education.

The USDA's Agricultural Biotechnology Research Advisory Committee (ABRAC) and its service unit, the Office of Agricultural Biotechnology (OAB), have under way an active program in bio-safety guideline development and case-by-case scientific review of individual projects. The ABRAC review process is modeled after the highly successful National Institutes of Health Recombinant DNA Advisory Committee which has for over a decade provided bio-safety assurances for contained laboratory experimentation with recombinant DNA. ABRAC will make use of NIH's existing network of Institutional Biosafety Committees (IBC). This IBC network will allow delegation of some types of review to the local level, or exemption from review for certain categories of experiments.

Another activity in the USDA has been the establishment by the Cooperative State Research Service (CSRS) of the National Biological Impact Assessment Program (NBIAP) to facilitate the safe field testing of genetically modified organisms. The NBIAP was first proposed by the National Association of State Universities and Land Grant Colleges' Committee on Biotechnology to address research needs in the public and private-sector research communities.

It is well recognized that agricultural research proceeds as a sequence of activities leading from the laboratory to eventual commercialization. A key step in this research process is the need to field-test genetically improved organisms to verify performance under realistic environmental conditions. This requirement is true not only for conventional research, but biotechnology as well.

The underlying assumption of the NBIAP is the well-recognized fact that there exists a tremendous body of knowledge on the methods and procedures of field testing from the conventional research of the past. This knowledge is applicable in many ways to the questions of safe field tests for genetically modified organisms. It is the purpose of NBIAP to facilitate the appropriate application of this body of knowledge to biotechnology and to develop new methods to assure safe and productive field tests with genetically modified organisms.

To accomplish its objectives, the Program staff have developed a strategic plan which identifies three areas of activity. Those areas are:

- Information networks
- Development of biological monitoring techniques
- Support for biosafety research

Details on how the Program is addressing these needs through specific activities are the subject of the remainder of this report.

Information Network

In collaboration with a number of institutions, NBIAP is working to develop and implement an information network to support biosafety needs in both the public- and private-sector research communities. The information network is accessible over telephone lines (an "800" number). This distribution system is intended to encourage maximum participation by being economical and as user friendly as possible.

Electronic Bulletin Board

The information network participants link via their computer terminal and modem to an electronic bulletin board that provides up-to-date information on biosafety-related research activity and announcements. Provision has been made for broad participation on the bulletin board and for individuals to communicate with each other via an electronic mail service. Thus, individuals can use the network to contact other scientists working in similar areas to obtain assistance when planning field research with genetically modified organisms.

Data Bases

The electronic bulletin board also serves as the gateway to 11 data bases designed to provide accurate and up-to-date information on a variety of topics. The purpose of the data bases is to assist principal investigators in securing authoritative and scientifically valid information.

The USDA's National Agricultural Library (NAL) is developing and supporting some of the NBIAP information systems data bases. The first one is called the "Yellow Pages." The types of information will include several bibliographic and nonbibliographic listings such as (1) directories of companies and information sources; (2) current videotapes that are available on techniques and field studies; and (3) data bases. A second data base contains current literature. This bibliographic file will provide information on currently available books, reports, and newsletters.

The NBIAP will directly support several data bases. One data base contains the current text of all federal laws, regulations, rules, guidelines, and executive orders pertaining to biotechnology and biosafety. The system also contains a current data base of all public and private Institutional Biosafety Committees (IBC). Information will include the names of contacts, addresses, telephone numbers and facsimile numbers. NBIAP will also maintain a current data base on all approved applications to federal agencies related to field-test permits, licenses, and scientific reviews. Proprietary information will not be included.

Other data bases are to be maintained in cooperation with the information network participants. These include information on topics such as new jobs and positions available, funding sources, and equipment and instruments that are related to field-test biosafety.

The University of Arizona has constructed a site data base. This data base records the locations where field tests with genetically modified organisms have been conducted. Similar collaborative projects are being planned to support organism data bases to make available to participants information on sources of germ plasm, clones, genetic tags, genomic sequencing information, and resource scientists.

Knowledge Base

The third component of the information network is the knowledge base that is designed to assist principal investigators in preparing applications for permits or licenses and/or scientific reviews. The knowledge base contains an expert system which identifies the responsible federal agencies under the Coordinated Framework to which the application(s) should be directed.

The knowledge base also has an educational component of information drawn from broad sources but focused on issues in biosafety, and field tests. We have identified an enormous amount of biosafety literature that has been organized into "hypertext." Hypertext is a hierarchal presentation of information ranging from general to specific explanations that can be interactively selected by the user. Through hypertext the reader can pursue information to his or her desired level of detail.

The third segment of the knowledge base is the "intelligent form generator." This unit of the knowledge base can be used to develop first drafts of applications for federal permits, licenses, and scientific reviews. The "intelligent form generator" will present to the user menus of experimental design options for their selection as appropriate to their proposed test. This information is assembled from existing knowledge (expert panels, literature, and previously approved applications). Users have access to previously written standardized text, technically precise descriptions, test-site information, and other data base resources. By combining data base information with extensive menus specific to the 79 organism categories that have been designated for agriculture (e.g., pond-contained fish, nightshades, nonpathogenic soilborne bacteria) specific methods, procedures, and protocols can be brought together to form a technically precise application. The system offers a word processing option that will allow redrafting of the materials for clarity and scientific accuracy.

The purpose of the intelligent form generator is to lift somewhat the burden of the principal investigator in complying with federal biosafety oversight requirements. This is important as some researchers now see the regulatory process as a deterrent to conducting field tests with genetically modified organisms. A national survey has been conducted to determine what factors are limiting the number of requests. Hopefully, by offering this biosafety information network safe field testing will be facilitated through assisting scientists in the exchange of biosafety information and by helping them in the application for permission to conduct field tests.

Biological Monitoring

Another activity of the National Biological Impact Assessment Program is to facilitate biological monitoring of genetically modified organisms tested outside of contained laboratories. To do this, the Program is supporting research in the monitoring of organisms to build a better scientific understanding of the dispersal of organisms into the environment. This activity is supported as a grants program that will continue to expand as funds become available for this type of research. NBIAP will also continue the development of plans for a research-based reporting system using the existing, extensive network of public- and private-sector scientists who now continuously watch over U.S. agricultural production. Scientists of the State Agricultural Experiment Stations, the Cooperative Extension Service, and others serve as sentinels in a national agricultural monitoring system. This warning system will need augmentation and support to monitor adequately agricultural biotechnology research and the commercial use of the products of biotechnology. NBIAP is identifying their needs and reaching out to make plans for implementation.

Biosafety Research

The National Biological Impact Assessment Program is also supporting specific biosafety research projects. Biosafety is a research topic unto itself. Although much useful information can be derived from conventional research, some aspects of biotechnology research raise unique biosafety questions. New biosafety methods must be developed for those needs; these methods will require new predictive models to understand the likely consequences of a field test with a genetically modified organism or the commercial release of a biotechnology product prior to that activity. Improved biosafety methods and accurate prediction models probably represent the best way to provide biosafety assurance in the long term. They will require a more thorough understanding of how organisms interact and how varying environmental conditions affect movement and dispersal patterns, how physical factors affect survival, fitness, and reproductive behavior and how all these come together to impact the environment or public health. More knowledge on these topics should lead to better biosafety assurances.

Conclusions

The present U.S. federal procedures for scientific review and regulatory permits and licenses are seen as an adequate interim step. Still, a more complete understanding of how a genetically-modified organism will perform, how it can be controlled, and how one is to recognize the unexpected are the key elements in a national bio-safety program if it is to have scientific as well as public acceptance. The National Biological Impact Assessment Program, by promoting the exchange of information, biological monitoring to detect the unexpected, and scientific research to predict the expected, will contribute to the safe and rapid application of biotechnology to agriculture.

Environmental Protection Agency Oversight of Microbial Pesticides

MICHAEL MENDELSOHN
AMY RISPIN
PHILLIP HUTTON

Abstract: The purpose of this paper is to provide an overview of the Environmental Protection Agency's oversight of microbial pest-control agents. The application of the Federal Insecticide, Fungicide, and Rodenticide Act and Sections 408 and 409 of the Federal Food, Drug, and Cosmetic Act are discussed. General data requirements for EPA registration of microbial pest-control agents are discussed along with the specific requirements for nontarget aquatic organism testing.

Background

This report presents an overview of the Environmental Protection Agency's (EPA) oversight of microbial pest-control agents (MPCAs). These agents are regulated under the Federal Insecticide, Fungicide, and Rodenticide Act (FIFRA) and Sections 408 and 409 of the Federal Food, Drug, and Cosmetic Act (FFDCA). MPCAs include living microorganisms such as bacteria, fungi, protozoa, algae and viruses that are introduced into the environment to prevent, repel, destroy, or mitigate the population or biological activities of another life form considered to be a pest under Section 2(t) of FIFRA.

The first of these microbial pest-control agents *(Bacillus popilliae)* was registered in 1948. At this writing, there are 21 microbial pesticide active ingredients used in several hundred products registered for use in agriculture, forestry, mosquito/blackfly

Table 1. EPA registered microbial pesticides.

Microorganism	Year Registered	Pest Controlled
Bacteria		
Bacillus popilliae/B. lentimorbus	1948	Japanese Beetle larvae
B. thuringiensis "Berliner"	1961	Moth larvae
Agrobacterium radiobacter	1979	*A. tumefaciens* (crown gall disease)
B. thuringiensis aizawai	1981	Wax Moth larvae
B. thuringiensis israeliensis	1981	Mosquito larvae
Pseudomonas fluorescens	1988	*Pythium, Rhizoctonia*
B. thuringiensis San Diego	1988	Coleopterans
B. thuringiensis tenebrionis	1988	Coleopterans
B. thuringiensis EG2348	1989	Gypsy moth
B. thuringiensis EG2371	1989	Lepidopterans
B. thuringiensis EG2424	1990	Lepidopterans/coleopterans
Viruses		
Heliothis Nuclear Polyhedrosis Virus (NPV)	1975	Cotton Bollworm, Budworm
Tussock Moth NPV	1976	Douglas Fir Tussock Moth larvae
Gypsy Moth NPV	1978	Gypsy Moth larvae
Pine Sawfly NPV	1983	Pine Sawfly larvae
Fungi		
Phytophthora palmivora	1981	Citrus Strangler Vine
Colletotrichum gloeosporioides	1982	Northern Joint Vetch
Trichoderma harzianum/ Trichoderma polysporum	1989	Wood rot
Protozoa		
Nosema locustae	1980	Grasshoppers

control and homeowner situations (Table 1). The Office of Pesticide Programs (OPP) formally recognized in 1979 that MPCAs are distinct from conventional chemical pesticides and made the commitment to develop appropriate testing guidelines for microbial pesticides. The guidelines for microbial and biochemical pesticides were published in 1982 as the Pesticide Assessment Guidelines, Subdivision M. (The microbial section of Subdivision M was updated and revised in July of 1989.) In 1984, data requirements for MPCAs were codified in the Code of Federal Regulations, Title 40, Part 158.170 (40 CFR Part 158.170).

FIFRA and FFDCA Requirements for Microbial Pesticides

The main aspects of the Federal Insecticide, Fungicide, and Rodenticide Act (FIFRA) as they relate to the pesticide registration process are summarized below. EPA's oversight extends from premarket testing, small-scale field-test notifications and Experimental Use Permits (EUPs), through full commercial use of a pesticide product, Section 3 Registrations. Any pesticide product that is to be used on a food crop to be distributed in commerce must have a tolerance (maximum legal residue level) or temporary tolerance under Section 408 or 409 of the FFDCA or must be granted an exemption from the requirement of a tolerance. All MPCAs registered to date for use on food crops are currently exempt from the requirement of a tolerance.

In the process of pesticide development, field testing is often necessary to evaluate the environmental fate and efficacy of a pesticide. Title 40 of the Code of Federal Regulations Part 172 (40 CFR Part 172) describes when it is necessary to obtain an Experimental Use Permit under Section 5 of FIFRA. Briefly, if the pesticide is to be used on a food crop that will be distributed in commerce or the size of the test acreage is greater than 10 acres on land or 1 surface acre of water, an EUP is required. For nonfood uses, it is generally presumed that an EUP is not required for field-tests under 10 acres on land or 1 surface acre of water. Other criteria used to determine when an EUP must be obtained are set forth in 40 CFR Part 172.3. An EUP is of limited duration and requires that the test be carried out under controlled conditions.

In 1984, EPA recognized that there may be potentially significant impacts from the use of genetically altered and nonindigenous MPCAs in the environment, even at the small-scale testing stage. To address this concern, EPA issued an interim policy statement that the presumption in the EUP regulation (40 CFR 172.3) that an EUP would not be required for small-scale testing would not be applicable to such tests involving genetically altered and nonindigenous microbial pesticides. This policy requires notification of EPA prior to small-scale field testing of geneticaly altered and nonindigenous MPCAs so that the Agency can determine whether an EUP is required. The 1984 interim policy was subsequently incorporated into a 1986 policy statement that retains the notification requirement. The Agency is currently working on revising the existing EUP regulation to codify the notification requirement.

As noted earlier, before a company can market a pesticide product, it generally must obtain a Section 3 registration; however, there are two additional means under FIFRA whereby a company may distribute a pesticide product in the absense of an experimental use permit. The first of these is pursuant to an emergency exemption under Section 18 of FIFRA. Under this section, federal or state agencies may request an unregistered use of a currently registered pesticide product or the use of an unregistered pesticide product. Such a request can only be granted when there is a potentially severe economic or human health impact and no other alternatives are available for pest-control. A Section 18 exemption usually allows use of the particular pesticide product for a year; however, the time for use allowed may be more or less. In addition to emergency exemptions under Section 18, cases exist where a particular pesticide product may be registered for one or more uses, but not for a particular use which is determined by the state as being a special local need. In these cases, the state may register that use or formulation needed for the special local need under Section 24(c) of

FIFRA. The EPA has 90 days to disapprove of such state registrations. If the Agency does not respond, then that use and/or formulation, heretofore not part of a federal registration, becomes part of either an existing or a new federal registration. (Refer to Section 24(c)(2) of FIFRA for specific details.)

Data Requirements

The recommended test methods provided in the 1989 revised Pesticide Assessment Guidelines Subdivision M and corresponding data requirements in 40 CFR Part 158 are set forth in four basic areas: product and residue analysis, environmental fate, nontarget organism testing, and human health effects. The testing schemes for human health and nontarget organism effects are tiered, i.e., certain testing is not required unless triggered by initial testing. Residue analysis and environmental fate requirements are usually triggered by human health effects data and nontarget organism data, respectively. An example is the nontarget organism/environmental fate tier testing scheme (Table 2). At the first tier, short-term testing utilizes maximum hazard dosing. If no adverse results are observed in Tier I, then further testing is not warranted nor is environmental fate data required. In the first tier of nontarget organism testing, avian oral, freshwater fish, freshwater aquatic invertebrate, and honeybee testing are required. In addition, tests to evaluate MPCA effects on wild mammals, plants and beneficial insects are required, depending on the proposed use site, target organism and degree of anticipated exposure.

Like the nontarget organism testing, the toxicology testing is also tiered (Table 3). Tier I consists of studies including oral toxicity/pathogenicity, dermal toxicity, pulmonary toxicity/pathogenicity, intravenous toxicity/pathogenicity, primary eye irritation, reporting of any observed hypersensitivity incidents, and cell culture tests with viral pest-control agents.

Pursuant to FIFRA Section 3 and within the specifications of 40 CFR Part 158 and Subdivision M, EPA has the authority to ask for data to address any additional questions regarding any of

Table 2. Nontarget organism testing.

Tier	Test
Tier I	Avian oral Wild mammal Freshwater Fish Freshwater Aquatic Invertebrate Estuarine and Marine Animal Nontarget Plant Nontarget Insect Honeybee
Tier III	Terrestrial Wildlife & Aquatic Organism Avian Chronic Pathogenicity and Reproduction Aquatic Invertebrate Range Fish Life Cycle Studies Aquatic Ecosystem Special Aquatic Nontarget Plant
Tier IV	Simulated and Actual Field Tests (Birds, mammals, aquatic organisms, insects)

Table 3. Toxicology Testing.

Tier	Test
Tier I	Acute Oral toxicity/pathogenicity Acute dermal toxicity Acute pulmonary toxicity/pathogenicity Acute intravenous toxicity/pathogenicity Primary eye irritation Hypersensitivity incidents Cell culture with viral pest-control agents
Tier II	Acute toxicity Subchronic toxicity/pathogenicity
Tier III	Reproductive and fertility effects Oncogenicity Immunodeficiency Primate infectivity/pathogenicity Nontarget Plant

the aforementioned data areas. EPA also has the authority to waive data requirements if (1) based on information provided by the applicant, (2) agency scientists determine that they are not applicable to the risk assessment or (3) they are inappropriate for the MPCA in question.

Nontarget Aquatic Organism Data Requirements

How does EPA assess the risk of an MPCA to nontarget aquatic organisms? In Tier I, several studies are required for all end-use products intended for outdoor application and all manufacturing-use products that legally may be used to formulate such end-use products. The first of these is the freshwater fish toxicity/pathogenicity study. If direct application to water is not expected from the use pattern of the product, then only one species of fish need be tested, preferably the rainbow trout. If direct application to water is expected, then bluegill sunfish is to be tested as well as the rainbow trout. As with all Tier I tests, maximum hazard dosing is required.

Besides freshwater fish studies, freshwater aquatic invertebrate toxicity and pathogenicity testing is required. Unless the pesticide product is to be applied directly to water, one species of aquatic invertebrate is to be tested. Products that are expected to have direct water application need two species tested, including one benthic and one planktonic species. Again, maximum hazard dosing is necessary. If no toxic or pathogenic effects are observed, then no further testing is warranted.

In addition to freshwater testing, estuarine and marine animal toxicity and pathogenicity testing are required when the end-use product is intended for direct application into estuarine or marine environments or is expected to enter this environment in significant concentrations due to the proposed use pattern or intrinsic mobility. Toxicity and pathogenicity are determined for one species of shrimp, preferably *Paleomonetes vulgaris*, and one estuarine or marine fish species. Again, maximum hazard dosing is utilized. If no toxic or pathogenic effects are observed then further testing is not required.

Nontarget plant testing may be required depending on the proposed use site, target organism and degree of anticipated exposure. The number of plant species tested depends on the host range of the MPCA and its similarity to known plant pathogens. For microbial pesticides that have aquatic uses or that may be expected to disseminate to, and survive in aquatic ecosystems, aquatic plants must be included within the testing regimen. As with plants, nontarget beneficial insect testing may also be required. The kinds and number of species to be tested depends on the host range of the MPCA and use sites.

OPP Organization

The mechanics of the Agency's review for field testing or registration involves the coordination of many people and offices. Once a submission has passed preliminary screens, the Registration Division routes it to the science-support divisions for technical review, namely the Health Effects Division and the Environmental Fate and Effects Division. The Registration Division Product Managers who are responsible for coordinating the registration process and who are involved in final decisions regarding registration or field test approval of MPCAs are Product Manager 17 and Product Manager 21. The coordination of the scientific reviews is handled through the Science Analysis and Coordination Staff within the Environmental Fate and Effects Division. In addition to the Office of Pesticide Programs, the Office of General Counsel is often involved in assisting with legal issues associated with MPCA registrations and field testing. The approval of field testing or registrations ultimately rests with the Assistant Administrator for Pesticides and Toxic Substances. However, it is referred to this level only in special situations. These decisions are usually delegated to the Director of the Office of Pesticide Programs or the Director of the Registration Division.

Shellfish Health and Protection

FREDERICK G. KERN
AARON ROSENFIELD

Abstract: This report discusses the movement of living organisms *de novo* into marine and estuarine ecosystems as a result of human actions, particularly from the viewpoints of how resource and environmental managers and decisionmakers cope, or at least attempt to contend, with the many problems and ramifications associated with the introductions and transfers of molluscan shellfish. Approaches and strategies to protect and provide for the health and continuing productivity of shellfish resource species are discussed.

Introduction

Considerable concern and anxiety are being voiced by shellfish resource management and regulatory agencies over findings that apparently link previously unobserved and undetected organisms (macro and micro) with adverse biological and environmental phenomena. Some of the undescribed, unreported biological agents (such as pathogens) now being observed have always been present within their ecological niches. They may have been present, however, in minute numbers, or as unrecognized cryptic life history stages, or as opportunists in their mode of action. Certain biotic or abiotic conditions may be necessary to stimulate them to exert their effects. It is also possible that these organisms were not detected because no one looked for them. Furthermore, we may have lacked the devices and systems necessary to observe them; we misinterpreted their symptoms, actions, and effects, or out of ignorance we have simply attributed effects to the wrong causes.

Alternatively, past history has often noted, and current history continues to record, incidents of the dispersal and spread of biotic agents from one ecosystem to another through actions by natural forces or by deliberate or accidental actions of man (Regier 1968; Lachner et al. 1970; Walford and Wicklund 1973; Rosenfield and Kern 1979; Rosenthal 1980; Carlton 1985; Carlton 1987). If these organisms establish self-sustaining, reproducing populations, i.e., colonize an ecological niche, this would account for their presence in areas and in tissues and cells of hosts in which they had not been detected previously.

There are several activities, some might call them mechanisms, whereby living organisms, for intended or unintended purposes, can find access into new aquatic ecosystems (Table 1). The National Marine Fisheries Service (NMFS), as the nation's lead agency for the conservation, management, and utilization of marine species, has frequently been asked to provide counsel and to comment on these activities — some already taken, others representing future planned or theoretical actions. In most instances, communications with NMFS have dealt with requests for details on what policy, regulations, and operational and informational needs have to be satisfied to ensure that risks of adverse effects resulting from introductions and transfers of marine shellfish species are minimal. In many cases, no counsel whatever is sought from NMFS or any other federal or state agency. Sometimes, counsel and information provided are simply ignored, inadequately applied, or abandoned after brief trials.

It should be emphasized that not all biological invasions, transfers, and colonizations by shellfish can be considered to have had adverse impacts. Indeed, many have proved to be very beneficial, mostly, however, from economic and not biological perspectives. Consequently, cooperative state-federal-industry programs must be developed and implemented to prevent, control, or otherwise reduce the risk of disease dissemination by transfers of shellfish from one aquatic ecosystem to another. Furthermore, measures must be made available to prevent the introduction and establishment of injurious aquatic animal and plant species into

Table 1. Activities representing potential routes of entry of genetic material into aquatic ecosystems.

- Introductions and transfers of species for aquaculture purposes, disease, pests, predator, competitor risks
  - Adults as brood stocks
  - Gametes, larvae, and post-larvae
  - Genetically manipulated — hybrids
- Current commercial practices
  - Growth enhancement and production
  - Direct consumption
  - Depuration and relaying
  - Product manufacture and processing
- Education, scientific study, and biomedical-bioveterinary research
  - Disease resistant or tolerant forms — bioremediation/biotechnology
  - Genetically engineered animals/plants — gene insertion, DNA alterations, and recombinants (micro- and macroorganisms), cell fusions
- Biological control agents and agricultural crop improvement
  - Genetically engineered microorganisms (pathogens)
  - Non-engineered microorganisms (pathogens)
- Aquarium systems, pets, aesthetics, recreation, escapes
  - Complete confinement
  - Open systems and combinations, ship hulls
- Overboard disposal
  - Bait, ballast water, refuse and waste, discards,
  - Ocean dumping, sludge disposal
- Farm and land runoff — point and non-point sources
  - Refuse and waste, storm events, sewer discharges
- Construction and waterway modification
  - Canals, connecting waterways
  - Harbors, boat basins, marinas, dredging, and sludge disposal
- Military and outerspace operations
  - Biological warfare
  - Outerspace returns — vehicles, materials

waters of the United States where they might adversely affect native fish and shellfish resource species or disturb environmental integrity.

State and Regional Activities

The ultimate control over deposition of shellfish into waters of the United States must abide with state(s) receiving the shellfish. It is in state waters that any negative or beneficial results of the introduction most likely will be realized. To that end, concerned regional groups of state, federal, university and industry representatives have met to address problems related to the introduction and transfer of marine organisms. In 1980, the Pacific Marine Fisheries Commission (PMFC) took a lead role and formed a shellfish committee to address these problems. This committee developed a cooperative agreement for the Pacific Coast and Hawaii that was signed by all of the member states. The agreement established protocols for the "safe" movement of molluscan shellfish and provided for official lines of communications between the member states and the Canadian province of British Columbia (see Appendix). In 1983, three regional groups representing the northeast, mid-and south Atlantic coastal states met and developed an east coast policy statement. The policy statement outlined the problems associated with the movement of shellfish, and suggested steps that should be taken to address these problems. The policy was presented to the Atlantic States Marine Fisheries Commission (ASMFC) later in 1983 which, after consideration of the needs of the Atlantic Coast states, then formed a shellfish transfer committee to address molluscan transport problems. This committee adopted a cooperative agreement, largely based on the earlier PMFC shellfish committee agreement.

The ASMFC Shellfish Transport Committee has moved ahead and is now in the process of developing a Fishery Management Plan (FMP) for shellfish that would standardize procedures used to examine and move shellfish from one geographic area to another to prevent the spread of diseases that are now causing major losses to the molluscan shellfish industry. Essentially, the Plan calls for establishment of standard protocols for gross and microscopic (histologic) examination of shellfish prior to their interstate transfer. It also calls for development and implementation of a training program to increase proficiency of researchers at state shellfish management agencies so as to readily detect pests, parasites and pathogens of animals to be transported from one location to another. Copies of the Plan are available from the ASMFC main office.[1]

The Plan should prove useful as a model to the Gulf states and PMFC in developing regional strategies and plans as they relate to disease control and prevention for molluscan resources in aquaculture programs. Of course, a coordinated, unified plan would be a desirable national objective. At present, however, the Gulf States Marine Fisheries Commission (GSMFC) has indicated that member states agree that sufficient legislation and supportive regulations exist to mitigate the risk of importing biological agents harmful to fishery resources (L. Simpson personal communication).[2] Consequently, except for the oyster *(Crassostrea virginica)*, no attention is being given to developing a multistate coordinated fishery management plan similar to that of the ASMFC. Rather, each state within the GSMFC is preparing independent aquaculture plans that will address management questions and operations as they apply to molluscan species.

National and International Activities

One of the most important and visible areas of NMFS involvement with introductions and transfers is related to the introduction and transfer of molluscan species from one location to another for mariculture purposes. If these movements involve in-

---

[1]Atlantic States Marine Fisheries Commission, 1400 16th St. N.W., Suite 310, Washington, D.C. 20036

[2]Larry Simpson, Executive Director, Gulf States Marine Fish-eries Commission. P.O. Bocx 726. Ocean Springs, Mississippi 39564.

ternational shipments, then the guidelines and codes of practice developed by international organizations such as the International Council for the Exploration of the Sea (ICES) and the European Inland Fisheries Advisory Commission (EIFAC) (Sindermann, this volume) are recommended.

Of the seafood currently consumed in the United States, 80% originates and is imported from foreign sources. A significant proportion consists of shellfish, including bivalve molluscs such as mussels, clams, oysters and scallops. In addition to the foreign origin of seafood for direct consumption, a number of aquatic organisms are shipped from distant locations intercoastally, interterritorially or intercontinentally for commercial, educational and research purposes, and even for use as bait. Many of these organisms are imported live, or processed incompletely; thus, they, and sometimes their associated milieu, have the potential to serve as vehicles for the import of exotic species, or of other agents that may be attached to their outer surfaces or within their cells and tissues. These latter organisms may in turn affect the health of domestic biota, including humans, as well as environments into which they may be placed (Rosenfield and Kern 1979).

Strict regulations are enforced to exclude live and processed agricultural products that do not meet inspection guidelines as established by the Animal and Plant Health Inspection Service (APHIS) of the U.S. Department of Agriculture. Obviously, there is a similar need for such a service with regulations as they apply to seafood (a Marine or Aquatic Animal and Plant Health Inspection Service). The National Marine Fisheries Service has established an in-house multiregional committee to examine the several problems associated with past and proposed introductions and transfers of aquatic species, including those that may have been accidentally translocated. The committee is charged with making contacts and gathering information appropriate for preparing a draft policy statement and devising strategies and plans for mitigating the adverse effects of introductions and transfers of marine species.

The Lacey Act as amended (18 U.S.C. 421, Part 16) partially addresses this problem by regulating specific import-transport actions into the United States that would result in harm to domestic fishery resources. The Act makes it unlawful to deliver, carry, transport or ship by any means for commercial or non-commercial purposes, or sell in interstate or foreign commerce, any fish, mollusc or crustacean, taken or sold, in violation of any law or regulation of any state or foreign country or in violation of federal law or regulation. It also prescribes package marking requirements, authorizes enforcement procedures and establishes both civil and criminal penalties. Currently, regulations imposed at the federal level deal primarily with salmon and their diseases.

A Presidential Order (11987) signed by President Carter addresses actions by federal personnel that would encourage them not to export or import exotic species, and directs the various department secretaries to develop regulations implementing this order. Several attempts have been made to develop federal regulations, but none have been implemented (Peoples et al., this issue). The American Fisheries Society has taken a strong stand in adopting a position on the introductions of aquatic species in U.S. waters (Kohler and Courtenay 1986; Kohler, this volume).

Congress most recently passed Public Law 101-646 on November 29, 1990 to be known as the "Nonindigenous Aquatic Nuisance Prevention and Control Act of 1990." The Act identifies four purposes: (1) to prevent unintentional introduction and dispersal of nonindigenous species into waters of the United States through ballast water management and other requirements; (2) to coordinate federally conducted, funded or authorized research, prevention control, information dissemination on the zebra mussel and other aquatic nuisance species; (3) to develop and carry out environmentally sound control methods to prevent, monitor and control unintentional introductions of nonindigenous species from pathways other than ballast water exchange; (4) to understand and minimize economic and ecological impacts of nonindigenous aquatic nuisance species that become established, including the zebra mussel; and (5) to establish a program of research and technology development and assistance to states in the management and removal of zebra mussels. Programs are just beginning to be developed.

The Food and Drug Administration (FDA), under an authority to assure that fresh shellfish are safe and wholesome for human consumption, administers the National Shellfish Sanitation Program (NSSP). Foreign governments wishing to ship fresh shellfish products to the United States have affirmed by Memorandum of Understanding (MOU) with FDA their intentions to cooperate to assure that fresh and frozen shellfish meet the guidelines set forth in the NSSP. Both FDA and NMFS recognize that these authorized introductions of foreign shellfish have the potential for certain environmental consequences resulting from the incidental importation of shellfish diseases, parasites, predators or other organisms (pest species) should these materials be placed in U.S. waters. Since there is a potential environmental consequence of the introduction of live shellfish, an environmental assessment of agency-initiated actions should be prepared.

Measures to mitigate the environmental consequences are being implemented on a case-by-case basis between NMFS, the foreign government and FDA. These measures include a review of existing research data, a two-year disease inspection and monitoring program by NMFS, and shipping labeling in the form of a "NOTICE TO RECIPIENTS" informing them not to relay shellfish in U.S. waters, not to hold in wet storage, and not to discard waste shell material where it may reach and contaminate the aquatic and/or marine environment. Should this program identify a potential environmental problem, then the need for preparing a new environmental impact statement will be reexamined by FDA, the permitting agency.

Safeguards must be developed to prevent problems that can, and will, arise from the examples given in Table 1. For the most part, the possible routes of entry are only just beginning to be covered by any of the above mentioned regulations. Obviously, however, there will never be enough funds or resources to adequately install meaningful efficient programs to prevent or control injurious or nuisance impacts that may result from all of the actions as listed in Table 1. Their importance to federal, state or local jurisdictional agencies and officials must be given priority according to their interpretation of legal and ethical mandates. Furthermore, the potential qualitative and quantitative impacts in terms of effects on resident biota, human health and environmental quality must be evaluated before actions are taken or projects implemented. At a minimum, environmental impact statements (i.e., risk assessments) should be a requirement on the part of those who would participate in these actions.

In summary, resource and habitat management decisions should not be made reactively; rather, they should be made in a proactive mode, if possible. Many laws, regulations, rules, action plans, operational strategies, guidelines, codes of practice, decision models and a plethora of recommendations produced by research and managerial organizations and planning committees are already available to those who could use them to good advantage. From the personal perspectives of the authors, it is time they do so!

Literature Cited

Carlton, J.T. 1985. Transoceanic and interoceanic dispersal of coastal marine organisms: the biology of ballast water. Oceanogr. Mar. Biol. Annu. Rev. 23: 313- 317.

Carlton, J.T. 1987. Patterns of transoceanic marine biological invasions in the Pacific Ocean. Bull. Mar. Sci. 31: 452-465.

Kohler, C.C. and W.R. Courtenay. 1986. American Fisheries Society position on introductions of aquatic species. Fisheries 11(2): 39-42.

Lachner, E.A., C.R. Robins and W.R. Courtenay, Jr. 1970. Exotic fishes and other aquatic organisms introduced into North America. Smithsonian Contr. Zool. 59: 1-29.

Peoples, R.A., Jr., J.A. McCann and L.B. Starnes. 1991. Introduced Organisms: Policies and Activiies of the U.S. Fish and Wildlife Service. This volume.

Regier, H. A. 1968. The potential misuse of exotic fish as introductions. Canadian Commission on Freshwater Fisheries Research, Ottawa, pp. 92-111.

Rosenfield, A. and F.G. Kern. 1979. Molluscan imports and the potential for introduction of disease organisms, p. 165-189. *In* R. Mann (ed.), Exotic species in mariculture. The MIT Press, Cambridge, Massachusetts.

Rosenthal, H. 1980. Implications of transplantations to aquaculture and ecosystems. U.S. Dept. Commerce, Nat. Mar. Fish. Serv., Mar. Fish. Rev. 42(5): 1-14.

Walford, L. and R. Wicklund. 1973. Contribution to a world-wide inventory of exotic marine and anadromous organisms. Food and Agriculture Organization of the United Nations, Fish. Tech. Pap. No. 121.

Appendix

Cooperative Agreement for Interstate Transfer of Shellfish

Because of the increasing danger of spreading shellfish pests, predators, and disease problems during the interstate transfer of shellfish (molluscan and crustacean), it is recognized that coordinated control is necessary.

Therefore, the states of Alaska, California, Hawaii, Oregon and Washington and the province of British Columbia, under aegis of the Pacific Marine Fisheries Commission (PMFC), agree to the following operating policies and procedures:

1. Primary control of imports lies with the importing state. The decision about adequacy of information and suitability of any evaluating agency or laboratory will be with the importing state. However, consistent with the foregoing, the states and province declare their intention to cooperate on a reciprocal basis in the conduct of inspections and issuance of permits.

2. In the event application for transfer is made to the exporting state, the appropriate agency of the importing state will be immediately notified. In the event a permit is to be issued by the exporting state, a minimum 10-day delay will be observed after notification of the appropriate agency in the recipient state prior

to the effective date of any permit or certification for transportation within the exporting state.

3. When feasible from both a legal and operational aspect, the exporting state will issue the same certification and/or inspection concerning shellfish pests, predators, pathogens or parasites that would be involved for a domestic shipment.

4. Every reasonable effort will be made to expedite applications and minimize impact on the applicant. However, unless the required information is routinely available, or can be readily obtained from recognized sources, it shall be the applicant's responsibility to underwrite costs of the needed evaluation. Also, this procedure could significantly lengthen the time for approval.

5. Available information will be utilized if deemed adequate. In the absence of such, new information will be sought. The impracticality of requiring complete or absolute information is recognized.

6. There shall be free communication and exchange of information among the states and province sharing knowledge or techniques, as well as disclosure of new problems or restricted areas. The available information is to be stored through PMFC and will be updated at least on a yearly basis.

7. It is recognized that considerable information and expertise exists within the federal government and other research institutions concerning the diseases of aquatic molluscs and crustaceans. It is the intent of the member states and province, working through PMFC, to utilize these resources.

Introduced Organisms: Policies and Activities of the U.S. Fish and Wildlife Service

ROBERT A. PEOPLES, JR.
JAMES A. MCCANN
LYNN B. STARNES

Abstract: This paper presents information about the statutory basis for U.S. Fish and Wildlife Service programs and policies regarding introduced species and then describes the development and current status of three Service activities related to introduced species. The paper concludes with a discussion of cooperative approaches that are being, or might be, taken by the fisheries community to resolve the growing, increasingly complex, problem of introduced species.

Introduction

Introduction of fish and other aquatic organisms into North American waters is a long-standing practice. Beginning three centuries ago (McCann 1984a) and continuing sporadically into this century, the introduction of exotic fish accelerated substantially after World War II and peaked in the early 1960s (Welcomme 1986). More than 100 exotic fishes — as used here, species not native to North American waters (Shafland and Lewis 1984; Kohler 1986) — are now known to occur in the United States, including at least 41 that have become established (Courtenay et al. 1984).

Professional involvement with the introduction of fish and other aquatic organisms is also long-standing; fishery managers were responsible for many of the early introductions of exotic fish into waters of the United States, as well as for transplants of native fish in North America. However, this practice has only recently become a concern of fishery and other professional resource managers.

Organized professional and governmental concern about the introduction of aquatic organisms into the United States, and U.S. Fish and Wildlife Service involvement in the problem, date from the late 1960s, (McCann 1984b). This interest resulted in a spate of state and federal policy and regulatory initiatives over several years to control or prohibit the importation or introduction of aquatic organisms.

Regulatory actions related to this issue generally stalled by the mid-1970s, when predicted ecological disasters failed to materialize, and in the face of strong assertions that tough preventive measures would result in significant impacts on the pet trade and other industries. In the absence of scientific evidence to the contrary, the potential benefits from introducing additional species overshadowed the often general and seemingly speculative concerns about the immediate, long-term, and cumulative impacts of introductions. Only in the last few years has research documenting the adverse effects of past introductions and an increase in the number of aquatic introductions, both intentional and unintentional, revived concern about the consequences of such actions and elicited interest in further development of policies and programs to confront this issue.

Until recently, U.S. Fish and Wildlife Service efforts directed toward introduced species problems and issues have been limited to three areas. First is a regulatory process, based on the injurious wildlife provision of the Lacey Act of 1900, which allows the Service to prohibit the importation of species that are potentially injurious to fish and wildlife and certain human activities. However, Service regulations (50 CFR Part 16) cover only a limited number of species — a list that has not significantly expanded for several decades. Second, the Service, along with all other Federal agencies, must abide by the provisions of Presidential Executive Order 11987 — Exotic Organisms, which prohibits those agencies, or activities they fund or authorize, from introducing exotic species or exporting native species into new ecosystems without first determining that such action will not adversely affect the receiving system. In 1978, the Service adopted regulations drafted, but never formally proposed, to implement the Executive Order as its policy.

Lastly, the Service is involved in research on exotic fishes research headquartered in Florida at the National Fishery Research Center — Gainesville. This research includes studies of the current status, distribution and biology of 43 species of fish already established in the nation's open waters as well as other aquatic species previously introduced or expected to be introduced. The Center supports federal, state and local efforts to prevent further introductions of harmful species and to evaluate species with beneficial characteristics to determine if they can be introduced in ways that do not have adverse environmental or other effects.

Given the inability of past actions to effectively address this issue, the central question now facing the Service and other interested entities is this: where do we go from here? Several new attempts to address introduced species problems have suggested that the pertinent question today is not so much *whether* action is needed, but rather *what* that action should be. As introductions continue, and even expand, solutions become correspondingly more complex difficult, and elusive, and will require far more substantial and costly — perhaps draconian — corrective measures.

To be effective, any policy or program must recognize and accomodate several constraints. One of the most important is that it will not be feasible, nor necessarily desirable, to prohibit all introductions of aquatic organisms. A strategy of managing the risks of intentional and unintentional introductions could provide the framework for developing equitable and effective solutions to the introduced framework for developing equitable and effective solutions to the introduced species problem. A second constraint is that, although the federal government shares responsibility for the introduced species problem, effective corrective action will depend on the full exercise of state prerogatives in a cooperative undertaking.

Statutory Basis for Federal Introduced Species Policies and Programs

Authority for federal activities in support of the introduction of fish and other aquatic organisms has not been a problem, because several of the Service's basic statutes have been interpreted as providing such authority. Authorities that have been used to introduce exotic fish and wildlife into the United States include the 1871 Act establishing the position of U.S. Fish Commissioner (16 Stat. 593) and the Fish and Wildlife Act of 1956 (16 U.S.C. 742a-742j). However, there is substantially less statutory basis for Federal efforts to control the introduction of aquatic organisms.

The most comprehensive federal authority for controlling the introduction of fish and other aquatic fauna is the 1948 injurious wildlife amendment of the Lacey Act of 1901 (18 U.S.C. 42). This authority prohibits the import of a few specified species of wildlife into the United States and its territories and possessions, and their transport between the continental United States and those territories and possessions. This provision also authorizes the Secretary of the Interior to prohibit by regulation the import of any other wild mammal, wild bird, fish (including molluscs and crustaceans), amphibian, reptile or their offspring or eggs, that are determined to be potentially injurious to humans, fish and wildlife, or to the interests of agriculture, horticulture or forestry in the United States.

Other authorities that might be used to control the import and introduction of non-native fish include the Fish and Wildlife Act of 1956 (16 U.S.C. 742a-742j), the 1972 Migratory Bird Convention with Japan (24 UST 3329), and the Endangered Species Act of 1973 (16 U.S.C. 1531-1543). With limited exceptions, however, these authorities have not been used to control such introductions.

The Fish and Wildlife Act of 1956 established a comprehensive national fish and wildlife policy and provided broad guidance about Service responsibilities. Some consider this statute to be the Service's "organic act." Among other things, the statute authorizes the Secretary of the Interior to take steps "required for the . . . conservation, and protection of fisheries resources."

Section (b) of Article VI of the Migratory Bird Convention between Japan and the United States requires that both signatories endeavor to take measures to control the importation of live animals and plants that are determined to be hazardous to migratory birds protected under the treaty. Under section (c) of the same Article, both signatories must also endeavor to take measures to control the introduction of live animals or plants that could disturb the ecological balance of unique island environments.

Sections 7 and 10 of the Endangered Species Act might also provide a vehicle for prohibiting the introduction of aquatic organisms if it can be determined *a priori* that the introduction is likely to jeopardize a listed species. For instance, consistent with the requirements of section 7, the U.S. Fish and Wildlife Service in the past has conditioned its fishery activities, especially fish stocking, to avoid any real or potential conflicts with threatened or endangered species.

Activities of the U.S. Fish and Wildlife Service Relating to Introduced Species

Until recently, Service involvement with introduced aquatic organism problems and issues involved only three activities: administration of injurious-wildlife regulations, Executive Order 11987 — Introduced Organisms, and research on exotic fishes.

Regulation of Injurious Wildlife Imports

Service injurious-wildlife regulations are based on a 1948 amendment of the Lacey Act of 1900 (18 U.S.C. 42). The current version of these regulations is included as Part 16 of Title 50 of the Code of Federal Regulations. As used in these regulations, the terms wildlife and wildlife resources include mammals, birds, fish, molluscs, crustaceans, amphibians and reptiles; the eggs and offspring thereof; and aquatic and land vegetation upon which such wildlife resources are dependent. Although this mechanism allows the Service to directly address the problem of importation of aquatic organisms, it has a number of shortcomings (discussed later).

The regulations begin with a general restriction prohibiting the importation into the United States or its territories of live wildlife, except for psittacine birds, and live or dead fish or eggs of the family Salmonidae. However, the regulations then provide a number of exceptions to these prohibitions that have the effect of limiting the import prohibitions to a few clearly undesirable species or taxa by entities other than federal agencies:

- Other than species or taxa listed in Table 1, all live wildlife may be imported, transported or possessed in captivity for scientific, medical, educational, exhibition or propagation purposes without a permit; a written declaration must merely be filed with United States Customs at the point of entry.

- The species or taxa listed in Table 1 may be imported into, and shipped within, the United States for zoological, educational, medical or scientific purposes under the terms and conditions of permits issued by the Director of the U.S. Fish and Wildlife Service.

- Imported wildlife can be released into the wild only by the State wildlife conservation agency with jurisdiction over the area of release or by persons with prior written permission from that agency.

Table 1. Import of wildlife prohibited by injurious-wildlife regulations.[1]

Live Wild Mammals (50 CFR 16.11[2])

"Flying fox" or fruit bat of the genus *Pteropus*.

Mongoose or meerkat of the genera *Atilax, Cynictis, Helogale, Herpestes, Ichneumia, Mungos,* and *Suricata*.

Any species of European rabbit of the genus *Oryctolaus*.

Any species of Indian wild dog, red dog, or dhole of the genus *Cuon*.

Any species of multimammate rat or mouse of the genus *Mastomys*.

Raccoon dog, *Nyctereutes procyonoides*.

Live Wild Birds or their Eggs (50 CFR 16.123[3])

"Pink starling" or "rosy pastor," *Sturnus roseus*.

Dioch, *Quelea quelea,* including its black-fronted, red-billed or Sudan subspecies.

Java sparrow, *Padda oryzivora*.

Red-whiskered bul-bul, *Pycnonotus jocosus*.

Eggs of wild nongame birds.

Live or Dead Fish. Molluscs and Crustaceans or their Eggs (50 CFR 16.13).[4]

Any live fish or viable eggs of the family Clariidae.

All live or dead fish or eggs of the family Salmonidae.[5]

Live Amphibians or their Eggs (50 CFR 16.14)

None.

Live Reptiles or their Eggs (50 CFR 16.15)

None.

---

[1]As of January 1989.

[2]Except live game mammals from Mexico, importation of which is governed by regulations in 50 CFR Part 14.

[3]Except live migratory birds and live bald and golden eagles, importation of which is governed by regulations in 50 CFR 16.21 and 16.22, respectively; and birds of the family Psittacidae (parrots, etc.), importation of which is governed regulations in 42 CFR Parts 71 and 72.

[4]Except shellfish and fishery products imported for purposes of human or animal consumption, importation of which is governed by regulations in 50 CFR 14.62.

[5]Except by direct shipment accompanied by appropriate certification that the importation is free of the protozoan *Myxosoma cerebralis* and the virus causing viral hemorrhagic septicemia or "Egtved disease."

- Federal agencies can import and transport live wildlife, except bald and golden eagles and migratory birds, solely for their own use upon filing a written declaration with United States Customs at the point of entry. Any such import, however, would be subject to Executive Order 11987 — Exotic Organisms.]

- Importation of dead natural history specimens of wildlife, except migratory birds, game mammals from Mexico, and bald and golden eagles, is allowed without a permit for museum or scientific collection purposes upon the filing of a written declaration with United States Customs at the point of entry.

- Import by direct shipment of live or dead fish and eggs of the family Salmonidae is allowed if accompanied by a certificate stating that they are free of the protozoan *Myxosoma cerebralis* and the virus causing viral hemorrhagic septicemia. The certification must be in a specified format, in English, and signed by a qualified fish pathologist acceptable to or designated by the Secretary of the Interior.

For a variety of reasons, the potential for effective regulation of the import of exotic organisms based on the injurious-wildlife provisions of the Lacey Act has not been realized. As a consequence, the current regulatory approach for implementing that statutory authority remains essentially reactive. Fundamental problems with this "exclusionary list approach" are the limited number of species and taxa that are currently regulated, and the difficulty of adding species to 50 CFR Part 16 — only five new species or taxa were added from 1966 to 1973, and only one more — the raccoon dog *(Nyctereutes procyonoides)* — through 1988. However, several other potentially injurious species are presently under consideration for listing in these regulations.

Although providing a desirable opportunity for public comment, the modification of federal regulations is a lengthy, involved process that is not effective in reacting to imminent importations. Timing problems associated with the regulatory process are compounded by the limited agency resources allocated to this task. For example, efforts to add mitten crabs of the genus *Eriocheir* to the list of injurious wildlife took more than two years (53 FR 45784). In the meantime, there is some evidence suggesting that mitten crabs have already been introduced into open waters in the United States. Furthermore, there has been little outside pressure, such as petitions from non-Service entities, to change or increase the species or taxa listed in the injurious-wildlife regulations. Under these circumstances, there is little incentive to give priority to reviewing the likely effects of species in anticipation of their import, although the Service has supported a minimal effort during the past several years to identify species with potential value to aquaculture that could provide a basis for such action.

By the early 1970s, it was clear that to be effective the implementation of the injurious-wildlife provisions of the Lacey Act of 1900 would have to be based on anticipation of the import of potentially harmful species. In recognition of the shortcomings of the existing approach, an ultimately unsuccessful attempt was begun in 1973 to shift the strategy for implementing the statute by adopting a proactive posture. Based on the assertion that all imported species of wildlife were or could be injurious to the interests defined in the Lacey Act's injurious-wildlife provision, it was proposed (38 FR 34970, December 20, 1973) that, with limited exceptions, the importation of all exotic species be prohibited. That proposed rule would have allowed import, under a permit system, of any species for scientific, educational, zoological, or medical purposes. In addition, the import of several hundred "low risk" species or taxa, predominantly freshwater fishes, that the Secretary had determined posed little threat to indigenous species, agricultural and forestry activities, and the other interests defined in the Act would have been allowed. Because of the large number of species or taxa exempted from the rule, this proposal came to be referred to as the "clean" list approach, and would have replaced the existing "dirty" or "exclusionary" list approach that prohibited the importation of a limited number of species known to be undesirable. One important effect of this proposed new approach would have been to shift the burden of proof of whether a species could be imported from the Federal Government to those proposing to introduce a species.

The proposed change in strategy quickly became highly controversial. The public comment period on the initial proposed rule eventually totalled nearly 9 months, during which four public hearings were held and more than 4,300 comments, predominantly critical, were received. This outcry led to the development of a revised proposal. Although the revision retained the same basic approach as in the initial proposal, many changes were included in response to the concerns raised. In particular, a substantial number of species or taxa — principally tropical fish — were added to the proposed "clean" list.

The revised proposal was republished on February 24, 1975 (40 FR 7935). By the end of the comment period on April 10, 1975, nearly 1,200 comments on this new proposal, identifying a variety of adverse impacts, were received. The pet industry and others contended there was insufficient proof that the importation of all wild animals and plants was inherently injurious. It was claimed that the proposed regulations would have been particularly disruptive to the tropical fish segment of the pet industry. Since past experience suggested that previously unknown tropical fish species would command high prices when they were discovered, their exclusion until proven harmless would have an adverse effect on this element of the pet industry.

Based on the predominantly adverse comments on the second iteration of the proposed rule, yet another approach was proposed on March 7, 1977 (42 FR 12972). The preamble to that proposed rule reiterated the belief that all wildlife outside its native habitat is potentially injurious to one or more of the interests designated in the Lacey Act provision. Recognizing, however, that the degree of risk varies from species to species, the new proposal added a number of species to those already listed in 50 CFR Part 16. These proposed additional species and taxa were determined to be injurious, as judged by nine criteria specified in the preamble to the proposed rule. Like the comments on the earlier proposals, however, the numerous comments received were mostly critical; consequently, all formal attempts to modify the injurious-wildlife regulations to enhance their effectiveness were abandoned. Hence, the current approach is still based on the "dirty" or "exclusionary" list strategy involving a limited number of prohibited species.

Because of lingering opposition to the concept of a "clean" list, the adoption of that strategy to reduce the risk of undesirable species introductions probably remains infeasible. Hence, if the Federal Government is to play a useful role in addressing the problem of harmful introductions of aquatic organisms without a new statutory mandate, several actions may be required. First, it would be desirable to improve internal Service procedures for modifying the list of injurious wildlife in 50 CFR Part 16 by establishing listing criteria and procedures and otherwise specifying how the Service will use the existing species-by-species listing mechanism. Second, species should be added to that list a few at a time as they are considered for introduction into the United States or otherwise brought to the Service's attention. Although past experience suggests that such actions might prove controversial and difficult, such problems should not be insurmountable. In this era of limited financial resources, one major hurdle to a more proactive posture may be the availability of appropriate staff to process petitions for listing, to assemble and interpret data on potential impacts, and to prepare necessary environmental and regulatory analyses.

Executive Order 11987: Restriction of Federal Activities

On May 24, 1977, President Carter signed Executive Order 11987 (42 U.S.C. 4321) which addressed the import and export of exotic species of plants and animals. That directive requires Federal agencies, to the extent permitted by law, to restrict three activities:

- Introduction of exotic species into land and waters under their jurisdiction.

- Importation of exotic organisms for introduction into any natural ecosystem of the United States.

- Export of native species for introduction into ecosystems outside the United States.

The Secretaries of Agriculture or the Interior may make exceptions to these restrictions if they find that such importations or exportations will have no adverse effect on natural ecosystems. Exotics were defined "as all species of plants and animals not naturally occurring, either presently or historically, in any ecosystem of the United States." In addition, the Secretary of the Interior, in consultation with the Secretary of Agriculture and the heads of other appropriate agencies, was directed to develop regulations implementing the Executive Order on a Government-wide basis.

The U.S. Fish and Wildlife Service, in fulfillment of the Secretary's responsibility prepared a draft of the required regulations. The regulations were intended to be in addition to, not in lieu of, current; federal restrictions and conditions on the introduction or importation of exotic and native animal species, including endangered and threatened insects. Under the draft proposed rule, executive agencies would be required to identify: (1) all proposed introductions into natural ecosystems; (2) any proposed importation for the purposes of introduction into a natural ecosystem; and (3) any proposed export of native species for introductions into natural ecosystems outside the United States that would be conducted, funded or authorized by the agency. Whenever such proposed activities were identified, the draft regulations required that the agency request a biological opinion, together with any available information, from the U.S. Fish and

Wildlife Service Regional Director in whose Region the proposed activity would be carried out. The agencies were to be responsible for conducting appropriate studies and providing the Service with enough biological information to establish the effects of a proposed importation or exportation on natural ecosystems.

Within 90 days after the receipt of a written request for a biological opinion from an executive agency that provided all necessary biological information, the Service would analyze the proposed action and issue its biological opinion. The opinion could recommend modification of the proposed importation or exportation to ensure that such actions would not result in any adverse effect on a natural ecosystem. In rendering its biological opinion, the Fish and Wildlife Service would have to ascertain whether receiving States or nations concur with a proposed introduction and that such actions would be in compliance with all applicable laws and regulations. Upon receipt of the Service's biological opinion, the agency proposing the activity would be responsible for satisfying the requirements of Executive Order 11987.

Although never published as a proposed rule, let alone made final and implemented, the Service adopted the draft regulations on December 14, 1978, as the guidelines for discharging its own responsibilities under the Executive Order. Since the draft of the proposed rule had not undergone full public and peer review that might have tightened and clarified its provisions, further scrutiny is warranted. For instance, previously introduced exotic species that had become established as viable, self-sustaining populations in a natural ecosystem of the United States were considered in the draft proposed rule as "naturally occurring," and therefore a "native species" not subject to its provisions. In addition, non-feral domesticated animals and plant cultivars were not covered by the draft of the proposed rule.

Due to the lack of government-wide implementation guidelines mandating U.S. Fish and Wildlife Service involvement in decisions regarding the intentional importation or exportation of exotics, only limited — essentially anecdotal — information is available concerning compliance with Executive Order 11987. Such information is of only limited value in assessing the effectiveness of the Executive Order in preventing problems associated with intentional introductions by Federal agencies. For instance, the Service has not been involved in, nor funded or authorized, the introduction of exotics for nearly three decades and now rejects requests that it export fish or eggs. On the other hand, little is known about introductions or exportations by other Federal agencies, let alone whether any such actions since 1977 were in compliance with the requirements of the Executive Order. Added study of the effectiveness of the mechanism proposed to implement Executive Order 11987 is warranted prior to any further attempt to promulgate implementing regulations.

Exotic Fish Research

The principal responsibility for exotic fish species research within the U.S. Fish and Wildlife Service is assigned to the National Fisheries Research Center —Gainesville, Florida, and its field stations in Stuttgart, Arkansas and Marion, Alabama.

In early 1977, due to increasing pressure by federal, state, private and professional organizations concerned about introduction of exotic fish species into our nation's open waters, the U.S. Fish and Wildlife Service established the National Fisheries Research Laboratory —Gainesville. In 1987, the laboratory was upgraded to the status of a National Fisheries Research Center.

The mission of the Center is to identify and determine the distribution, status and impacts of exotic fish species already established in the nation's waters and to evaluate the exotic species under consideration for introduction or likely to be released into open waters. It serves as the major national information exchange center on exotic fish species. Working closely with other components of the Service and other federal, state and private organizations, the Center supports national policy prohibiting Federal actions that result in the introduction of additional exotic species without a full evaluation of their impact on the receiving environments. The Center is also responsible for promoting beneficial exotic species when they pose little or no threat to the Nation's waters.

Construction of a $5.5 million maximum security installation to house the Center at the University of Florida in Gainesville was completed in 1988. The facility is isolated from all major rivers and their drainage systems and there are no permanent bodies of standing surface water within 5 miles. The facility is double fenced and its 12 acres of ponds are enclosed by an earthen berm system that will retain up to three times the maximum recorded 24-hour rainfall of 9 inches in Gainesville. The watershed in which the Center is located is small and discharges into a sink hole one mile away. Here the surface water mixes with ground water with an oxygen level near zero and then flows underground for 60 hours before surfacing in the Santa Fe River system. All water exchanged between ponds and the laboratory is double filtered and any water exchanged between ponds must pass through three screens and a 100 µm filter system. No live exotic fish can be moved out of the facility without the Center Director's approval.

The laboratory funded and participated in the publication of an *Atlas of North American Freshwater Fishes* (Lee et al. 1980), which contained summary accounts of the status and distribution of all native freshwater fish species, and highlighted endangered species and exotic fish species that have become established (i.e., breed in open waters). A manuscript summarizing published and unpublished data about the impacts of exotics species was developed, as was a procedures manual for in-house species evaluations.

In 1980 the Service contracted with the American Fisheries Society to identify the exotic fish species of economic importance to United States interests. Over 2,000 species were identified as being imported into the United States or of particular interest or concern to North Americans. A special publication of the Society entitled *World List of Fishes Important to North Americans Exclusive of Species from Continental Waters of the United States* (Robins et al.

1991), summarizes this information.

Concerns in recent years about the import of several new exotic fish species and the expanded use of several others led to the development of biological synopses for the grass carp *(Ctenopharyngodon idella)*, bighead carp *(Hypophthalmichthys nobilis)*, and the Mozambique tilapia *(Oreochromis mossambicus)*. In addition, an in-Service review of the rudd *(Scardinius erythrophthalmus)* has been prepared. Established and expanding populations of blue tilapia *(Oreochromis aurea)* in the St. Johns River system and the blackchin tilapia *(Sarotherodon melanotheron)* in the Indian-Banana River system along the east coast of Florida have been investigated in the field to determine their distribution, status, biology and impact on native fish populations. Studies in the laboratory have centered on determinations of critical environmental factors such as temperature and salinity, which control the survival and reproductive potential of exotic species.

Ecological and laboratory studies on grass carp and its hybrid and triploid forms have also been funded. The ecological studies have assisted local aquatic plant control agencies in developing appropriate stocking rates and management systems. Laboratory studies have centered on the development of techniques to produce triploid grass carp. The Center's Fish Farming Experimental Laboratory, Stuttgart, Arkansas, had conducted some of the earlier research on triploidy in grass carp. It subsequently developed an effective and practical protocol for ensuring that grass carp are functionally sterile (i.e., have triploid rather than diploid chromosomes).

To encourage and facilitate the use of sterile grass carp, the Stuttgart station began providing triploid certification inspections as a service to state conservation agencies in September 1985. It now certifies, for interested state agencies, that grass carp shipped from large Arkansas producers are in fact triploid. To date 26 states have adopted regulations that allow only triploid grass carp to be imported. Many of those states require inspection by the Fish and Wildlife Service to ensure triploidy. Consequently, the number of triploidy inspections conducted by the Stuttgart station increased from 2 in 1985 to 216 in 1988. As this technique is now operational, responsibility for certifying triploidy in grass carp will shift to the Service's fishery component on October 1, 1989.

The Southeastern Fish Cultural Laboratory has conducted studies on the use of redbelly tilapia *(Tilapia zilli)* to control nuisance vegetation in striped bass production ponds. Two contracted studies are developing baseline data on the morphometric, meristic (Cichochi et al. In review) and electrophoretic (Phelps In review) characteristics of most populations of tilapia now in the United States. Involved are detailed comparative analyses of 60 samples representing different tilapia populations from United States and foreign sources.

Another major U.S. Fish and Wildlife Service installation, the National Fisheries Research Center — Great Lakes in Ann Arbor, Michigan, also conducts research on exotic fish species. In recent years, significant numbers of exotic plants and animals have been introduced and become established in the Great Lakes due to introductions that were either intentional (e.g., Pacific salmon) or unintentional [e.g., the ruffe *(Gymnocephalus cernuum)*, zebra mussel *(Dreissena polymorpha)* and water flea *(Bythotrephes cederstromi)* introduced when ships from outside the Great Lakes dumped fresh or brackish water ballast before loading export goods]. The Great Lakes Center is involved in a coordinated effort with other federal and Canadian, state and private agencies to identify and determine the status, distribution and impacts of these exotics. Efforts to stop the dumping of these large volumes of biologically contaminated water are also under way.

Where Do We Go from Here?

Time has not diminished the concern over unrestricted introductions of aquatic organisms that resulted in the initiatives of the 1960s and 1970s to address this problem. If anything, concerns are far more substantial, being based on a more complete understanding of the breadth and scope of the problem and a growing awareness that past efforts to address this issue have proven largely ineffective, particularly with respect to unintentional introductions. These concerns are compounded by the realization that introductions, particularly unintentional ones such as those that occurred recently via ballast water discharges in the Great Lakes, are still occurring. Furthermore, there seems to be a significant potential, in the absence of effective control mechanisms, for additional introductions. Waterborne commerce continues to grow as the economies of the world become increasingly interdependent. In addition, as noted previously, more than 2,000 species of non-native fishes and shellfishes have been identified as having potential for use in aquaculture in North America.

A concrete indication of this growing concern is the renewed effort to address the problem being made by a number of governmental entities: Canada's Department of Fisheries and Oceans, the North American Commission of the North Atlantic Salmon Conservation Organization, the Council of Lake Committees and the Fish Disease Committee of the Great Lakes Fishery Commission, and the U.S. Fish and Wildlife Service.

As introductions continue, and even expand, solutions will become correspondingly more complex, difficult and elusive. There are no easy answers. Furthermore, continued inaction based on denial of the problem, on explicit choice or on other reasons can only result in the problem becoming more severe and intractable. This increase in severity will, in turn, require more substantial and costly corrective measures, perhaps bordering on the draconian, if a major human health or agricultural problem is attributed to introductions. Although the need for immediate action sees obvious to us, the central — and far more difficult — question becomes: what should — and can — be done about this situation?

To be effective, any policy or programmatic initiative must recognize and accommodate several constraints. One of the most important is that it will not be feasible, nor necessarily desirable, to prohibit all introductions of aquatic organisms. This is due to the nature of the problem (especially with regard to unintentional introductions) and the potential benefits to be derived from intentional introductions. In addition, the fisheries and aquatic communities have been unable to agree on the magnitude of the threat of introduced organisms, let alone corrective actions, due (perhaps in large part) to the fundamental dichotomy between those interested in maintaining the viability and productivity of established ecosystems and fisheries and those who believe that human intervention (i.e., introducing non-native species) can diversify and enhance capture fisheries and aquaculture. Means must be found for these conflicting interests to equitably share the burden of corrective actions if this standoff is to be resolved.

We suggest that a strategy of managing, i.e., minimizing, the risks of unintentional introductions as well as those associated with intentional introductions could provide the framework for developing equitable and effective solutions to this problem. This presents the difficult challenge of balancing the potentially substantial benefits for both aquaculture and fishery resource management from the introduction and exploitation of new species against the risks of ecological and financial effects of what could turn out to be ill-advised introductions. Although such a strategy is likely to be more effective, it would require a comprehensive, cooperative implementation effort significantly more complex than a simple prohibition on introductions. In addition, such a risk minimization approach would be costly to develop, administer and enforce.

A second constraint is inherent in the federal system of government in the United States. Although the federal government shares responsibility for this problem, effective corrective action will depend on the full exercise of state prerogative in a cooperative initiative. Because of this interjurisdictional nature and the potentially large number of management entities involved, a substantial effort to educate resource managers and legislators will be required. Undoubtedly, substantial inertia will also have to be overcome before such an initiative can proceed.

Although less global than the two constraints discussed previously, some questions remain about the appropriate scope of any cooperative action to resolve the introduced species problem. There seems to be a fairly complete consensus that all aquatic species are of concern. There may be somewhat less agreement that the focus must be on unintentional, in addition to intentional, introductions. The major subject of disagreement, however, concerns whether any introduced species program should focus on only exotic species (i.e., those not native to North America) or include transplants of native North American species outside their natural range into waters where they are not already established It will be important to resolve these remaining policy dichotomies, as well as any that develop later, but such essentially scientific and technical issues should not be allowed to hinder progress on the overall effort.

Being mindful of such constraints, two general areas of action leading toward development of a comprehensive initiative can begin immediately and be undertaken with little controversy:

- Broadening and intensifying research and educational activities aimed at developing a professional consensus on the need for effective action and resolving remaining technical issues.

- Better coordination of recent efforts to resolve specific introduced species problems and development of additional cooperative efforts to ensure a comprehensive, timely, and effective response to the overall problem.

Enhancing Professional Awareness and Information

Resolution of contentious, multifaceted technical issues, such as whether and how to establish more effective control over new introductions of aquatic species, depends on two crucial ingredients. First, professionals with the most direct responsibility for and intimate involvement with the problem, including scientists, field biologists, aquaculturists and resource managers, must become fully conversant with the issues involved. Second, additional scientifically valid information from the biological disciplines as well as the policy disciplines, including the emerging field of risk assessment, is essential to focus the debate and facilitate the development of rational, effective policy choices.

Scientific, professional, resource management and even trade organizations are naturally well adapted to the task of informing their members about such complex technical issues. For instance, the fisheries profession has been in the forefront of efforts to keep this issue before the scientists, biologists and resource managers involved in or potentially affected by this issue over the past 20 years. Organizations such as the American Fisheries Society have established special sections dealing with introduced species and have sponsored a number of symposia as forums where research information is presented and subjected to peer review as well as to provide opportunities to identify and discuss emerging policy issues. In addition, the American Fisheries Society has developed a protocol for evaluating the feasibility and desirability of intentionally introducing additional aquatic species (Kohler and Courtenay 1986).

The symposium from which this book derives, which is sponsored primarily by aquacultural interests, builds on and extends this important role. Much new information is being presented on many issues. There is a balanced emphasis on unintentional versus intentional introductions and on introductions of plants and non-vertebrate animals as well as vertebrate animals. Although directed principally toward the biological community, information presented is also relevant to the policy choices that will probably have to be made in the not too distant future. One paper appropriately addresses the concepts of risk assessment and how those concepts might be applied to the problem of introduced species. We commend the organizers of the symposium, particularly Dr. Aaron Rosenfield, for their significant contribution raising our collective consciousness about this issue.

Although significant strides have been made, much more needs to be done to increase professional and public awareness of the problem of introduced species to create the foundation or common ground for efforts to resolve the important issues involved and develop acceptable and effective corrective actions. In recognition of this need, the introduced species issue was included in a comprehensive set of environmental initiatives submitted to then President-Elect Bush by the conservation community in its *Blueprint for the Environment* (Haize 1988). A later publication (Anonymous 1989), supporting the summary document, recommended that one of the key actions for implementing any introduced species initiative should be the convening of a national meeting to, among other things, reach a consensus on the problem and the policy issues involved.

Important, if limited, research related to the nature and extent of the introduced species problem, including its policy dimensions, has been conducted since this issue surfaced in the 1960s and 1970s. This research and related policy and management implications have been summarized in several papers and discussed at a number of national and international symposia. This book is one example. McCann (1984b) summarized the major works on exotic fishes in the United States during the period 1969-1982. The proceedings of a symposium on the distribution, biology and management of exotic fishes in North America, held in Albuquerque, New Mexico, in September 1981, have been published (Courtenay and Stauffer 1984). In August 1984, the Introduced Fish Section of the American Fisheries Society sponsored a special session at the Society's Annual Meeting in Ithaca, New York, entitled Strategies for Reducing Risks from the Introduction of Aquatic Organisms. Many papers from that session have been published (American Fisheries Society 1986). A symposium entitled Role of Exotic Species Introductions in Fisheries Management was held at Lake of the Ozarks, Missouri, in March 1985 (Stroud 1986). An unpublished symposium entitled Quantitative Effects of Introduced Organisms was held at the American Fisheries Society's Annual Meeting in Toronto, Ontario, Canada, in September 1988 (American Fisheries Society 1988).

Because understanding of the problems created by introductions is so central to their resolution, we will not belabor the need for additional research about introduced species, except to reiterate that additional study of these issues is essential. Courtenay and Robins (1989) discussed some research needs and priorities, along with several protocols for evaluating species before introduction and the successes and failures of some recent fish introductions.

Program Development

In response to the growing concerns about both specific and general introduced species problems, several governmental entities have initiated actions to address them. In response to proposals to introduce additional Pacific salmonids in the Great Lakes and along the Atlantic coast of North America, the North American Commission of the North Atlantic Salmon Conservation Organization established a scientific working group to study potential adverse impacts. That group recommended that the Commission acknowledge the potential for adverse effects from, and develop protocols governing, future intentional introductions to reduce the risks of European Atlantic salmon and Pacific salmonids being introduced (Bilateral Scientific Working Group on Salmonid Introductions and Transfers 1987).

Canada's Department of Fisheries and Oceans is increasingly concerned about unrestrained species introductions in general, a concern reinforced by proposals to introduce the zander (*Stizostedion lucioperca*) into United States waters, from which the fish would eventually reach Canadian waters (D.G. Wright personal communication). In response, they are moving toward formal adoption of the species introduction protocol included in the American Fisheries Society Position on Introductions of Aquatic Species (Kohler and Courtenay 1986).

The Great Lakes Fisheries Commission has also been active in addressing this problem in response to the recent unintentional introductions (via ballast water) of at least three aquatic organisms identified previously that have spread rapidly and are viewed with alarm. Further, its Council of Lake Committees and Fish Disease Committee have expressed concern about the possibility of additional undesirable introductions into the Great Lakes Basin as a result of fishery management and aquaculture activities (M. Dochoda personal communication). From extensive experience with the full range of consequences of numerous previous introductions, they believe additional introductions could have significant ecological and disease impacts on long-term programs for the restoration of lake trout (*Salvelinus namaycush*) and other native fishery resources of the Great Lakes. To address these concerns, it has been suggested that a "Model Program for Management of Introductions in the Great Lakes Basin" be developed.

The Fish and Wildlife Service has also become more active with regard to this issue in recent years. In 1987, the Service's fishery staff completed an analysis of policies for reducing the risks associated with the introduction of aquatic organisms. For that analysis, introductions were defined as involving both the import of exotics and the transplant of species native to North America outside their native range. However, the analysis focused only on intentional introductions. Several likely policy constraints, such as the infeasibility of prohibiting all introductions, the interjurisdictional and national and international nature of the problem, and the need for coordinated and universal action were identified. Four illustrative strategies that could be the basis for renewed initiatives to address this problem and several implementation options were discussed.

That analysis concluded that perhaps the most prudent course of action, given the still substantial opposition to Federal action, is to adopt an evaluation protocol approach. Although acknowledging that under no circumstances should this approach be considered a panacea, the analysis pointed out that that approach is the least threatening and provides a means of building a consensus and momentum for more aggressive initiatives. In addition, this approach could result in the generation of useful information about the likely effects of a variety of species.

One immediate result of the Service policy analysis was the drafting of a new internal Service policy on introduced aquatic organism problems and issues that took into consideration the emerging understanding of those problems and the near-term limits on the Service's ability to act on them. This new policy was intended to be in addition to, not a replacement for, existing Service policies — including policies for implementing Executive Order 11987. The draft policy was generally well received within the Service, but has not been finalized. However, growing concerns about the efficacy of current policies and programs and recognition of the significance of unintentional introductions suggest that further broadening of Service policy might be warranted and desirable.

In developing policy recommendations on natural resource and environmental issues for the President-Elect in 1988, the Nation's conservation community recognized the importance of reducing risks associated with new introductions of aquatic organisms (Anonymous 1989). Recommended was the development, based on existing and new statutory authority, of effective strategies and cooperative programs to ensure systematic biological evaluation of the desirability and feasibility of potential introductions of aquatic organisms. The initiative would focus on new transplants of native species outside their present range as well as on the introduction of species native to waters outside North

America. Primary responsibility for implementation of this co-operative venture by the states and other appropriate jurisdictions, rather than by direct federal action, was emphasized.

Given recent initiatives addressing introduced species problems, the pertinent question today is not so much whether action is needed, as what action is needed. Because a number of efforts to address this problem are under way, the immediate challenge is how best to coordinate and diligently pursue those efforts to ensure that they are effective and make wise use of limited resources. In addition, it would be desirable to address how best to fill the gaps in current activities to ensure that a comprehensive solution to the introduced species problem is developed. This effort should result coherent and clearly stated goals and objectives and strategies built around concepts of risk reduction; it should include a balancing of the benefits of additional intentional introductions against the risk of impacts and resultant monetary and non-monetary costs.

Literature Cited

American Fisheries Society. 1986. The good, the bad, the ugly: Introduced species. Fisheries 11(2):2-42.

American Fisheries Society. 1988. Quantitative effects of introduced fishes. 118th Annual Meeting, Symposia Sessions. Fisheries 13(3):62-63.

Anonymous. 1989. Blueprint for the environment: The environmental community's recommendations to President-Elect George Bush. Howe Brothers, Salt Lake City, Utah.

Bilateral Scientific Working Group on Salmonid Introductions and Transfers. 1987. Report of activities on salmonid introductions and transfers to the North American Commission. Annex 13 to the Report of the Fourth Annual Meeting of the North American Commission [NAC(87)20]. 1987.

Cichochi, F. P., E. L. Garcia and W. R. Courtenay, Jr. Morphometrics and meristics of exotic introduced and native populations of tilapine fishes (Cichlidae). manuscript in preparation.

Courtenay, W. R., Jr., D. A. Hensley, J. . Taylor and J. A. McCann. 1984. Distribution of exotic fishes in the continental United States. p. 41-77. *In* W. R. Courtenay, Jr. and J. F. Stauffer, Jr. (eds.), Distribution, biology, and management of exotic fishes. Johns Hopkins University Press, Baltimore, Maryland.

Courtenay, W. R., Jr. and C. R. Robins. 1989. Fish introductions: Good management, mismanagement, or no management. CRC Critical Reviews in Aquatic Sciences 1(1):159-172. 1989.

Courtenay, W. R., Jr. and J. R. Stauffer, Jr. editors. 1984. Distribution, biology, and management of exotic fishes. Johns Hopkins University Press, Baltimore, Maryland.

Kohler, C. C. 1986. Strategies for reducing risks from introductions of aquatic organisms. Fisheries 11(2):2-3.

Kohler, C. C., and W. R. Courtenay, Jr. 1986. American Fisheries Society position on introductions of aquatic species. Fisheries 11(2):39-42.

Lee, D.S., C.R. Gilbert, C.H. Hocutt, R.E. Jenkins, D.E. McAllister and J.R. Stauffer, Jr. 1980 et seq. Atlas of North American Freshwater Fishes. North Carolina State Museum of Natural History, Raleigh, North Carolina.

Maize, K.P., editor. 1988. Blueprint for the environment: Advice to the President Elect from America's environmental community. Blueprint For the Environment, Washington, D.C.

McCann, J. A. 1984a. Preface p. ix-x. *In* W. R. Courtenay, Jr. and J. R. Stauffer, Jr. (eds.), Distribution, biology, and management of exotic fishes. Johns Hopkins University Press, Baltimore, Maryland.

McCann, J. A. 1984b. Involvement of the American Fisheries Society with exotic species, 1969-1982. p. 1-7. *In* W. R. Courtenay, Jr. and J. R. Stauffer, Jr. (eds.), Distribution, biology, and management of exotic fishes. Johns Hopkins University Press, Baltimore, Maryland.

Phelps, S. R. An electrophoretic investigation of the genetic relationships of eight species of tilapia and species composition of populations in the U.S. and Puerto Rico. Manuscript in preparation.

Robins, C.R., et al. 1991. World list of fishes important to North Americans, exclusive of species from continental waters of the United States and Canada. American Fisheries Society, Bethesda, Maryland. AFS Supplement Special Publication 20.

Shafland, P. L. and W. M. Lewis. 1984. Terminology associated with introduced organisms. Fisheries 9(4):17-18.

Stroud, R. H., editor. 1986. Fish culture in fisheries management. American Fisheries Society, Bethesda, Maryland.
Welcomme, R. L. 1986. International measures for the control of introductions of aquatic organisms. Fisheries 11(2):6-9.

Legal and Regulator Citations

CF	—	Code of Federal Regulations
E.O.	—	Executive Order
FR	—	Federal Register
Stat.	—	United States Statutes At Large
U.S.C.	—	United States Code
UST	—	United States Treaties

Effective Application of Aquaculture Disease-Control Regulations: Recommendations from an Industry Viewpoint

RALPH A. ELSTON

Abstract: Transportation of aquatic animals, organisms and their tissues is inevitable and results from a variety of activities including movement of fishery commodities, shipment of eggs, larvae and brood stock for aquaculture or research purposes, transfer of ship ballast water, and transport of aquatic animals by the public. Such transfers can have serious pathological and ecological effects. However, the aquaculture industry is frequently and unfairly targeted as the only or prime practitioner of aquatic animal transfers. Regulations to protect natural and farmed resources must recognize the inevitability of a certain level of animal transfer, and formulation and enforcement of these regulations must involve the user groups causing the transfers and affected by the regulations. Without voluntary compliance with animal-transport regulations, legislation aimed at reducing risks is doomed to failure.

Risks can be reduced. Education of the public, researchers and the aquaculture industry regarding the risks of animal transfers is a key area needing attention. Transfer and availability of technical information can help alleviate serious effects of animal transfers. The introduction of the oyster parasite, *Bonamia ostreae*, to Europe could have been easily prevented by adequate review of technical information since its existence and significance were recognized at least ten years before it was introduced into Europe.

Resource managers must work with aquaculturists and other user groups to develop workable and effective policies, while recognizing that these cannot be zero-risk policies. Technical information regarding infectious disease distribution is needed as well as information on the actual numerical significance of known diseases so that appropriate risk assessments can be attached to these diseases. Researchers must be careful not to overstate the significance of diseases in the quest for research funds and recognition. Often lacking sufficient technical information in the decision-making process, resource managers and regulators must recognize that philosophy often overrides technology and determines regulatory implementation.

Introduction

The following perspectives are derived from working both with commercial aquaculturists and in the aquatic animal health research field. These ideas do not encompass every facet of disease control regulation or all the details of proposed approaches to this issue, such as that of the International Council for the Exploration of the Seas (ICES) Working Group on Introductions and Transfers of Marine Organisms. Others in this volume address different aspects of this important issue. I recount some general views as shaped by my experiences in working with both industry and government. It is essential to view this problem from an aquaculture industry standpoint if we are ever going to be able to implement the concepts of animal health management in aquaculture.

Transportation of aquatic animals for commodity distribution, research purposes, by the general public, and for aquatic animal husbandry purposes is inevitable within North America and, I believe, between continents. History demonstrates that introduced pathogens can have catastrophic results. Although the risk is not necessarily proportional to the quantity and frequency of movement of a given species, it is likely to increase with the diversity of species movements. As a sophisticated aquaculture industry continues to expand, it is vital to address the problem and solutions realistically.

Most individuals, whether from industry, government or academia, agree that we need some level of control on the transport of fish and shellfish in order to prevent the damaging effects which infectious diseases can have on both husbanded and natural populations of aquatic animals when they spread into uninfected populations. It is important to recognize that workable regulations can reduce, but never eliminate, the risks of such diseases. Ineffective regulatory control of infectious disease can result from either no regulation on the one hand or, on the other hand, from an attempt to eliminate the risk posed by infectious diseases by a too conservative and unrealistic approach to the problem of disease control. Overzealous regulation, without a substantial technical base and without recognition of the realities of animal transports, simply encourages individuals and companies to disregard the law. There is no practical way that animal-transport regulations can be effective without voluntary and active support by the user groups.

Animal Transports Are Inevitable

The transportation of aquatic animals throughout the continent of North America and between North America and other continents is inevitable. Often the aquaculture industry is regarded as the primary practitioner of this activity. In fact, the transport of aquatic animals or their fresh tissues, which may contain viable infectious agents, is practiced by several other user groups. These include commodity distribution of harvested or husbanded fishery products, the transport of aquatic animals for research purposes, transfer of marine waters and organisms in ship ballast water, and transfer of fish and shellfish by the general public. Catastrophic damage which can result from the introduction of an infectious disease does not necessarily occur from the transfer of large numbers of a single species. I have observed some cases in which researchers and resource-management biologists somehow rationalize that animal transport regulations do not apply to them, even though the greatest potential damage could result from the movement of a small number of animals which carry nonindigenous pathogens. I suspect that a larger diversity of aquatic animals is moved by the research community than by the aquaculture industry. The control of the spread of infectious disease organisms through commodity distribution activities is not usually covered by the same, if any, regulations which pertain to aquaculture products; and, although regulations may forbid the transfer of aquatic animals by the public, these regulations are difficult, if not impossible, to enforce while the transfer of ballast water by circumglobal shipping is largely unregulated.

Another phenomenon I have observed is that while individuals in government agencies may pronounce the need for elimination of aquatic animal transports between or even within continents, such movements seem to continue frequently in the aquaculture industry. It sometimes seems that industry and government are operating in two entirely different worlds. There are great differences in how governments handle these issues. In Washington State, which has a large aquaculture industry, the Department of Fisheries implements a careful case-by-case policy which, so far, has been effective in preventing the introduction of any exotic diseases with catastrophic consequences. In the United States, there is really no national legislation currently implemented which deals with this issue. Several European countries have relatively strict legislation in place forbidding the transport of aquatic species in many cases. History shows that legislation alone fails to solve the problem of the introduction of exotic pathogens and may indirectly exacerbate the problem.

Reducing the Risk

What can be done to reduce the risk of spreading infectious aquatic animal diseases? Education is a key area needing attention. We can target education of the aquaculture industry relatively easily. My experience is that the industry will act responsibly when it recognizes that disease control is in its own interest and that such education will encourage self-enforcement efforts — the most effective means of enforcement of disease control regulations. In fact, if the aquaculture industry is not aware of and supportive of animal-movement regulations, the regulations are certainly doomed to failure. The expansion and economic imperative for commerce in aquaculture products encourages aquatic animal transfers within and, in fact, between continents. Furthermore, there are simply too many avenues for such distribution to occur for animal transfers to be prevented by regulatory enforcement. This does not mean that some regulation is not needed. It does mean that ignoring the realities of commerce in aquaculture will not be effective. It also means, I believe, that we need to initiate a new emphasis in the support of the industry through educational efforts by government on the risks of animal transfers.

Those of us in the fish and shellfish health field are the most effective professionals to educate the industry through workshops, publications, and taking the opportunity to speak on this subject where appropriate. As fish and shellfish health professionals, we should begin a dialogue on how to effectively extend this education to the general public and other user groups involved in commodity distribution.

Transfer of technical information may help reduce the risks of the introduction of catastrophic disease. *Bonamia ostreae* is a parasite of the European flat oyster which was introduced into Europe in the latter part of the 1970s. The disease and the parasite had been observed and descriptions published from research conducted in California, the site of origin of the European introductions, before 1970. If the prevention of any exotic aquatic animal disease has ever been possible, this surely stands as the signal case. The solution is the effective dissemination of technical information — another aspect of the need for education of both industry and government officials.

In the application of effective disease control in an age of incomplete technology, it is incumbent, I believe, on resource managers charged with controlling animal diseases to adopt reasoned and workable policies. These cannot be zero-risk policies. Failure of policies has often resulted from an overzealous attempt to control animal transfers and the failure to recognize the multiple avenues of animal transfers as well as from the lack of implementation of animal transfer controls.

We should strive toward enacting regulation which is based on substantial technical information rather than incomplete information. If we are going to forbid transfer of a particular animal species when it carries a particular disease it should be because this disease is truly exotic to certain areas. While it is not a good animal husbandry or resource-management practice to move animals which are sick, we cannot, from a regulatory point of view, reasonably attempt to control the movement of diseased organisms from one enzootic area to another. Regulations should be formulated with a view toward protecting both natural resources and the aquatic animal husbandry industry. Finally, and perhaps most importantly, we need to devise ways to address all avenues of risk for the introduction of aquatic animal diseases, not just the most visible, easily targeted avenues of risk, such as the aquaculture industry.

Technical Needs

There are certainly some outstanding technical needs if we are going to manage and prevent the spread of infectious aquatic animal diseases effectively. We need to develop complete regional inventories of diseases. This is not the most glamorous research problem nor, necessarily, the highest priority of fisheries-management agencies, but a disease inventory is one of the key categories of information needed. We cannot rationalize excluding a disease or species from a region if we do not know of its presence or absence from that region and, conversely, we cannot know the risk of moving an animal population if we do not have a good idea of the diseases it harbors.

We also need information on the numerical significance of a given disease to all of the life stages of fish or shellfish which it affects. As fish health professionals, we know there is a great variety of pathogens and parasites found on host animals. Obviously some of these are much more important than others in their effects on the host. As well, certain categories of infectious agents, e.g., viruses, are typically regarded as highly significant, *de facto*, without consideration of their potential for pathogenicity. In fact, some virus-host relationships represent a high degree of adaptation, like other categories of infectious agents, in which the virus apparently has little effect on the host. We must strive to be more quantitative regarding the effects of microorganisms on hosts. As a result we will be able to prioritize the diseases according to their importance and apply appropriate regulations to each disease depending on its importance. We can thus also prioritize the expenditure of our limited resources for fish and invertebrate health research.

Another need in the technical area is for researchers to be very careful not to overstate the importance of particular infectious diseases. This happens all too frequently in the quest for the acquisition of research funds or in the aim of establishing the importance of a disease on which one may be currently working. Published technical information finds its way into many uses and that which is loosely interpreted or not substantially supported can be very misleading when used to formulate public policy. The establishment of a list of diseases by their numerical significance to survivability, growth or other criteria will help alleviate this problem.

Philosophy Often Overrides Technology

Finally, as all of those in regulatory roles in government know, decisions must usually be made in the face of insufficient technical information. Even as we strive to shore up our technical information base, resource managers will be faced with this state of affairs. Thus it is of utmost importance to recognize that one's philosophy toward animal transports will often determine the character of regulations and their implementation as much or more so than supporting technical information. Therefore, it is incumbent on those of us in resource management to adopt a reasonable and workable philosophy on aquatic animal transports, recognizing the need for a stronger technical information base and for the education of all user groups.

Acknowledgments

The Battelle Marine Sciences Laboratory is part of the Pacific Northwest Laboratory, which is operated for the Department of Energy by the Battelle Memorial Institute under contract DE-AC06-76RLO 1830.

California's Approach to Risk Reduction in the Introduction of Exotic Species

ROBSON A. COLLINS

Abstract: California now takes a conservative approach to the introduction of new species to its lands and waters. Although there have been some notable successes with the introduction of exotic species, the introduction of others, both with official permission and without, has resulted in problems that are still prevalent today. California is especially concerned with the possible introduction of disease organisms to already existing populations, and with the displacement of native species by introduced species.

Introduction

California has a fairly extensive history of the introduction of exotic species. Various species of oysters were first planted in our waters before the beginning of the twentieth century, and there is an established commercial oyster industry in the state that is dependent on introduced species. Most people are aware of the successful introduction of striped bass *(Morone saxatilis)* into the Sacramento-San Joaquin delta, which produced a substantial recreational fishery. However, we have also had our share of bad experiences as a result of exotic introductions, a recent example being the spread of white bass *(Morone chrysops)*, originally introduced experimentally in a single reservoir, to several lakes and reservoirs in central California where they are in competition with native species. Tilapia (*Tilapia* sp.), originally introduced as an aquaculture species, has also spread into the Salton Sea and other waters in southern California where it has been able to survive much colder water temperatures than originally expected and competes with native species.

As a result of experiences like this, a much more conservative view of exotic introductions has been taken by California officials and the aquaculture industry in recent years. This conservative view is expressed in the official policy of the California Fish and Game Commission, which requires: (1) that all proposals for the introduction of exotic species are thoroughly evaluated for their potential impacts on native species and that a negligible or positive impact is determined; (2) the initial introduction of an exotic species must be made under conditions that will permit the action to be reversed; and (3) clear need for the action exists and the need cannot be satisfied through improved management to enhance native species or previously established nonnative species.

This policy also states that the introduction of an established nonnative species into areas of the state where it has not been previously established will be permitted only after it has been determined that there will be no significant negative impacts on native plant and animal species in the new area.

This Fish and Game Commission policy impacts aquaculturists in two significant ways. First, for freshwater aquaculture where we can expect that the introduced organism may be reasonably confined to the aquaculture facility, it requires that any application to import an exotic species be evaluated for the potential of the organism to become established in state waters. This evaluation may result in a requirement that the receiving facility either not discharge into state waters or that the waters leaving the facility are appropriately treated. Second, for the marine environment, where isolation is not practicable, initial introductions of exotic species will be allowed to take place only after an extensive investigation of the potential impact of the introduction on native species. Once the decision has been made to proceed, the state will require that procedures recommended by the International Council for the Exploration of the Sea (ICES) will be followed. Very briefly, this means complete isolation of the parent stock and release of the first-generation progeny only if no diseases or parasites become evident in the parent stock. (Sindermann, this volume.)

In the case of a species that has been previously introduced into the state, the Fish and Game Commission Policy allows more latitude. For instance, importation of giant Pacific oyster *(Crassostrea gigas)* seed to California from Japan and other west coast states has been a commercial practice for most of the twentieth century. Currently, importation of Pacific oyster seed is allowed only from one prefecture in Japan and a single hatchery in Washington, and then only when accompanied by a health certificate from appropriate authorities at the origin. Each shipment is also inspected upon arrival at the planting location. The combination of an extensive health history for the export zone and regular pathology examination limits the risk associated with this practice.

California has attempted to encourage the development of its own certified disease-free brood stock for shellfish species and most growers in California can now obtain Pacific oyster larvae and seed from a source within California. The California Fish and Game Department believes that development of our own disease-free broodstock for all commercial shellfish species is preferable to continued importation of animals or seed, and we would like to see sources of seed for other oyster species such as eastern and European oysters developed as well because of the continuing possibility of the introduction of disease organisms originating outside California.

The only other importation of marine organisms allowed by California outside ICES guidelines is the interstate movement of species within their established range from other west coast states. It has been established commercial practice to ship some organisms, most notably ghost and mud shrimps (Genera *Callianassa* and *Upogebia*) and marine annelid worms (Phylum Annelida), between west coast states for use as marine baits. These animals are naturally distributed along most of the U.S. west coast, and no adverse effects of this practice have been noted to date.

Finally, I would like to note that the west coast states made attempts in the late 1970s and early 1980s to develop a unified approach to importation policies under the auspices of the Pacific Marine Fisheries Commission (PMFC), but currently each state has gone its own way in developing and implementing these policies. There is clearly still a need for this kind of interstate cooperation, and I am hopeful that a unified west coast policy on the importation of exotics will yet be developed.

CHAPTER 6

International Activities and Programs

Role of the International Council for the Exploration of the Sea (ICES) Concerning Introductions of Marine Organisms

CARL J. SINDERMANN

Abstract: The North Atlantic nations, functioning through the International Council for the Exploration of the Sea, have made good progress during the past 15 years in drawing attention to problems associated with transfers and introductions of marine species, and have developed a basic code of practice which member countries have endorsed. A critical ingredient of the code is a recommendation that only F1 individuals derived from quarantined adults should be introduced, and not the adults themselves. Original concern about introductions of exotic pathogens has been augmented in recent years by concerns about ecological disturbances and genetic modifications. Because of the virtual irreversibility of successful introductions to marine waters, the problem is particularly acute, and calls for concerted international response.

Some of the ICES objectives include effective communication at appropriate levels, adoption of codes of uniform practices (insofar as national capabilities permit), and attempts at international uniformity in inspections and regulations. Proposed strategies to reduce risks from deliberate introductions include the development of governmental awareness of the potential effects of such actions; the establishment of regional and international committees to discuss problems related to introductions and to develop mutually acceptable procedures; and the inclusion of considerations of introductions on the agendas of international regulatory bodies concerned with living resources.

Introductions

Ten years ago, at an aquaculture meeting in a delightful Puerto Rican coastal city, a proposal was made for an "International Decade of Indiscriminate Ocean Transfers (IDIOT)." The proposal grew out of frustration over what seemed to be a rising tide of introductions and transfers of marine species from country to country and ocean to ocean. The core of project IDIOT was to be a decade of deliberate unrestricted movements of animals and plants from one place to another, whether for aquaculture, ornamental purposes, or any other reason (or for no reason at all). Then, after the expected great ecosystem disruptions and epizootics subsided (which might take half a century or longer) there would be no need for concern about future introductions of marine species, no oppressive regulations, no inspections at any border for diseases or pests.

Somehow the proposal elicited only minimal enthusiasm from attendees at that half-forgotten meeting in Puerto Rico. It was too extreme a concept for that age, and it may be too radical today. We need, therefore, to look for more rational strategies to reduce risks from transfers and introductions of marine species.

Transfers and introductions of marine species have occurred and are occurring on a worldwide basis, largely in response to perceived needs of expanding aquaculture industries. Greatest interest is in salmon (cage rearing and ocean ranching), shrimp, and bivalve molluscs, although other organisms are being considered. One important consequence of the global expansion of marine aquaculture has been the deliberate and often large-scale movement of plants and animals. Introductions of marine species to hydrographic provinces where they have not previously existed have been increasing, and will probably continue to do so. Thus far, the control of such movements has been variable in the extreme from country to country. Some have no restrictions; some have poorly enforced regulations; and a few have strict inspection and licensing laws. There is a growing perception that practices in coastal waters of one country may affect adjacent countries. This perception has led, haltingly, to interest in uniform standards for importation of living marine plants and animals.

An important development within the past two decades has been enunciation and endorsement by countries bordering the

North Atlantic of an international policy concerning introductions. Under the leadership of the International Council for the Exploration of the Sea (ICES), a "Code of Practice" has been formulated and then approved by member countries. ICES is the oldest and one of the most respected marine scientific organizations in the world, and its views — though they do not have the force of law — are taken very seriously by member nations, especially in Europe. The ICES Code is somewhat idealistic, but it does provide an international uniform policy concerning introductions of marine species.

The quintessence of the Code can be given in one simple paragraph:

"The species proposed for introduction should be studied in its native habitat. The study should include known diseases, pests and predators, food habits, and biotic potential. To be included would be consideration of pathological, environmental, and genetic implications of the introduction. The study should extend over several years, and the results should be examined by a committee of specialists. If a decision is made to proceed, then a brood stock should be established in quarantine in the recipient country. Only the F1 generation should be introduced to open waters, provided that no problems emerge."

A schematic plan for introductions of marine organisms, following the Code of Practice, is presented in Figure 1 and, because the ICES Code of Practice is widely cited but not yet generally available in the world literature, it is reproduced in its entirety as an appendix of this paper.

To ensure continuing scrutiny of actions by member nations with respect to introduced species, the Council has designated a permanent working group, with representation from each country, which meets annually to review the status of introductions and to recommend appropriate measures.

It would be reasonable to ask how the system has worked and is working. The answer is "marginally," with little progress except as a consequence of reactions to crises. Because of recognized problems resulting from importation of live marine animals,

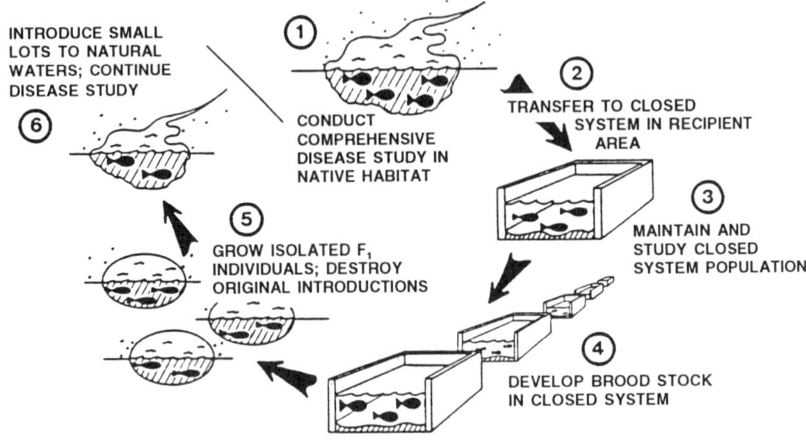

Figure 1. Recommended steps in the introduction of a new species, following the ICES Code of Practice.

some countries (e.g., United Kingdom, Ireland, Netherlands, Canada) have been sensitized, and have enacted conservative legislation governing such imports. Other countries have encountered problems, but have not reacted. Still other countries deny the existence of import-related problems. In general, though, because of ICES involvement, there does seem to be an increased awareness of potential difficulties — pathological, genetic, and environmental — associated with introductions.

All this emphasis on problems tends to obscure instances in which introductions have had positive outcomes from a human perspective. The usual examples cited are introduction of striped bass, *Morone saxatilis,* to the United States west coast, and the introduction of Japanese oysters, *Crassostrea gigas,* to the west coast of the United States and Canada. Other examples of successful introductions are easier to find in fresh water.

A few case histories can illustrate some of the positive and negative effects of deliberate introductions. These are: (1) introductions of Pacific salmon to Atlantic waters (largely a failure in terms of establishing self-sustaining populations), (2) the global dissemination of a pathogenic virus (IHHNV) in cultured penaeid shrimp stocks as a consequence of movements of brood stocks of several species, and (3) the large-scale introduction of Pacific oysters on the coast of France beginning in 1967 — an introduction which has led to establishment of the species there as a major aquaculture resource.

So the principal contribution of ICES concerning introductions of nonindigenous marine species has been the development of awareness among nations that the importation of nonindigenous species in quantity, without adequate controls, can have serious effects on native stocks and ecosystems, and that it is in the best economic interest of every country to have an effective regulatory mechanism in place. Such a system must have the flexibility to accommodate to new situations, but enough rigidity to resist political manipulation.

A recent informal internal critique of successes and failures in the activities of the ICES working group on introductions during the two decades of its existence resulted in the following assessment:

- "The *rate* of introductions and transfers of marine organisms seems to be accelerating."

- "Some ICES member countries are actually making a sincere effort to *follow the code* — at least by quarantining brood stock and introducing F1's only."

- "*Communication* among intergovernmental groups concerned with introductions is improving, but as it does we all come to realize the true extent of the activity."

- "*Newly recognized introductions* of pathogens (the eel nematode, *Anguillicola,* and the shrimp virus, IHHNV) reaffirm the extent of potential disease problems inherent in introductions and justify ICES concerns."

- "*Ecological and genetic problems* are being discussed, and examples identified (ballast water introductions of invertebrates would be a good example). More data from long-term studies are clearly needed."

- "Faint outlines of some *concepts* concerning introductions and transfers are emerging:

  - escapes from confined culture environments are inevitable;

  - diseases which can affect native species severely *are* imported with related nonindigenous species; and

  - actions by governments and intergovernmental bodies such as ICES are worthwhile and productive."

The ICES working group has proposed, in addition to the Code of Practice, a set of general operating principles useful in evaluating introductions of marine species:

1. Proposed introductions should have clearly stated and demonstrated rational bases. Proposals which are without adequate rationale, poorly planned, or unnecessarily risky, should not be approved.

2. Decision-makers should be aware of, and sensitive to, the practical, economic, social, and political aspects of introductions, but should evaluate proposals principally on the basis of the available scientific data. Relevant scientific implications and viewpoints include but are not limited to:

a. *Ecological considerations:* including competition, predation, and community characteristics of species (diversity, carrying capacity).

b. *Genetic considerations:* including the potential for hybridization, change in gene frequency (genetic diversity), and change or modification in disease and/or parasite resistance.

c. *Behavioral considerations:* including interactions between native and nonnative species.

d. *Pathological considerations:* including the potential for unintentional introduction of diseases and parasites.

3. Risks from introductions (e.g., diseases, parasites, predators, pests, environmental modification) are never zero.

4. The development of *native* species or of species stocks, through scientific management and aquaculture practices (including selective breeding and genetic manipulation), should be encouraged whenever and wherever feasible, as an alternative to introducing nonnative species.

5. Consideration should be given to nonmigratory species rather than migratory species, because of the potential of the latter for uncontrolled straying and subsequent colonization. Of course, it is also necessary to take into account the fact that sessile or sedentary forms (mussels, some reef fishes) may colonize distant areas by dispersion of eggs or larvae with ocean currents.

6. All importations should be made under adequate national control and surveillance supported by an adequate legal framework of laws and regulations, including those focused on mandatory and standardized inspection, quarantine, and certification procedures.

7. The outcome of an introduction or transfer cannot be fully predicted, so that any importation is an exercise in risk-assess-

ment and risk-taking, where positive and negative factors, insofar as they can be determined, are weighed.

8. All proposed introductions should be accompanied by full and adequate procedures and provisions for post-importation (follow-up) monitoring.

A final contribution of ICES is that its representatives form a thin line of rationality and conscience in each member country, ready to speak out against abuses and to publicize the positive aspects of regulating introductions. Their role is rarely a popular one; they can be accused of obstructionism, misplaced enthusiasm, manufacturing problems, and causing economic disaster for aquaculture ventures. Their existence and persistence can provide, however, a measure of reason and caution in a time of indiscriminate ocean transfers.

Appendix

ICES Code of Practice to Reduce the Risks of Adverse Effects Arising from Introduction of Non-Indigenous Marine Species

At its Statutory Meeting in 1973, the International Council for the Exploration of the Sea adopted a "Code of Practice to Reduce the Risks of Adverse Effects Arising from Introduction of Non-Indigenous Marine Species." At its Statutory Meeting in 1979, the Council adopted a revised code as follows:

1. Recommended procedure for species prior to reaching a decision regarding new introductions. (This does not apply to introductions or transfers which are part of current commercial practice.)

    a. Member countries contemplating any new introduction should be requested to present to the Council at an early stage information on the species, stage in the life cycle, area of origin, proposed place of introduction and objectives, with such information on its habitat, epifauna, associated organisms, potential competition to species in the new environment, etc., as is available. The Council should then consider the possible outcome of the introduction, and offer advice on the acceptability of the choice.

b. Appropriate authorities of the importing country should examine each "candidate for admission" in its natural environment, to assess the justification for the introduction, its relationship with other members of the ecosystem, and the role played by parasites and diseases.

c. The probable effects of an introduction into the new area should be assessed carefully, including examination of the effects of any previous introductions of this or similar species in other areas.

d. Results of b. and c. should be communicated to the Council for evaluation and comment.

2. If the decision is taken to proceed with the introduction, the following action is recommended:

a. A brood stock should be established in an approved quarantine situation. The first-generation progeny of the introduced species can be transplanted to the natural environment if no diseases or parasites become evident, but *not the original import*. The quarantine period will be used to provide opportunity for observation for diseases and parasites. In the case of fish, brood stock should be developed from stocks imported as eggs or juveniles, to allow sufficient time for observation in quarantine.

b. All effluents from hatcheries or establishments used for quarantine purposes should be sterilized in an approved manner (which should include the killing of all living organisms present in the effluents).

c. A continuing study should be made of the introduced species in its new environment, and progress reports submitted to the International Council for the Exploration of the Sea.

3. Regulatory agencies of all member countries are encouraged to use the strongest possible measures to prevent unauthorized or unapproved introductions.

4. Recommended procedure for introductions or transfers which are part of current commercial practice.

a. Periodic inspection (including microscopic examination) by the receiving country of material prior to mass transportation to confirm freedom from introducible pests and disease. If inspection reveals any undesirable development, importation must be immediately discontinued. Findings and remedial actions should be reported to the International Council for the Exploration of the Sea.

b. Inspection and control of each consignment on arrival.

c. Quarantining or disinfection where appropriate.

d. Establishment of brood stock certified free of specified pathogens.

It is appreciated that countries will have different attitudes towards the selection of the place of inspection and control of the consignment, either in the "country of origin" or in the "country of receipt."

Canadian Strategies for Risk Reductions in Introductions and Transfers of Marine and Anadromous Species

David J. Scarratt
Roy E. Drinnan

Abstract: Canada, with a huge land mass, extreme climatic variation, commercial fishery activity on three oceans, and a growing aquaculture industry, presents a diversity of risks associated with introductions or transfers: introductions or spread of pest organisms or diseases; genetic impacts from stocks transferred within the species range; and negative ecological impacts of introduced species by direct competition.

"Introduction" includes imports to the country, and internal "transfers" are of both native and exotic species between geographically and biologically separated areas. The unit-control area varies with the perceived risk. Basically, for constitutional and administrative reasons, Canada utilizes provincial boundaries; for more subtle control, and given sufficient knowledge, smaller discernible systems and biological subunits such as watersheds may be specified.

Control is effected by mandatory government approval of virtually all introductions and transfers to a province or smaller area. For salmonids, national legislation specifies requirements for approval. For other species groups, regional authorities operating within broad guidelines assess risk and potential impact, and, where approval is granted, specific procedures and requirements, including holding in or breeding from quarantine, monitoring of biological conditions, and criteria for release.

Introduction

Evidence is clear that the ill-considered or indiscriminate introduction or transfer of both exotic and native stocks can be accompanied by the introduction of diseases, pathogens, and parasites to which indigenous stocks have little or no resistance, or by other free-living species which may significantly affect the receiving ecosystem. Examples in Canada include the introduction of Malpeque disease of oysters to Prince Edward Island, possibly from southern New England, between 1910 and 1915, and the transfer of enteric redmouth disease in certified trout from Idaho in the 1970s. A local issue includes the transfer of salmon infected with bacterial kidney disease to two sea-cage sites in New Brunswick, which severely constrained the growth of the salmon industry in that province while disease-free brood-stocks were established on other farms. A classic example of a "stowaway" introduction is that of the slipper limpet, *Crepidula fornicata,* which was introduced to Europe in a shipment of American oysters in the 1880s.

Similar concerns have been expressed for the protection of the genetic integrity of native stocks, particularly from dilution by closely related stocks of the same species, which may be at a competitive disadvantage in a remote environment and, by interbreeding, reduce the success of the native stock.

Finally, native stocks may be displaced, or put at competitive disadvantage by the willful introduction of closely related species or other species occupying a similar ecological niche. An example is the potential for competition between brown trout and other salmonids in eastern Canada and elsewhere. At least one stream in Nova Scotia is now supporting five species of salmonids, of which only two, Atlantic salmon and brook trout, are native, while the other three, rainbow and brown trouts and coho salmon, are the result of introductions or escapes from early aquaculture experiments. The coho presumably escaped from New England, as none have been introduced to the Maritimes, and illustrate the potential for interactions between neighboring states, and the need for cooperation, especially in the case of highly mobile species.

Against these risks must be set the economic and social advantages that can be realized by the judicious exploitation of nonindigenous species, particularly where there can be human intervention in, and control of, one or more parts of the species life cycle in, say, a hatchery or grow-out farm.

The Canadian experience has been such that both the potential benefits and pitfalls of transferring and introducing new species are well known and, while progress has been significant in realizing the benefits and addressing the pitfalls, it is fair to say that progress has been irregular. While we are quite advanced in some areas, others leave much room for improvement. While there are national standards in Fish Health Protection Regulations, transfer of fishes within provinces is based on guidelines which differ from region to region and are not everywhere backed by effective legislation. So far, shellfish in Atlantic Canada, with one or two exceptions, are specifically excluded, except that a review of these regulations is imminent, and many shellfish growers are prudently requesting inspections and advice on shellfish movements in the absence of any legal requirement. In British Columbia, transfer of major commercial species is regulated.

Some of the instruments and procedures used in Canada to control introductions and transfers, drawing largely on Atlantic Coast examples, may be regarded as illustrative of the country as a whole, recognizing that regional differences reflect regional needs and priorities.

Fish Transfers

The Fish Health Protection Regulations (FHPR)

The FHPR was set up in 1977 specifically to control the spread and introduction of diseases of fish, principally among salmon and trout. They apply only to fish crossing provincial boundaries from other provinces or outside the country. They are comparable in intent with, but possibly more stringent than, regulations in the United States (Title 50) and in several European and other countries. A series of protocols described in the Manual of Compliance outlines procedures by which fish hatcheries and farms may receive and maintain certification as being free from specified pathogens. In general, four representative samples of fish taken from each site at 6-month intervals must be pathogen-free. Certification is lost if subsequent annual samples are shown to harbor any of the named pathogens, or if fish from a noncertified source are introduced to the site.

These protocols work extremely well in shore-based hatcheries, particularly those using ground-water sources. They are less effective for farms using river water where wild fish may carry pathogens, and are impossible in ocean farms where small wild fish may freely enter the cages. Protocols exist where wild fish may be certified for brood-stock purposes using statistical sampling techniques. A protocol for brood-stock certification from sea-cages has been developed and subjected to experimental verification, but has not yet been approved for routine use. This protocol requires the lethal sampling of all brood fish at the time of stripping, the identification and isolation of the fertilized eggs, and the subsequent destruction of any lots coming from parents shown to have harbored pathogens. Approved lots of eggs may be transferred to quarantine units in other provinces and the fry may be released four months after first feeding if bacteriological and virological examination shows them to be pathogen-free.

Mandatory examinations of market stock to control the spread of pathogens in dead fish sold as meat have been used to provide a profile of diseases in fish immediately prior to marketing and some indication of potential problems in brood-stock on the same farm. Salmon at this size are extremely valuable ($50-$100 each), so any comprehensive examination for disease is best combined with routine harvesting programs.

Provincial Fishery Regulations

Notwithstanding their name, Provincial Fishery Regulations are issued under the authority of the Fisheries Act for Canada, in respect of each province. In general, these regulations prohibit the importing or transfer of fish to or between waters of any province. Ministerial approval may be given if it is determined there is little risk due to disease, genetic or ecological threats. Unfortunately, the exact wording differs from province to province, so the legal force of the prohibition in certain provinces is somewhat reduced. Also, unfortunately, in the Maritime Provinces, the regulations specifically exclude shellfish. (See below.)

Regional Fish Health Guidelines

These guidelines have been developed in the Maritimes and are designed to support the provincial regulations by establishing protocols and criteria for determining which fish may be moved where and under what circumstances. The inspection criteria for fish are largely drawn from the FHPR Compliance Manual, but include experimental protocols developed for broodstock evaluation and locally developed "carrier" tests. The diseases of particular concern in eastern Canada are furunculosis, enteric redmouth, and bacterial kidney diseases. Several weeks prior to the transfer of smolts, samples are submitted to the Fish Health Unit at the DFO, Halifax Laboratory, and stress tested for the presence of the pathogens causing furunculosis and ERM, and examined for BKD. Fish carrying pathogens, but not showing disease symptoms, may receive standard therapeutic treatment before transfer.

We recommend that fish with active epizootics not be transferred under any circumstance, and that fish harboring pathogens not be transferred unless there is already a history of that disease in the receiving waters. Compliance with these guidelines is voluntary except in provinces where approval is specified by regulation, but the policy has met with remarkable success among the members of the salmon culture industry in eastern Canada. Many hatchery operators have slaughtered diseased smolts rather than sell them to sea-cage operators, and several sea-cage owners have destroyed fish which have proved to be diseased after transfer.

Provincial Aquaculture Acts

Several of the provinces have established legislation which includes powers to control the movement of fish between licensed aquaculture sites, and hence control unwanted diseases and genetic or ecological threats; however, the paramount authority is derived from the British North America Act and rests with the federal government.

Introductions Committees

A number of regions have introductions committees which, *inter alia,* rule on the genetic advisability of introducing a particular stock or species, and specify protocols to be followed in all introductions. Canada subscribes to the ICES protocol on introductions, which states that wherever possible native stocks shall be used to extend the range of a species, be it in captivity or for release into the wild. We have in recent years discouraged the import of foreign strains of Atlantic salmon into east coast waters where the species is native, although European Atlantic salmon have been introduced into farming operations on the Canadian west coast. The discouragement has often been for vaguely stated uncertainties about disease, or the disease-inspection protocols of the country of origin of the proposed consignment, and in many cases arrangements have been made for access to local stocks. Less firmly stated have been the genetic implications, and the concern for ecological competition by escaped farmed salmon with native wild runs, particularly in small streams. The federal government does have a policy on the introduction of rainbow trout, and certain streams are off limits because of potential competition with native species.

Shellfish

Disease

To date in the Maritimes, shellfish have been excluded from the regulations dealing with the transfer of fish, with the specific exception of Cape Breton Island, Nova Scotia. The reason for the exclusion has been to allow the traditional free movement of live shellfish (lobsters, oysters, clams, etc.) to markets elsewhere. The exception has been to prevent the spread of Malpeque disease to oysters in Bras d'Or Lake on Cape Breton Island. Canadian officials believe that Malpeque disease was introduced to the Gulf of St. Lawrence with oysters from southern New England. The fact that it took some 30 years to affect all the stocks in the Gulf is attributed partly to deliberate attempts to contain it and partly to the water circulation characteristics of the Gulf. No epizootics have been observed on Cape Breton Island, and oysters from there are still susceptible to the disease, in contrast to all other stocks which are resistant.

Notwithstanding the obvious implications of the uncontrolled movement of shellfish, and the more recent experiences of the French oyster industry, the Fish Health Protection Regulations apply only to salmonids. Shellfish introductions to the waters of a Province, as opposed to normal traffic for food, are covered in regulations requiring a Ministerial Permit for all plants or animals. Proposed introductions are reviewed and, if approved, appropriate conditions are written into the permit. In the absence of established protocols, responsible scientists engaged in experimental transfers and introductions have established experimental protocols by which species transferred or introduced out of their normal ranges, have been maintained in quarantine for varying lengths of time or for a number of generations. Currently, the protocol for European oysters and Bay scallops from U.S. stocks is that introduced brood-stock will be maintained in quarantine, and only F1 and subsequent generations will be released if, on examination, they are shown to be pathogen-free. This requirement is made by regional fisheries officials, and has not been challenged. Similar requirements are made on the west coast.

In addition, newly established shellfish hatcheries and shellfish farmers are approaching fish-health officers and requesting advice and guidance on the transfer of brood-stock and spat around the regions. The Fish Health Unit at Halifax, for example, is prepared to examine specimens and advise on the disease status of the sample. Similarly, the Fisheries Inspection Officers responsible for ensuring that shellfish are wholesome and nontainted are advising when live, in-shell shipments are entering the region, in case there is any possibility that effluent from live holding facilities will contaminate local waters.

To some degree this shellfish control policy smacks of shutting the door after the event, given the decades in which shellfish have been shipped back and forth without hindrance. In any case, precious little evidence exists of which diseases, pathogens, viruses and parasites are critical, although ignorance, in itself, is no reason to allow a laissez-faire attitude to jeopardize an important industry.

Genetics

To date there has been little movement of native shellfish for selection purposes, although recent experiments with blue mussels, *Mytilus edulis*, have shown that some strains have superior performance over others when transplanted. In Prince Edward Island, one harbor in particular is proving popular as a source of mussel spat for planting out in a number of other locations, so possibly there will be increased pressure to address the wisdom of this practice, and the conditions under which such transfers should be approved.

Given that the shellfish culture industry in eastern Canada is largely dependent upon mussels and native oysters, there is a need for diversification. To some degree the quahaug, *Mercenaria mercenaria*, offers potential, but the industry sees European oysters, *Ostrea edulis*, Bay scallops, *Argopecten irradians*, and possibly other species as having a significant future role to play. There are already stocks of these species in eastern Canada, although the genetic base of Bay scallops is rather impoverished. Potential importers of new strains, including government-sponsored programs, will be obliged to provide sound reasons to justify the risk of importing exotic diseases and parasites with the stock, notwithstanding the disease control protocols already in place.

Conclusion

While Canada has made substantial progress on controlling the introduction and spread of exotic and indigenous species, including their pests and diseases, it is equally clear that some significant legal and logistic loopholes remain, particularly in the detection of movements and enforcement of the regulations. To a degree, these are being addressed by the careful application of commonsense and policy choices, but some opportunity for legal challenge remains, and in some instances instruments are not available to ensure adequate control. In general, these loopholes are recognized and will be addressed; however, delays always occur between the recognition of a problem and the development of legislation to tackle it. The Fish Health Protection Regulations are currently being rewritten and will cover shellfish. The proposed amendments to the Provincial Regulations are already part of an "omnibus" revision.

We recognize that controlling the spread of exotic diseases, and reducing the risk of deleterious effects of introductions and transfers must be part of a continental, and worldwide, understanding, and so will collaborate, as in the past, with officials from the United States and other countries to ensure the health and safety of wild and cultivated fish and shellfish stocks around the world.

The Status of the U.S.-Japan Cooperative Program in Natural Resources (UJNR) Policy on the Introduction of Exotic Species for Aquaculture

JAMES McVEY

Abstact: The United States–Japanese Natural Resources Panel on Aquaculture (UJNR) consists of scientists from both countries who are interested in the exchange of information on the developing field of aquaculture. Meetings are held once a year on an alternating basis in each country to discuss new developments in aquaculture and opportunities for cooperation between the two countries. Topics for discussion are chosen well in advance and leading scientists from each country are asked to give presentations on their areas of expertise within the topic.

There has been a long-standing concern and discussion on the hazards and opportunities associated with the use of exotic species in aquaculture for both countries. The UJNR Aquaculture panel recognizes the need for establishing protocols concerning the movement and transport of exotic species; however, in the most recent meeting of the UJNR it was the general opinion that it is premature to formulate such guidelines until there is more input from the U.S. and Japanese aquaculture industries. Additional technical information will also be needed before such guidelines are established.

The U.S. agencies that are responsible for aquaculture are developing a coordinated policy, through the Joint Subcommittee on Aquaculture (JSA), which will clarify the U.S. federal policy in this area. Once this is done, the UJNR will reconsider the issues and make recommendations in line with the JSA and the existing Japanese policies on the use of exotics in aquaculture.

Introduction

The U.S.-Japan Cooperative Program in Natural Resources (UJNR) has provided for a technical exchange in aquaculture through the Aquaculture Sub-Panel since 1969. The panel currently includes specialists drawn from the federal departments most concerned with aquaculture in both governments. The panel is charged with exploring and developing bilateral cooperation which could be of benefit to both countries. Past activities and accomplishments have included increased communications and cooperation among technical specialists; exchanges of information, data and research findings; annual meetings of the panel; administrative staff meetings; exchanges of equipment, materials and samples; several major technical conferences and beneficial effects on international relations.

During the course of the cooperative program there has been a constant interest by the Aquaculture Panel to develop guidelines for the introduction of exotic species to new geographic locations. However, the issue is a difficult one because many exotic introductions have already occurred, resulting in both new industries and economic and recreational benefits as well as the introduction of diseases and parasites to new areas and the establishment of unwanted populations in the wild. In most instances, the positive aspects of the introductions have outweighed the negative ones on a financial and economic basis, but the fact that there are both positive and negative consequences of an introduction makes the formation of an overall policy very difficult.

In the United States, the federal government has not set forth policy on the introduction of exotics for aquacultural purposes. The U.S. Fish and Wildlife Service has policies regulating the introduction of exotics for enhancement purposes but not for aquacultural activities. The legal authority relating to the introduction of exotics is with the States. These policies differ substantially from State to State depending on past experiences, geographical location, species involved, the perceived future potential for the development of an industry, and many other considerations. Obtaining a general consensus is, therefore, very difficult.

In Japan, the development of aquaculture is more closely tied to the federal government and the aquaculture industry is more developed and thus more important to the overall economy of the nation. The federal agencies responsible for aquaculture are concerned that any exotic introductions are safe for the existing industry, but they want to maintain their options for the introduction of desirable exotics in order to develop the aquaculture industry to its full potential.

During the last few years, the UJNR has worked actively to develop guidelines for the introduction of exotic species. Guidelines, based primarily on those that have already been adopted by ICES (International Council for the Exploration of the Seas), have been proposed by Dr. Carl Sindermann in cooperation with Dr. Ryo Suzuki of Japan and have been circulated to members of the UJNR. In the most recent meeting of the UJNR Aquaculture Panel held in Japan, these guidelines were discussed in detail. After lengthy deliberation the Aquaculture Panel decided to table the issue indefinitely for the following reasons:

1. Each state in the U.S., as well as the federal government, has separate and often different policies regarding introduced species. The complexity of this situation as well as the lack of common agreement among all U.S. federal agencies makes it impossible to devise a uniform set of guidelines at this time, except in very broad terms. The U.S. government needs to consolidate its opinions first before establishing guidelines for others.

2. The proposed guidelines were too restrictive. Many of the very successful transplants, such as the introduction of salmon to the Great Lakes, Chile and New Zealand, would never have occurred had the guidelines been in force.

3. The panel felt that one set of policies for all species and all places would not work because of the diverse considerations in different situations. Therefore, guidelines, especially those that

could be construed as laws, should not be promulgated until more technical information is available on a species-by-species basis.

4. Guidelines, even though not laws, tend to become law when interpreted by government agencies. The guidelines, once adopted, will be very difficult to change. It is outside the sphere of the UJNR panels to set such international policy.

5. The guidelines, should they become policy, might pose too much of a hardship on existing U.S. and Japanese industry participants while other countries would be free to operate as usual. In addition, the U.S. and Japanese aquaculture industry that would be most affected by the guidelines has not had an opportunity to provide input.

6. The aquaculture industry is not the worst offender when it comes to exotic introductions; as long as the tropical aquarium industry, scientific researchers and the general public are allowed to import organisms with essentially no controls, it is unfair to burden the aquaculture industry with restrictive measures. In truth, the aquaculture industry has more to gain by proper controls to avoid the introduction of disease than most other groups; if a realistic procedure is identified, they will be among the first to adopt it.

7. Guidelines that are too restrictive would lead to companies and individuals disregarding the law. There is no practical way that animal transport regulations can be effective without voluntary and active support by the user groups. Enforcement of guidelines and laws that are too restrictive is impossible, especially when there is no funding for enforcement.

In summary, the UJNR Aquaculture Panel recognizes the need for establishing protocols concerning the movement and transport of exotic species. However, it was the general opinion of the Panel that it is premature to formulate such guidelines until there is more input from the U.S. and Japanese aquaculture industry. Modifications of the proposed guidelines will probably have to be made in order to allow the existing aquaculture industry to survive and new developments to occur.

Additional technical information is important to the formulation of future guidelines, and work should continue on the catalog of aquatic diseases which is now being worked on by the UJNR Aquaculture Panel.

The U.S. Joint Subcommittee on Aquaculture, which is composed of representatives of all the major federal agencies involved in aquaculture, has formed a working group to review U.S. policy and procedures with regard to the introduction of exotics. In addition, the outcome of the workshop "Human Influences on the Dispersal of Organisms and Genetic Materials into Aquatic Ecosystems" (which was the basis of this book) reflected a good cross-section of both industry and the research community and will be important to the establishment of U.S. aquaculture policy. The UJNR will be watching these developments closely and will take action when more information is available for decision-making.

Toward a Reasoned Approach to Introduced Aquatic Organisms

CHRISTOPHER C. KOHLER

Abstract: The American Fisheries Society (AFS) has taken a leadership role on the global issue of fish introductions dating back to 1969. A formal "Position of the American Fisheries Society on Introductions of Exotic Aquatic Species" was en-dorsed by the membership in 1972. The AFS Exotic Fish Section was formed in 1980 and has initiated a number of activities addressing exotic fish issues. The section name was changed to be the Introduced Fish Section in 1985 to broaden its scope to include transplanted species. The section, upon the request of the AFS Environmental Concerns Committee, updated the AFS position statement which was subsequently published and adopted in 1986. Voluntary compliance with the intent of the position statement within the aquaculture industry is recommended. Non-compliance could lead to legislation being enacted that would not provide the same flexibility in imports/exports currently being enjoyed by the industry.

Introduction

"Be careful and think" are the words which Hubbs (1977) used to summarize the meaning of guidelines emanating from the "Invitational Conference on Exotic Fishes and Related Problems" held on 18-19 February 1969, Washington, D.C., which was sponsored by the American Fisheries Society and the American Society of Ichthyologists and Herpetologists (Lachner et al. 1970). If that simple and eloquent message was always heeded during the process of planning and effectuating introductions there would be little need for formal guidelines and regulations. Such has not been the case and numerous introductions have been made, often ill-fated or of dubious benefit. Courtenay and Robins' (1989) recent review article entitled "Fish Introductions: Good Management, Mismanagement, or No Management?" aptly describes the situation with respect to introductions of aquatic organisms, and along with a number of other reviews (e.g., Taylor et al. 1984; Kohler and Courtenay 1986a; Courtenay and Kohler 1986) provides examples of all scenarios expressed in the above title. Thus, a given introduction might fall anywhere along the continuum of highly beneficial to disastrous. Accordingly, all potential benefits and risks should be carefully weighed whenever an introduction is being contemplated.

A number of international organizations have adopted or are considering adopting formal "codes of practice" for regulating the introduction of aquatic organisms (see Kohler and Courtenay 1986a; Sindermann 1986; Welcomme 1986). Implementation of such codes (protocols, guidelines, etc.) can ensure that decisions regarding future introductions are based on sound ecological evidence, and that introductions effectuated are properly evaluated (Kohler and Courtenay 1986b). In the sections that follow the approach being taken by the American Fisheries Society with respect to introduced aquatic species is reviewed, a protocol for evaluating potential introductions is described, and relative merits of guidelines versus regulations concerning introductions are presented.

American Fisheries Society and Introduced Aquatic Species

A review of past initiatives of the American Fisheries Society (AFS) with respect to regulating introduced aquatic species is presented in Kohler and Courtenay (1986a). The major activities are repeated here along with more recent developments.

Courtenay and Robins (1973) reviewed the issue of exotic aquatic organisms in Florida and made several recommendations as to the methods that should be employed for considering introductions. That article was approved by the Florida Chapter of the American Institute of Fisheries Research Biologists as constituting an official position statement of that organization. In addition, some of the recommendations put forth were adopted by the AFS Committee on Exotic Fishes in a position statement which was submitted to and approved by the membership of AFS at the 1972 annual meeting of the Society in Hot Springs, Arkansas. That statement appeared in *Transactions of the American Fisheries Society*, (Volume 102, Number 1, pages 274-276).

For reasons not fully known, the above position statement received little recognition and rarely was followed. Many fisheries professionals may have confused the term "exotic" with "ornamental" and thus perceived the position statement as pertaining only to the pet fish industry. Others may have chosen to ignore it, while perhaps the majority were unaware of its existence.

The AFS Exotic Fish Section was formed in 1980 and has since initiated a number of activities addressing exotic fish issues. The Section sponsored a symposium entitled "Distribution, Biology and Management of Exotic Fishes" held in 1981 during the 110th Annual AFS Meeting at Albuquerque, New Mexico. The papers appear in published form in a book of the same title, edited by Courtenay and Stauffer (1984). Nearly all aspects concerning exotic fish introductions were addressed, including a suggested protocol for evaluating proposed exotic fish introductions in the United States (Kohler and Stanley 1984a). That protocol was subsequently revised (Kohler and Stanley 1984b) and presented at the European Inland Fisheries Advisory Commission (EIFAC) Symposium on Stock Enhancement in the Management of Freshwater Fisheries (see Welcomme et al. 1983) to include all of North America and Europe.

One of the first accomplishments of the AFS Exotic Fish Section was to refine the terminology associated with introduced organisms (see Shafland and Lewis 1984). They define "introduced" as a plant or animal moved from one place to another by man (i.e., an individual, group, or population of organisms that occur in a particular locale due to man's actions). They define "exotic" as an organism moved outside its native range but within a country where it occurs naturally (i.e., one whose native range includes at least a portion of the country where found).

Recently, the AFS Exotic Fish Section changed its name to be the Introduced Fish Section. The name change was brought about by a desire of the membership to broaden its scope to include transplanted species. Transplanting native species (e.g., Pacific salmon to the Atlantic Ocean) carry the same inherent risks as exotic introductions. Although caution is advised with respect to movements of any species beyond its native range, the AFS Introduced Fish Section has taken the approach that with proper planning and evaluation, the odds in what Magnuson (1976) has described as "a game of chance" can be improved. Towards that end, the section and the parent society have taken a leadership role on the global issue of fish introductions. The first major step has been the adoption of an AFS Position on Introductions of Aquatic Species (Kohler and Courtenay 1986b). The specific statement is as follows:

Position of American Fisheries Society on Introduced Aquatic Species

A. Our purpose is to formulate a broad mechanism for planning, regulating, implementing, and monitoring all introductions of aquatic species.

Some introductions of species into ecosystems in which they are not native have been successful (e.g., coho salmon and striped bass) and others unfortunate (e.g., common carp and walking catfish).

Species not native to an ecosystem will be termed "introduced." Some introductions are, in some sense, planned and purposeful for management reasons; others are accidental or are simply ways of disposing of unwanted pets or research organisms.

It is recommended that the policy of the American Fisheries Society be:

1. Encourage fish importers, farmers, dealers and hobbyists to prevent and discourage the accidental or purpose-

ful introduction of aquatic species into their local ecosystems.

2. Urge that no city, county, state, province or Federal agency introduce, or allow to be introduced, any species into any waters within its jurisdiction which might contaminate any waters outside its jurisdiction without official sanction of the exposed jurisdiction.

3. Urge that only ornamental aquarium fish dealers be permitted to import such fishes for sale or distribution to hobbyists. The "dealer" would be defined as a firm or person whose income derives from live ornamental aquarium fishes.

4. Urge that the importation of fishes for purposes of research not involving introduction into a natural ecosystem, or for display in public aquaria by individuals or organizations, be made under agreement with responsible governmental agencies. Such importers will be subject to investigatory procedures currently existing and/or to be developed, and species so imported shall be kept under conditions preventing escape or accidental introduction. Aquarium hobbyists should be encouraged to import rare ornamental fishes through such importers. No fishes shall be released into any natural ecosystem upon termination of research or display.

5. Urge that all species considered for release be prohibited and considered undesirable for any purposes of introduction into any ecosystem unless that species shall have been evaluated upon the following bases and found to be desirable:

> *a. Rationale.* Reasons for seeking an import should be clearly stated and demonstrated. It should be clearly noted what qualities are sought that would make the import more desirable than native forms.
>
> *b. Search.* Within the qualifications set forth under *Rationale*, a search of possible contenders should be made, with a list prepared of those that appear most likely to succeed, and the favorable and unfavorable aspects of each species noted.

c. Preliminary Assessment of the Impact. This should go beyond the area of rationale to consider impact on target aquatic ecosystems, generally effect on game and food fishes or waterfowl, on aquatic plants and public health. The published information on the species should be reviewed and the species should be studied in preliminary fashion in its biotope.

d. Publicity and Review. The subject should be entirely open and expert advice should be sought. It is at this point that thoroughness is in order. No importation is so urgent that it should not be subject to careful evaluation.

e. Experimental Research. If a prospective import passes the first four steps, a research program should be initiated by an appropriate agency or organization to test the import in confined waters (experimental ponds, etc.).

f. Evaluation or Recommendation. Again publicity is in order and complete reports should be circulated amongst interested scientists and presented for publication.

g. Introduction. With favorable evaluation, the release should be effected and monitored, with results published or circulated.

Because animals do not respect political boundaries, it would seem that an international, national and regional agency should be involved at the start and have the veto power at the end. Under this procedure there is no doubt that fewer introductions would be accomplished, but quality and not quantity is desired and many mistakes might be avoided.

B. The Society encourages international, national, and regional natural resources agencies to endorse and follow the intent of the above position.

C. The Society encourages international harmonization of guidelines, protocols, codes of practice, etc., as they apply to introductions of aquatic species.

D. Fisheries professionals and other aquatic specialists are urged to become more aware of issues relating to introduced species.

Protocol for Assessing Planned Introductions

A protocol for assessing planned introductions of aquatic organisms was developed by Kohler and Stanley (1984a) for the United States and subsequently modified (Kohler and Stanley 1984b) to include much of the northern hemisphere.

Four categories are considered in the evaluation:

1. Feasibility, which deals with the validity of the proposed use, the status of the organism in the native range, the location and type of system into which it would be introduced, disease control measures, and various legal restrictions;
2. Acclimation potential of an organism, which is based on habitat requirements, reproductive viability and migratory behavior;
3. Control potential, which deals with methods that could be used to eliminate organisms introduced but later deemed undesirable or to prevent (limit) reproduction; and
4. Prediction of impact, which is defined as the balance between perceived benefits and risks.

The model is highly flexible and is comprised of five levels of review and five "decision boxes" (Figure 1). Although each level of review mandates progressively greater scrutiny of the proposed introduction, decisions can often be rendered during early stages of the evaluation because the more basic criteria for analyzing introductions are considered at the outset. The review and decision model contains five decision points for approval and seven for rejection of an introduction. An opinionnaire (Table 1) is used to generate the data base for the review and decision model. The opinionnaire consists of ten questions designed to evaluate any proposal to introduce an aquatic organism.

The protocol is an effective mechanism for considering progressively more complex and uncertain information to arrive at decisions to approve or reject proposals for introductions of aquatic organisms. The protocol has been adopted by the European Inland Fisheries Advisory Commission in a slightly modified form and is currently undergoing testing.

Guidelines or Regulations?

It is one thing to have formal position statements, protocols, guidelines, etc., with respect to introduced aquatic organisms but it is quite another to have enforceable regulations. There is also the question as to which is more desirable. There exist individuals who prefer no guidelines or regulations and would introduce any species they desire, others willing to give some lip service to guidelines but who will manipulate them to meet their own objectives, while still others that will follow guidelines even if it means they have to put their own best interests aside. If most individuals fell in the latter category then sufficient peer pressure would likely exist to keep the others in line. Unfortunately, this has never been the case, though there is more movement in that direction than the opposite. Does this mean there is a need for stringent regulations and severe legal penalties for those not complying? On the surface of the issue there appears to be a simple affirmative answer; however, nothing dealing with introduced species is ever simple, and that is particularly true when the issue concerns regulations. Stringent regulations would likely result in endless litigation with attorneys being the primary beneficiaries. Conversely, no or ineffective regulations set the stage for countless introductions and eventual environmental chaos. Any

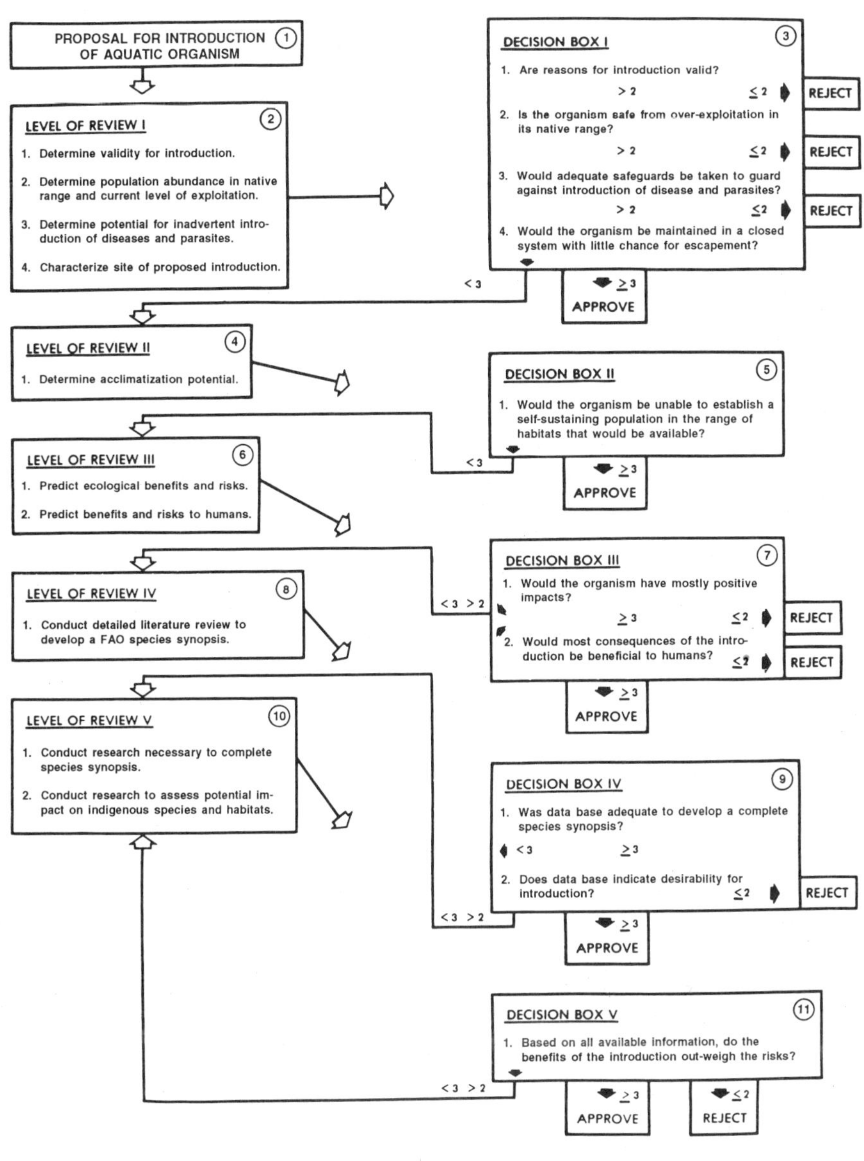

Figure 1. Review and decision model for evaluating proposed introductions of aquatic organisms. Mean opinionnaire values (see Table 1) are used at decision-making points. (Taken from Kohler and Stanley 1984b).

Table 1. Opinionnaire for appraisal of introductions of exotic aquatic species. Each member of an evaluation board or panel of experts circles the number most nearly matching his or her opinion about the probability for the occurrence of the event. If information is unavailable or too uncertain; "don't know" is marked. (Taken from Kohler and Stanley 1984b).

		Response					
Variable	Question	*No*	*Unlikely*	*Possible*	*Probably*	*Yes*	*Don't Know*
Valid	1. Is the need valid and are no native species available that could serve the stated need?	*1*	*2*	*3*	*4*	*5*	*X*
Status	2. Is the exotic species safe from over-exploitation in its native range?	*1*	*2*	*3*	*4*	*5*	*X*
Disease	3. Are safeguards adequate to guard against importation of disease/parasites?	*1*	*2*	*3*	*4*	*5*	*X*
Escape	4. Would the exotic species be limited to closed systems?	*1*	*2*	*3*	*4*	*5*	*X*
Sustain	5. Would the exotic species be unable to establish a self-sustaining population in the range of habitats that would be available?	*1*	*2*	*3*	*4*	*5*	*X*
Impact	6. Would the exotic species have only positive ecological impacts?	*1*	*2*	*3*	*4*	*5*	*X*
Hazard	7. Would all consequences of the exotic species be beneficial to humans?	*1*	*2*	*3*	*4*	*5*	*X*
Synopsis	8. Is there a species synopsis and is it complete?	*1*	*2*	*3*	*4*	*5*	*X*
Desired	9. Does data base indicate desirability for introduction?	*1*	*2*	*3*	*4*	*5*	*X*
Benefit	10. Would benefits exceed risks?	*1*	*2*	*3*	*4*	*5*	*X*

regulations that are enacted should be coupled with education programs on the role that introduced species can and should play within the context of aquaculture and aquatic resource management. In the final analysis, establishment of specific regulations is an admission on the part of the environmental community that it has failed to adequately educate that sector involved in introducing aquatic species. The end result may be the same, but, at least for those in education, it is just not quite as satisfying.

Literature Cited

Courtenay, W.R., Jr. and C.C. Kohler. 1986. Review of exotic fishes in North American fisheries management, p. 401-413. *In* R. Stroud (ed.), Role of fish culture in fishery management. Symposium held on 31 March-3 April 1985 at Lake Ozark, Missouri.

Courtenay, W.R., Jr. and C.R. Robins. 1973. Exotic aquatic organisms in Florida with emphasis on fishes: a review and recommendations. Trans. Am. Fish. Soc. 102:1-12.

Courtenay, W.R., Jr. and C.R. Robins. 1989. Fish introductions: good management, mismanagement, or no management? Reviews Aquat. Sci. 1:159-172.

Courtenay, W.R., Jr. and J.R. Stauffer, Jr. (eds.). 1984. Distribution, biology, and management of exotic fishes. Johns Hopkins University Press. Baltimore, Maryland.

Hubbs, C. 1977. Possible rationale and protocol for faunal supplementations. Fisheries 2(2):12-14.

Kohler, C.C. and W.R. Courtenay, Jr. 1986a. Regulating introduced aquatic species: a review of past initiatives. Fisheries 11:(2)34-38.

Kohler, C.C. and W.R. Courtenay, Jr. 1986b. American Fisheries Society position on introduction of aquatic species. Fisheries 11:(2)39-42.

Kohler, C.C. and J.G. Stanley. 1984a. A suggested protocol for evaluating proposed exotic fish introductions in the United States, p. 387-406. *In* W.R. Courtenay, Jr. and J.R. Stauffer, Jr. (eds.), Distribution, biology, and management of exotic fishes. Johns Hopkins University Press, Baltimore, Maryland.

Kohler, C.C. and J.G. Stanley. 1984b. Implementation of a review and decision model for evaluating proposed introductions of aquatic organisms in Europe and North America, p. 541-549. *In* European inland fisheries advisory commission documents presented at the symposium on stock enhancement in the management of freshwater fisheries, 2. EIFAC Tech. Pap. 42.

Lachner, E.A., C.R. Robins and W.R. Courtenay, Jr. 1970. Exotic fishes and other aquatic organisms introduced into North America. Smithsonian Contrib. Zool. 59:1-29.
Magnuson, J.J. 1976. Managing with exotics — a game of chance. Trans. Amer. Fish. Soc. 105:1-9.
Shafland, P.L. and W.M. Lewis. 1984. Terminology associated with introduced organisms. Fisheries 9(4):17-18.
Sindermann, C.J. 1986. Strategies for reducing risks from the introductions of aquatic organisms: a marine perspective. Fisheries 11:(2):10-15.
Taylor, J.N., W.R. Courtenay, Jr. and J.A. McCann. 1984. Known impacts of exotic fishes in the continental United States, p. 322-373. *In* W.R. Courtenay, Jr. and J.R. Stauffer, Jr., (eds.), Distribution, biology, and management of exotic fishes. The John Hopkins University Press, Baltimore, Maryland.
Welcomme, R.L. 1986. International measures for the control of introductions of aquatic organisms. Fisheries 11:(2):4-9.
Welcomme, R.L., C.C. Kohler, and W.R. Courtenay, Jr. 1983. Stock enhancement in the management of freshwater fisheries: a European perspective. N. Am. J. Fish. Mgmt. 3:265-275.

CHAPTER 7

Factors that Affect Management

Model Seafood Surveillance Project

G. Malcolm Meaburn
Lloyd W. Regier
E. Spencer Garrett

Abstract: The continuing concerns of the public, media and Congress that the seafood supply of the United States may present unacceptable health hazards due to pollution and/or mishandling resulted in congressional action. The Model Seafood Surveillance Program (MSSP) study was authorized by Congress in the 1987 fiscal year budget to have NOAA design "a program of certification and surveillance to improve the inspection of fish and seafood consistent with the Hazard Analysis Critical Control Point system." The National Marine Fisheries Service is proceeding with the study utilizing a three-pronged approach: product safety, plant hygiene, and economic fraud. The product-safety issue will be addressed primarily through a contract to the National Academy of Science. Plant hygiene and economic fraud issues will be addressed using the Hazard Analysis Critical Control Point (HACCP) concept during specific industry by industry workshops conducted in conjunction with National Fisheries Institute and other trade associations through Saltonstall-Kennedy grants.

The major health-related problems associated with seafood are ciguatoxin, scombrotoxin (histamine), paralytic shellfish poison(PSP), and viruses. Ciguatoxin is a naturally occurring toxin in the flesh of some reef fishes in tropical areas that has its source in certain dinoflagellates. Ciguatoxin production has not been shown to be related to man's activities. Histamine poisoning occurs when fish, which are high in natural concentrations of the free amino acid histidine, are subjected to abusive conditions after harvest. The PSP and viruses in molluscan shellfish arise from the contamination of the growing waters. Only the viruses have been shown to be definitely related to pollution from human activities. Each of these hazards requires a different part of an overall strategy to establish effective controls.

The final product which NOAA intends to deliver to Congress will be a surveillance system design for seafood products which provides for reasonable consumer protection in the consumption of fishery products, and treats imported, domestic and exported products equably.

Introduction

There is a strong public and congressional perception that the consumption of fishery products in the United States may represent an unacceptable public health risk to consumers since seafoods are not subject to mandatory federal inspection in a manner comparable to that for meat and poultry products.

In 1985, the National Fisheries Institute (NFI) conducted a survey of its members to determine their views of consumer-protection issues. Respondents to the survey expressed the opinion that, other than in the area of molluscan shellfish, seafood-safety problems were not significant. The NFI concluded that actions should be taken, including legislation, to correct deficiencies in the inspection of fishery products.

Legislation (H.R. 1483) has since been introduced to require a mandatory program of continuous fishery-product inspection. Testimony at hearings by both Senate and House Agriculture Committees on current meat and poultry inspection problems has emphasized the fact that fishery products do not receive similar inspectional scrutiny. It should also be noted that a philosophical change away from the traditional continuous surveillance concept developed for meat and poultry inspection is rapidly evolving, largely as a result of recent legislative action and more innovative risk-management approaches under consideration by the United States Department of Agriculture (USDA).

In both fiscal years (FY) 1987 and 1988, Congress appropriated $350,000 to National Oceanic and Atmospheric Administration to design an improved system of inspection and certification of seafood in the United States. The design study is also being augmented through redirection of agency effort and other resources.

The National Oceanic and Atmospheric Administration/National Marine Fisheries Service study, known as the Model Seafood Surveillance Project (MSSP), is based upon the Hazard Analysis Critical Control Point (HACCP) concept and will provide for equitable treatment of domestically produced and imported fishery products. Implicit in the HACCP concept is that only absolutely necessary, i.e., critical, control points of a food-processing operation are monitored. This is in contrast to all operational steps under a traditional inspection approach. The HACCP approach has been successfully employed by USDA and FDA in the low-acid canned-food industry for a number of years and is also the basis of recommendations to USDA by the National Research Council (NRC) concerning the design of more efficient and cost-effective inspection programs for meat and poultry products. The MSSP operational approach to the congressional charge is described in a detailed Plan of Operations first issued in September 1987, and updated in January 1989 (Anonymous 1987).

HACCP Considerations

In order to apply the HACCP concept systematically, MSSP is following the recommendations of the National Academy of Sciences (NAS) in a 1985 report on the role of microbiological criteria for foods and food ingredients (NAS 1985). In that report, NAS recommended that industry take the lead role in the development of the technical details of a HACCP-based food-control program. The report defines procedures for utilizing HACCP in food-processing plants and emphasizes the crucial role of industry groups in the design, testing, and implementation of a surveillance system. After such an industry-driven system is developed, the regulatory authorities should determine the adequacy of the HACCP plans, including the selection of critical control points, monitoring procedures, and records to be maintained. The National Academy of Sciences further recommended that, in order to be effective, such an HACCP program should be made mandatory through appropriate regulation. The need for a close working relationship between industry and regulatory agency, and the acceptance of their respective roles in the design and implementation of a HACCP program, was stressed throughout that report.

The design of a HACCP surveillance system is based upon analysis of the hazards associated with each step of a food-processing or food-handling operation and their importance relative to the end use of the product. The critical control points for significant hazards and preventative measures to minimize these hazards are identified, and monitoring procedures to be used for compliance purposes are established. In following the recommendations of the NAS report, the MSSP has classified potential consumer hazards into three categories: (1) product safety; (2) plant/food hygiene; and (3) economic fraud.

Product-Safety Issues

The initial operational approach of the MSSP focused on a quantitative analysis of *documented* product safety issues associated with consumption of seafoods. A summary review of data on seafood-borne illnesses reported during the years 1978-82, in the Annual Reports published by the Center for Disease Control (CDC) (Freeman 1987, 1988), indicates that safety problems are not as widespread as are often perceived. Most of the reported illnesses (approximately 87%) were attributable to consumption of ciguatoxic or scombrotoxic fish, or molluscan shellfish contaminated with microbial pathogens or toxins. However, it is recognized that the CDC statistics may be biased because of the large differences in reporting efficiency between the various states and territories.

The CDC, with assistance from NMFS personnel, has compiled data on outbreaks of illness attributable to consumption of all animal protein foods, including seafood, for the period 1983-86 which was not possible with CDC budgets. The data was received by MSSP in December 1988 for review and analysis.

The U.S. General Accounting Office (GAO) released a report in late September 1988 on its investigation of consumer-related safety problems with seafood (GAO 1988). While the GAO made no recommendations, it did make several observations based on the information gathered and the views of experts interviewed during the course of the study. They concluded that there was no "compelling case at this time for implementing a comprehensive, mandatory federal seafood inspection program similar to inspections used for meat and poultry." The GAO expressed support for several initiatives, including the development of a model surveillance system for domestic and imported seafoods (the MSSP), to provide a basis for mandatory inspection in the future.

The safety of fishery products available to the U.S. consumer from both commercial and recreational sources will be evaluated by NAS under a contract from NMFS which was negotiated in September 1988.

The study will address public health problems that are directly attributable to consumption of fish and shellfish, and other fishery products derived from marine and freshwater sources. The efficacy of systems currently in place for documenting and reporting cases of seafood-borne illness will also be examined, as will the prevalence and significance of imports as causative agents of illness. The expert committee appointed by the NAS/NRC through its Food and Nutrition Board met with representatives of NOAA/NMFS, FDA, and USDA, on January 30, 1989, to review its charge and discuss the major issues. The NAS also held a meeting on January 31, 1989, to introduce the study to the public formally.

Plant/Food Hygiene and Economic Fraud

The MSSP implemented this aspect of the study by following the NAS recommendation that industry should take a lead role in the design of specific HACCP-based food-control programs. The NMFS is further supporting this cooperative approach through several successive Saltonstall-Kennedy (S-K) grant awards to the National Fisheries Education and Research Foundation Inc., (NFERF) to develop HACCP models for specific seafood commodities, and to identify regulatory and/or research needs associated with each model. The series of workshops supported by the S-K funds was initiated in October 1987 to meet these objectives.

NFERF/NMFS has conducted more than 16 HACCP industry workshops covering 39 seafood commodities. The commodities include breaded shrimp, cooked shrimp, raw shrimp, fresh and frozen fish, tropical fish (Pacific species), molluscan shellfish, blue crab, smoked and cured fish, breaded fish and specialty items, scallops, lobsters, West Coast crabs (king, snow and Dungeness), and imports. With the exception of imports, HACCP inspection models have been developed for all commodities. Corresponding documents for evaluating plant hygiene have been generated and critical control points in food-processing operations for each commodity have been identified. Included in these considerations are controls to prevent unwholesome products and eliminate fraudulent practices such as species substitution and short weights. A HACCP model for imports, which represent 65% of the volume of seafood consumed in the United States, was developed following a second workshop in May 1989. A series of HACCP workshops for vessels was also be conducted in 1989.

The workshop on molluscan shellfish processing considered the hazards related to the safety of the product as related to end use. The industry participants identified pathogenic viruses and paraly-tic shellfish poison (PSP) and related toxins as major items of safety concern but recognized that these contaminants are related to the growing water conditions as they may affect raw materials received for processing. While the methods for monitoring shellfish for the preformed PSP toxin are adequately refined to provide protection of consumers of shellfish from controlled waters, other toxins and viruses are not yet adequately identifiable or measurable. While there are bacteriological safety hazards that need to be considered in the processing of shellfish, most of the effort or concerns must still be placed in the management of the harvesting of molluscan shellfish.

The harvesting and distribution parts of the industry have not yet been covered by the workshops. From the processing operations analyses, the industry workshop participants identified 11 to 16 process steps of which 5 to 7 were considered critical and in need of recorded evidence of monitoring. Most of these critical control points were process steps involving receipt of the raw material, final inspection of the product, or shipment.

Model Testing

The HACCP models for each commodity are being tested at processing plants that are statistically stratified according to production volume. In-plant testing of the HACCP models for the breaded and cooked shrimp industries has been completed. A final report on the breaded shrimp model, including the test results, was prepared by NFI/NFERF in October 1988 (Anonymous Oct. 1988). The final report on the cooked shrimp model was released in January 1989 (Anonymous Jan. 1989).

Only one plant visit remains to be made in order to complete the testing of the raw shrimp model. The HACCP model for fresh and frozen fish, developed to accommodate important regional differences within the industry, is currently being tested at processing plants around the country. The schedule for in-plant testing of the molluscan shellfish models is being prepared.

A start has been made on the economic analysis of the HACCP models undergoing in-plant testing. The work is being conducted as part of the current NFERF S-K award. The general approach of the economic analysis is to evaluate the gross costs of compliance associated with a new HACCP-based surveillance program for the seafood industry. Estimates of economic impact are to be based primarily on data collected at the individual plant level for each commodity model and extrapolated to all plants processing that commodity, either singly or in combination with other commodities.

Report to Congress

In view of the intense public interest in seafood safety issues the MSSP issued an interim report to Congress on its activities in the Fall of 1989. The shrimp and fresh and frozen fish HACCP models, including the associated economic analyses, were included in the interim report to Congress.

The MSSP process includes the continuing input from steering committees of industry members who have participated in the workshops. Since there are wide differences in the processes and products and related hazards, separate manuals will be prepared for each commodity.

The overall report and the recommended system design was prepared in December 1990 and delivered to Congress shortly thereafter.

While the problems of providing greater consumer protection from safety, wholesomeness and economic hazards are complex, they appear soluble. The hazards are controllable with the HACCP approach, which should minimize costs to both the industry and the responsible government agency.

Literature Cited

NAS. 1985. National Research Council (U. S.), Food Protection Committee, subcommittee on Microbiological Criteria (1985). An evaluation of the role of microbiological criteria for foods and food ingredients. National Academy of Sciences Press.

Anonymous. 1987. Plan of Operations, NMFS Model Seafood Surveillance Program. Office of Trade and Industry Services, National Marine Fisheries Service, National Oceanic and Atmospheric Administration, U.S. Department of Commerce, Washington D. C.

Freeman, M. 1987, 1988. Reported foodborne disease outbreaks attributed to the consumption of seafood as summarized from Centers for Disease Control 1978 through 1982 annual reports. Memorandums to Project Leader E. Spencer Garrett, NMFS/NOAA.

GAO. 1988. U.S. General Accounting Office (August 1988), Seafood safety: Seriousness of problems and efforts to protect consumers. GAO/RCED 88-135 Report to the Chairman, Subcommittee of Commerce Consumer and Monetary Affairs, Committee on Government Operations, House of Representatives.

Anonymous. October 1988. Seafood industry hazard analysis critical control point workshop and model tests — Breaded shrimp final report, National Fisheries Institute, Washington, D.C.

Anonymous. January 1989. Seafood industry hazard analysis critical control point workshop and model tests — Cooked shrimp final report, National Fisheries Institute, Washington, D. C.

Economic Pressures Driving Genetic Changes in Fish

NICK C. PARKER

Abstract: The demands for fish and fishery products in this country and throughout the world are expected to continue to expand faster than the supply of fish. The new tools of genetic engineering — gene insertion, cloning, androgenesis, gynogenesis, transgenetic production, ploidy manipulation — and other techniques will be used in conjunction with selective breeding and hybridization to produce fish tailored for selected environments or with exceptional traits. Data from the Food and Agricultural Organization in Rome indicates that the global supply of fish — the catch from the ocean, inland waters, and all farm-reared aquatic products — increased 11% from 1982 through 1988. During this time, the volume of fish traded among 162 nations increased 16% indicating a more rapid increase in demand than in supply.

Introduction

Imports of fish and fishery products into the United States were valued at $365 million in 1960 and $8.8 billion in 1987, when the imports consisted of $3.1 billion worth of non-edible products (animal feeds, industrial products, etc.) and $5.7 billion for edible fish-ery products. Recently, imports have expanded at an average rate of $860 million per year from 1982 to 1987, while exports increased at only $120 million per year. The annual per capita consumption of fish increased over 20% from 1975 to 1987, when the per capita rate reached 7.0 kg; it is expected to be 13.6 kg by the year 2020. Based on data from the Food and Agricultural Organization, the world's catch of fish (millions of metric tons) was 27 in 1954, 57 in 1966, 74 in 1976, 83 in 1984, and 90 in 1986. The catch has increased with the demand only because previously unused resources — those formerly classified as "trash" fish — are now being captured and processed into consumer-acceptable forms such as imitation lobster, shrimp and scallops. The ocean's resources are recognized as finite, having an estimated maximum sustainable yield of about 100 to 120 million metric tons. The expansion of demand in a market with limited supply is expected to continue to drive prices up and to make fish farming even more lucrative than it is today, when more than 11% of global fish landings are produced by aquaculture. The forecast is for the global yield from aquaculture to increase to 22 million metric tons by the year 2000 when farm-raised species will represent 25% of the world's harvest of aquatic organisms.

Demand for Game Fish Increases

According to a survey conducted by the U. S. Fish and Wildlife Service in 1985, 58 million anglers spent $28.2 billion in 987 million angler days. (U. S. Fish and Wildlife Service 1988). Sport fishing is projected to double by the year 2030. The fishing pressure on public waters is expected to increase much more rapidly than the ability of the resource to produce. Even today some anglers have abandoned public waters to fish in more productive private waters. Public waters often yield less than one legal-size bass in 10 hours of fishing, whereas 30 to 40 bass can be taken from some privately owned and managed lakes in only 3 or 4 hours. Many U.S. citizens are willing to buy catchable-size fish to be stocked in private ponds for recreational purposes. Others are willing to pay sizable fees to enjoy quality fishing in private waters.

Aquaculturists are producing hybrids of striped bass and white bass, Florida strain largemouth bass and other warmwater species for food and sport fish. Some anglers have paid $900 per day for the opportunity to catch 3-kg trophy-size bass in private waters and other fishermen routinely pay $90 in the off-season and $165 per day in the peak season to fish for 1- to 2-kg fish

(Parker 1988). Adoption of the user-pay philosophy has long been evident in private hunting clubs and on game management areas requiring special licenses and fees. The growing popularity of feefishing operations offers expanded opportunities for selling farm-raised fish for recreational purposes.

The catch from privately owned fee-fishing operations is frequently so high that it resembles a supermarket activity. In a single day anglers have harvested over 500 kg of channel catfish from a 0.l-ha pond. Activities such as these may relieve the fishing pressure and certainly provide recreational opportunities in excess of those available on public waters. In many locations public waters are already being managed as catch and release fisheries. If fishing pressure increases as projected, a greater percentage of the commercially captured fish will be redirected to recreational fisheries. Farm-raised fish are expected to become increasingly important for food and recreational purposes.

Expected Contributions of Genetics

Domestic beef, poultry and swine have evolved through selective breeding over thousands of years and are quite different today from their ancestral stocks. By contrast, man has had little influence on genetic selection in most fish. Man's activities to domesticate and improve livestock were directed to a limited number of species whereas the gene pool of fishes resides in thousands of freshwater and marine species of which only a few — predominantly goldfish *Carassius auratus*, common carp *Cyprinus carpio* and rainbow trout *Oncorhynchus mykiss* — have been selectively bred in excess of 100 years.

In recent years, a few other species have been selectively bred as tropical or hobby fish or for the recreational and food-fish markets; the likelihood is that well under 1% of the estimated 20 to 25 thousand species of finfish have been spawned in captivity. The percentage of molluscan and crustacean species that have been spawned in captivity is probably similar to that of finfish.

Geneticists are increasingly turning their attention to fish and in conjunction with aquaculturists have produced common carp and channel catfish *Ictalurus punctatus* which carry the human gene for growth hormone (Dunham et al. 1987). Others are justifiably concerned about the influence of hatchery stocks on the population structure of indigenous stocks (Ryman and Utter 1987). It seems likely that aquaculturists, managers of natural stocks, and those interested in fish for recreation and the commercial fisheries will turn to geneticists to "improve" their stocks. Fish will be selected and developed to thrive in modified environments, to tolerate extremes in temperature, dissolved oxygen, pH, and other water quality variables. Sport fish records will fall in every category as genetically improved fish are stocked in public and private waters. Androgenesis and gynogenesis coupled with reversal of genetic males and females into functional individuals of the opposite sex and then mated back with their own genotype, will produce monosex populations posing little threat to native species. Hybrid and polyploid individuals will become increasingly more popular and widespread due to unique characteristics which may include rapid growth, disease resistance, trophy size, and their ability to fill niches unoccupied by other fishes. The genetic tools being developed and perfected for human medicine and animal sciences will be increasingly applied to fish due to growing economic incentives.

Proposed Action

Aggressive positive programs must be established quickly if the demands for food fish and sport fish are to be met. According to U. S. Department of Commerce figures, the 1987 trade deficit for fish and fishery products ($7.1 billion) was 4.1% of the total U. S. trade deficit ($171.2 billion) and, excluding manufactured goods, was second only to the deficit for petroleum and petroleum products ($16.2 billion) — 9.5% of the total. By comparison, the top five agricultural products imported, listed by actual value and as a percent of the total deficit, were vegetables and fruits $4.3 billion, 2.5%; coffee $2.8 billion, 1.6%; crude rubber $1.2 billion, 0.7%; cocoa $1.1 billion, 0.6%; and sugar $0.4 billion, 0.2%.

The export value of several other agricultural commodities exceeded their import value, resulting in a positive trade balance for these few items. These agricultural trade surpluses in billions were as follows: soybeans $4.3, corn $3.3, wheat $3.0, cotton $1.6, rice $0.5, and tobacco $0.5. Each of these commodities produced in surplus have benefitted from strong government support programs, including research, extension, loans and even price support. Similar positive action programs for aquaculture, if funded at just a small percentage of the trade deficit for fish and fishery products, would produce almost immediate benefits. These benefits would reduce the deficit, establish new jobs for U. S. citizens, provide additional food fish for consumers and sport fish for anglers, and better prepare American fish farmers to compete in international markets. Even with this positive action, the trade deficit in fish and fishery products is expected to continue its upward spiral as the demand grows and the supply shrinks. The economic incentive to produce speciality fish or "improved" fish will draw more geneticists into the fisheries field. Natural resource managers may face their greatest challenge in trying to balance the public's demand for genetically improved fish with the preservation of native stocks.

Literature Cited

Dunham, R. A., J. Eash, J. Askins and T. M. Townes. 1987. Transfer of the metallothionein-human growth hormone fusion gene into channel catfish. Trans. Am. Fish. Soc. 116:87-91.

Parker, N. C. 1988. Aquaculture — natural resource managers' ally? Trans. 53rd N. Am. Wild. and Nat. Res. Conf. 53:584-593.

Ryman, N. and F. Utter, editors, 1987. Population genetics and fishery management. University of Washington Press, Seattle.

U. S. Fish and Wildlife Service. 1988. 1985 National survey of fishing, hunting and wildlife associated recreation. U. S. Department of Interior, Fish and Wildlife Service, Washington, D. C.

A Decision Framework for Managing the Risks of Deliberate Releases of Genetic Materials

ROBIN GREGORY

Abstract: Releases of genetically altered organisms into the environment constitute a concern because of their potential for adverse consequences to human health and the natural environment. There is a need for a defensible framework to assess the risks of releases and to structure a meaningful dialogue with the public about this information. This paper examines risk-management needs from the perspective of a potential regulator, emphasizing the integration of technical information and public values in developing a decision-making framework for evaluating the net social benefits of deliberate releases of genetic material to the aquatic environment.

Introduction

Releases of genetically altered organisms into the aquatic environment constitute a concern because of their potential for adverse consequences to human health and the natural environment. Recent surveys have shown that the public believes biotechnology and genetic engineering will create social benefits but also will lead to costs, including risks to health and the environment, of unknown magnitude (McGarity 1985). Thus, the starting point for risk-management is to think about strategies for reducing this risk; i.e., for reducing the costs associated with the achievement of the anticipated benefits, and for choosing among alternative options that yield different types or amounts of benefits and risks.

This paper outlines a framework for making defensible decisions concerning the risks of deliberate releases of genetically altered marine organisms. The comments are intended to focus attention on a general approach to thinking about risks and risk-management rather than on a particular proposed introduction. The basic analytic structure is derived from an approach known as decision analysis (Keeney and Raiffa 1976), which has been called ". . . a formalization of common sense for decision problems which are too complex for informal use of common sense" (Keeney 1982, p. 806). Decision analysis draws on the theories and methods of psychology, statistical decision theory, economics and management science to provide a means for thinking about decision problems that combines technical information with the value judgments of affected individuals or groups.

Expert and Public Views of Risk

What do we know about risk-assessment and risk-management procedures that may be useful in the context of potential releases of genetically altered organisms to the marine environment? The short answer is: quite a bit, but not nearly enough.

Risk assessments focusing on the expected damages of established and novel technologies have become commonplace, and much of this general structure is relevant for assessments of deliberate genetic releases (Fiksel and Covello 1986). However, there are specific problems in evaluating releases of genetically altered organisms that require special attention. First, both dispersal and exposure models for genetically altered aquatic organisms typically will be highly uncertain because of the relatively short track record of genetic manipulation. This creates special difficulties in terms of how the analysis is done (e.g., the assessment of cumulative impacts) and in terms of how information is communicated (e.g., problems of unintentional bias). Second, risk-management policies that encompass low probability-high consequence events need to combine technical assessments of potential physical damage with social and economic assessments of psychological and sociological considerations. Yet analysts have little experience with such evaluation frameworks, particularly in cases where the benefits are likely to be highly uncertain as well (Gregory 1987). Third, technologies that are new and perceived to present possibly catastrophic consequences tend to be highly feared by large segments of the public. This has important implications for the conduct of risk-management decision-making; for example, in terms of the identification of stakeholder groups and the types of tradeoffs likely to be acceptable.

The general goal of the risk manager is to make better decisions involving potentially hazardous technologies. This requires good science to complete an accurate and comprehensive assessment of the expected risks based on an integrated model that combines information about the risk source, exposure channels, and dose-response. A clear presentation of the basis for physical (health and environmental) effects, combined with treatment of the uncertainties underlying the occurrence of consequences and the scientific knowledge base, can be accomplished by trained experts (see, for example, Morgan and Henrion 1989).

However, a framework for managing the risks of deliberate genetic releases must be responsive to concerns of the public as well as experts: understanding values, as well as facts, is essential to the risk-management decision framework. This means that a socially acceptable decision approach must pay attention to any differences in how experts and laypersons perceive risk, how information about risks is communicated, and the social processes by which acceptable risks are defined. It must also pay attention to how individuals and groups respond to information they are given about risks; for example, how well do people understand the meaning of a low-probability event? Failure to take account of public values and public response might result in an analysis that ignores key factors; it certainly is likely to result in a decision that will be strongly opposed by affected parties who feel that concerns important to them have not been adequately considered.

The following six-stage decision process provides a general basis for managing the risks of deliberate genetic releases to aquatic ecosystems:

1. Identify the key groups of people, or *stakeholders*, who might be affected by a management decision and decide on a means to communicate with representatives of these groups. In the usual case, stakeholders will include representatives of the developer, members of key public groups, and representatives of local, state, or federal regulatory agencies.

2. Determine the technical and managerial *alternatives* that are possible: these must be set out clearly so as to encourage consideration of the full range of options.

3. Identify the *consequences* of each alternative in terms that capture the most salient effects and that are sufficiently precise so as to address important distinctions. For example, separate estimates of morbidity and mortality may need to be made for different species or stocks of fish.

4. Incorporate the *likelihood* of these effects explicitly in the form of probability estimates or frequency distributions.

5. Obtain information from each of the key stakeholder groups regarding the *values* they place on the different consequences, in terms of their expressed objectives (i.e., what matters) and attributes (which measure the degree to which these objectives are met). This information enables comparisons to be made across impacts that differ over time, between geographic regions, or across social groups.

6. With this information in hand, the decision-maker can link values to the consequences of different project alternatives. This leads to a more insightful choice among alternatives and may lead to the creation of new alternatives or to the collection of new information that helps to distinguish between alternatives.

A key to this decision process is distinguishing successfully between issues of fact and issues of value (von Winterfeldt and Edwards 1986). The effects associated with releases of genetically altered organisms are complex and multidimensional. Thus, it is to be expected that stakeholders will disagree as to the importance of the various attributes that characterize the designated set of consequences. Discussions among stakeholders should be expected to clarify, but not resolve, differences in the values (or weights) given to different consequences and their attributes. On the other hand, there may be close agreement regarding the likelihood of specific consequences. Consultation between experts should help to resolve these factual differences; for example, relating to the numbers of fish that might be affected by a particular accident scenario or the magnitude of indirect effects on social and community indicators.

This basic distinction, between risk facts and risk values, is related to a second distinction between technical and perceived estimates of risks. Technical risk refers to quantitative measures of physical risks predicted by experts. Perceived risk refers to qualitative concerns about risks expressed by laypersons. A large number of studies conducted over the past decade (summarized in Slovic 1987) have demonstrated that experts typically think about the risks of a technology in terms of its impact on human health and environmental risks, as defined by changes in statistical mortality and morbidity. These concerns are important to the public, but so too are a number of additional factors that include (1) the possibility that a catastrophic accident will occur, (2) the extent to which exposure is voluntary, and (3) the extent to which the risk source is familiar.

Most current approaches to studying risk perception employ psychometric techniques to produce quantitative representations or "cognitive maps" of these multidimensional risk attitudes and perceptions. In the usual case, people are asked to make judgments about the riskiness of technologies, products, or activities in terms of ratings across (1) characteristics hypothesized to account for risk perceptions, such as dread or voluntariness or familiarity, (2) the benefits provided to society by the technology, (3) the environmental mortality and morbidity expected to be caused by the hazard. Investigation of these relations by means of psychophysical scaling and multivariate analysis techniques (e.g., statistical procedures that examine the intercorrelations among variables and then seek to define two or three underlying, explanatory "factors") have shown that the broad domain of characteristics can be condensed to a small set of higher-order characteristics. Two factors generally emerge as most important. One is dread risk, which includes characteristics such as fear and disgust as well as the perceived lack of control over a technology's effects, its catastrophic potential, and the ambiguity of consequence estimates. The higher a technology's dread rating, the more people want to see its risks reduced and the more likely they are to seek strict regulations designed to reduce the risk. A second factor is unknown risk, which includes characteristics such as the extent to which a technology is observable, known, and new.

Significant variations often exist in the risk perceptions of different stakeholder groups and in how experts and members of the public view the risks of a technology. In most cases, the characterization of a risk by the public will be considerably more broad than that by experts because it takes into account a wider variety of considerations. This means that assurances relating to the relative safety of a technology, calculated in terms of the expected probability and severity of likely accident scenarios, may fail to address a stakeholder's concerns: people may not be worried just about the number of deaths from a possible accident but about the way in which these deaths occur, their geographic incidence, or their latency period. The imaginability of an unfamiliar negative impact also plays a role here: if people believe that a valued fish stock could be entirely lost due to a proposed genetic manipulation, their concerns may focus on the catastrophic consequences of this event and assurances that the associated probability is very low may do little to reduce their fears.

Lessons for the Risk Manager

The predicted physical impacts of a risk source on the natural environment and human society are clearly important. Risk perceptions also are important, and public concerns about a technology and its consequences should play an important role in decisions between competing management strategies. Risk managers need to account for public values in making social decisions about uncertain initiatives.

Acknowledgment of public concerns about a risky option does not mean that a risk manager should go along with everything the public wants (Lichtenstein et al. 1990). One reason for caution is that the conceptual links between the perception of risks and the evaluation of risks remain tenuous; we still know little about the extent to which changes in the physical risks of technologies will affect their psychological assessment. A second reason is that some public attitudes concerning risks are likely to be related to socio-political considerations that have little to do with environmental or human health (Furby et al. 1988). For example, some attitudes may be influenced by the role of the government in an industry, by its relative use of capital and labor, or by the identity of the ultimate beneficiaries of a technology. The more that these various factors can be distinguished, the easier it will be for decision-makers to know which considerations are regarded as legitimate (e.g., which should serve as the basis for mitigation or compensation policies) and what might be done to encourage acceptance of a technology, product, or facility.

A third reason for exercising caution when integrating physical and perceptual measures of risk comes from research in behavioral decision theory which suggests that the framing of a decision can have a major influence on choice. In particular, it appears that risks and benefits are not assessed in isolation but as part of a mental balancing process in which individuals' understanding of the benefits of a technology can affect their estimate of the accompanying risks (Keller and Sarin 1988). Weighing benefits against costs is an everyday task but, as much experimental work demonstrates, it is not one people tend to do particularly well (at least, in terms of the traditional criteria for rational decisions; see Kahneman and Tversky 1984). One implication is that changes in the presentation of costs and benefits that are formally irrelevant can result in dramatic reversals of judgment. Thus, the same people who express support for a transgenic organism based on a colorful depiction of its benefits (e.g., its ability to feed starving people or to destroy oil and toxic substances) may voice their strong opposition to it when low-probability but highly adverse consequences are emphasized instead. In such cases, questions of order, salience and reference (e.g., whether impacts are expressed in terms of benefits gained or costs avoided) may dramatically affect the acceptability of a technology.

Risk-management decision-making will prove to be far easier if a basic structure for thinking about risks is clearly understood by the key parties involved in a dispute. Knowing ahead of time about some of the more common fallacies in thinking about risks may help to anticipate the basis for controversy and diminish the heated arguments that often accompany debates about hazardous technologies and management initiatives. The following five lessons, drawn freely from a lengthier list developed by Ralph Keeney (Keeney 1988a), are particularly important.

1. Zero risk is an illusion. No risk-free alternatives are available: all options under consideration, including the option of doing nothing (e.g., build no hatchery, release no fish), involve risks. These risks may vary importantly in their form; for example, the risks associated with release of a genetically altered organism may be perceived very differently from the risks associated with a reduced catch. Nevertheless, eliminating one alternative always means that something else will be done instead and that something else also entails a risk. Identification of the acceptable (nonzero) level of risk therefore will vary depending on circumstances and will be a function of the specific type of activity, its benefits and costs, and the social history of its development and use.

The myth of zero risk is important to remember in the context of public debates that often make the impossible demand that managers (e.g., regulators) select an alternative that will avoid risk. A preferred goal is rather to make use of risk to achieve social objectives: to recognize the inevitability of trial and error within acceptable social limits. Taking too narrow a view can in fact lead to increased risk, in cases where the small harm that is avoided forms an integral part of substantially larger forgone benefits (Wildavsky 1988).

2. ***Risk decisions involve conflicting objectives.*** Decisions about risk also involve decisions about other implications of the same activities: their economic and environmental costs and benefits, their effects on quality of life, their impacts on social structure (Keeney 1988b). Except in the case of dominating alternatives, multiple objectives will always conflict. This means that difficult choices will need to be made, because more of one objective implies less of some other objective. As a result, a key to successful risk-management decision-making is likely to rest with the ability of stakeholder groups to maintain flexibility in negotiating positions and to recognize the gains that are possible from trades across consequence types.

3. ***Risk decisions involve statistical rather than individual effects.*** Individual effects impact particular individuals or groups, whose identity is known in advance. Statistical effects impact a class of people or groups, with the identity of the affected individuals not known in advance. In most cases, risk-management decisions involve statistical effects: we know about the type of impact that may occur (e.g., the possibility that the harvest of some aquaculturalists may decline) but we do not know to whom or when it will occur. The value tradeoff between costs and statistical impacts usually will be different (and, in most cases, substantially smaller) than the value tradeoff between costs and identifiable impacts.

4. ***Risks to life must be traded off against other considerations.*** Developmental options involving the release of genetically altered organisms often have, as one category of effect, a low-probability risk of injury or death to marine organisms or to humans. In nearly all cases, each of the alternatives under consideration could be made safer at an additional economic cost through

the provision of additional safety features or additional testing. However, a tradeoff is required because funds used in this way will not be used in other ways and these other uses include options that also could promote health and safety. Sometimes the relevant tradeoff is between overall safety and equity, so that decision-makers must balance what is best for society as a whole against the achievement of a fair distribution of benefits and risks across groups that differ in location, age, or some other characteristic. The general point is that management decision problems typically involve trading health risks against other attributes: slogans to the contrary, none of us operates as if health and safety considerations, whether our own or those of other individuals and species, really are priceless.

5. ***The analysis of risks is never objective.*** The analysis of risks starts with framing the problem and structuring objectives (what is important) and alternatives (what project options are possible). Defining and integrating this information to obtain a decision requires value judgments. Subjectivity is therefore an essential element of risk-management decision-making and must be acknowledged as such. Key judgments therefore should be made explicitly and sensitivity analyses conducted so that the significance of assumptions for the bottom line of an analysis can be discerned.

Case studies of early experiments in genetic engineering suggest that each of these five lessons has been repeatedly ignored. The struggle to field-test the frost-resisting ice-minus bacteria provides one highly visible example: a four-year adversarial process, dominated by lawsuits and public hearings, succeeded in approval to go ahead with field tests but did little to address fundamental questions of conflicting objectives and the public's desire for a zero-risk alternative (Krimsky and Plough 1988). This suggests that regulatory success in genetic engineering may well hinge on the ability of risk managers to develop a process that effectively addresses some of the more general issues noted above before launching into discussions about any specific planned introduction of transgenic organisms into the environment.

Risk-Management in Practice

The decision-making approach outlined above focuses on a process for structuring a dialogue about management options and their consequences. It provides a means for making defensible decisions about highly complex social situations characterized by multiple objectives and high degrees of uncertainty. Perhaps most importantly, it provides a tested mechanism for encouraging dialogue between the potentially affected parties.

This last point is important when discussing genetic changes in aquatic resources because of the nature of the proposed initiatives. There are many reasons why genetic changes might be considered: faster growth rates among commercially valued species, improved disease resistance, lower age of individuals at maturity, increased thermal tolerance, or improved ability to utilize a food source. These changes focus on the creation of incremental benefits: something that now occurs could be done faster, or cheaper, or better. The accompanying risks, on the other hand, are characterized by a nonmarginal quality: a fish stock is or is not genetically altered; the introduction of a genetically changed organism to the marine environment is or is not accidental.

An increase in growth rates of 20% may mean the difference between profit and loss for an aquaculturalist, but for a member of the public the distinction is unlikely to be important. On the other hand, the image of a catastrophic event, whereby a single accident involving genetically altered fish leads to the extinction of an entire species, is highly salient. Terms like biotechnology or genetic engineering may have quite precise definitions for the scientist, but to many members of the public they convey images of change that are met by fear and that are highly influenced by mention of the existence of potentially adverse consequences. Paying attention to public understanding and criteria for public acceptance therefore is a key consideration in the development of any risk-management framework. Nuclear power is a prime example of a technology whose fate in the past decade has been profoundly affected by its failure to achieve public support.

In such a highly charged environment, it is important that a risk-management framework focus not only on the outcomes of a decision but also on the process that is followed (Gregory et al. In press). The approach discussed in this paper, which provides a tool for representing complex value tradeoffs and for explicitly incorporating the value preferences of concerned parties into the decision process, is particularly appropriate in such controversial social-risk problems (Edwards and von Winterfeldt 1987). At minimum, the identification of key stakeholders and the elicitation of information about their objectives will provide a mechanism for fostering dialogue. Both the public and experts are given the opportunity to (1) clarify their values in the context of a structured elicitation process and (2) learn about how others view the same problems.

Furthermore, this process is both iterative and interactive. As values information is obtained from different stakeholder groups, each is given the opportunity to assess its understanding in light of how others view the same problem. This helps to focus the discussion of management alternatives, by emphasizing what is most important about the decision problem, and provides information to experts about additional information that should be collected (e.g., regarding the range of impacts, their likelihood, or the variance of consequence estimates). Perhaps most significantly, disagreements about the consequences of technically feasible alternatives can be separated from disagreements about their relative desirability. Experts then will know where to place scarce financial and intellectual resources to address the most important technical questions and to work more effectively toward social agreement concerning the adaption of new technologies.

Acknowledgments

The content of this paper owes a special debt to conversations the author has had with Ralph Keeney about risk-management and the role of decision analysis in assessing public values.

Literature Cited

Edwards, W. and D. von Winterfeldt, 1987. Public values in risk debates. Risk Analysis 7: 141-158.

Fiksel, J. and J. Covello, 1986. Biotechnology risk assessment: Issues and methods for environmental introduction. Pergamon, New York.

Furby, L., P. Slovic, B. Fischhoff and R. Gregory. 1988. Public perceptions of electric power transmission lines. J. Environ. Psychol. 8:14-43.

Gregory, R. 1987. Nonmonetary measures of nonmarket fishery resource benefits. Trans. Am. Fish. Soc. 116:374-380.

Gregory, R., H. Kunreuther, D. Easterling and K. Richards. In press. Incentive policies to site hazardous waste facilities. Risk Analysis.

Kahneman, D. and A. Tversky. 1984. Choices, values and frames. Am. Psychol. 39:341-350.

Keeney, R. 1982. Decision analysis: An overview. Oper. Res. 30:803-838.

Keeney, R. 1988a. Facts to guide thinking about life-threatening risks. Proceedings of the 1988 IEEE international conference on systems, man, and cybernetics. Pergamon Press, Oxford, p.326-329.

Keeney, R. 1988b. Structuring objectives for problems of public interest. Oper. Res. 36:396-405.

Keeney, R. and H. Raiffa. 1976. Decisions with multiple objectives, preferences, and value tradeoffs. Wiley, New York.

Keller, R. and R. Sarin. 1988. Equity in social risk: Some empirical observations. Risk Anal. 8:135-146.

Krimsky, S. and A. Plough. 1988. Environmental risks: Communicating risks as a social process. Auburn House, Dover, Massachusetts.

Lichtenstein, S., R. Gregory, P. Slovic and W.A. Wagenaar. 1990. When lives are in your hands: Dilemmas of the societal decision-maker. *In* R.M. Hogarth (ed.), Insights in decision-making: Theory and applications. (A tribute to Hillel J. Einhorn.) University of Chicago Press, Chicago, Illinios

Morgan, G. and M. Henrion. 1989. Uncertainty. Cambridge University Press, New York.

McGarity, T. 1985. Regulating biotechnology. Issues Sci. Tech. 4:41-56.

Slovic, P. 1987. Perception of risk. Science 236:280-285.
Wildavsky, A. 1988. Searching for safety. Transactions Books, London, England.
Winterfeldt, D. von and W. Edwards. 1986. Decision analysis and behavioral research. Cambridge University Press, New York.

Genera and Species Index

D

E

F

G

H

I

K

L

M

N

O

P

Q

R

S

Geographic Index

D

E

F

G

H

I

J

K

L

M

N

O

P

Q

R

S

T

U

V

W

Y

General Index

B

D

G

H

S

U

V

W

X

Y

Z